Differential & Difference Equations: Theory, Techniques and Practice

Differential & Difference Equations: Theory, Techniques and Practice

Edited by **Charlin Chester**

Willford Press
New York

Published by Willford Press,
118-35 Queens Blvd., Suite 400,
Forest Hills, NY 11375, USA
www.willfordpress.com

Differential & Difference Equations: Theory, Techniques and Practice
Edited by Charlin Chester

International Standard Book Number: 978-1-68285-235-4 (Hardback)

Printed in China.

Contents

Preface

Differential and difference equations are applicable in a wide variety of fields. From mechanics in physics to predator-prey equations in biology, equations are one of the most effective methods of measuring rates of change. This book discusses various concepts like stochastic differential equations, delayed differential equations, differential algebraic equations, etc. This book is a valuable compilation of topics, ranging from the basic to the most complex advancements of differential and difference equations. It will prove beneficial for students of mathematics, physics and associated branches of science and engineering.

This book has been the outcome of endless efforts put in by authors and researchers on various issues and topics within the field. The book is a comprehensive collection of significant researches that are addressed in a variety of chapters. It will surely enhance the knowledge of the field among readers across the globe.

It gives us an immense pleasure to thank our researchers and authors for their efforts to submit their piece of writing before the deadlines. Finally in the end, I would like to thank my family and colleagues who have been a great source of inspiration and support.

Editor

1

A Riccati-Bernoulli sub-ODE method for nonlinear partial differential equations and its application

Xiao-Feng Yang[1*], Zi-Chen Deng[2,3] and Yi Wei[1]

*Correspondence: yangxiaofeng@nwsuaf.edu.cn
[1]Department of Applied Mathematics, Northwestern Polytechnical University, Xi'an, 710072, P.R. China
Full list of author information is available at the end of the article

Abstract

The Riccati-Bernoulli sub-ODE method is firstly proposed to construct exact traveling wave solutions, solitary wave solutions, and peaked wave solutions for nonlinear partial differential equations. A Bäcklund transformation of the Riccati-Bernoulli equation is given. By using a traveling wave transformation and the Riccati-Bernoulli equation, nonlinear partial differential equations can be converted into a set of algebraic equations. Exact solutions of nonlinear partial differential equations can be obtained by solving a set of algebraic equations. By applying the Riccati-Bernoulli sub-ODE method to the Eckhaus equation, the nonlinear fractional Klein-Gordon equation, the generalized Ostrovsky equation, and the generalized Zakharov-Kuznetsov-Burgers equation, traveling solutions, solitary wave solutions, and peaked wave solutions are obtained directly. Applying a Bäcklund transformation of the Riccati-Bernoulli equation, an infinite sequence of solutions of the above equations is obtained. The proposed method provides a powerful and simple mathematical tool for solving some nonlinear partial differential equations in mathematical physics.

MSC: 35Q55; 35Q80; 35G25

Keywords: Riccati-Bernoulli sub-ODE method; Bäcklund transformation; traveling wave solution; solitary wave solution; peaked wave solution

1 Introduction

Nonlinear partial differential equations (NLPDEs) are known to describe a wide variety of phenomena not only in physics, but also in biology, chemistry, and several other fields. The investigation of traveling wave solutions for NLPDEs plays an important role in the study of nonlinear physical phenomena. In recent years, many powerful methods were used to construct traveling wave solutions of NLPDEs. For example, the inverse scattering method [1], the Bäcklund and Darboux transformation method [2], the homotopy perturbation method [3], the first integral method [4–6], the $(\frac{G'}{G})$-expansion method [7–9], the sub-equation method [10, 11], Hirota's method [12], the homogeneous balance method [13–15], the variational iteration method [16, 17], the tanh-sech method [18], the Jacobi elliptic function method [19], the modified simple equation method [20–23], the $\exp(-\Phi(\xi))$-expansion method [24], the alternative functional variable method [25], and so on.

Many well-known NLPDEs can be handled by those traditional methods. However, there is no unified method which can be used to deal with all types of NLPDEs. Moreover, we always encounter the fractional NLPDEs, the NLPDEs which have nonlinear terms of any order or peaked wave solutions. It is significant to construct traveling wave solutions of NLPDEs by a uniform method. Based on those problems, the Riccati-Bernoulli sub-ODE method is firstly presented.

In this paper, the Riccati-Bernoulli sub-ODE method is proposed to construct traveling wave solutions, solitary wave solutions, and peaked wave solutions of NLPDEs. By using a traveling wave transformation and the Riccati-Bernoulli equation, NLPDEs can be converted into a set of algebraic equations. Exact solutions of NLPDEs can be obtained by solving the set of algebraic equations. The Eckhaus equation, the nonlinear fractional Klein-Gordon equation, the generalized Ostrovsky equation, and the generalized Zakharov-Kuznetsov-Burgers (ZK-Burgers) equation are chosen to illustrate the validity of the Riccati-Bernoulli sub-ODE method. A Bäcklund transformation of the Riccati-Bernoulli equation is given. If we get a solution of NLPDEs, we can search for a new infinite sequence of solutions of the NLPDEs by using a Bäcklund transformation.

The remainder of this paper is organized as follows: the Riccati-Bernoulli sub-ODE method is described in Section 2. In Section 3, a Bäcklund transformation of the Riccati-Bernoulli equation is given. In Sections 4-7, we apply the Riccati-Bernoulli sub-ODE method to the Eckhaus equation, the nonlinear fractional Klein-Gordon equation, the generalized Ostrovsky equation, and the generalized ZK-Burgers equation, respectively. In Section 8, our results are compared with the first integral method, the $(\frac{G'}{G})$-expansion method, and physical explanations of the obtained solutions are discussed. In Section 9, some conclusions and directions for future work are given.

2 Description of the Riccati-Bernoulli sub-ODE method

Let there be given a NLPDE, say, in two variables,

$$P(u, u_t, u_x, u_{xx}, u_{xt}, \ldots) = 0, \tag{1}$$

where P is in general a polynomial function of its arguments, the subscripts denote the partial derivatives. The Riccati-Bernoulli sub-ODE method consists of three steps.

Step 1. Combining the independent variables x and t into one variable

$$\xi = k(x + Vt), \tag{2}$$

with

$$u(x,t) = u(\xi), \tag{3}$$

where the localized wave solution $u(\xi)$ travels with speed V, by using Eqs. (2) and (3), one can transform Eq. (1) to an ODE

$$P(u, u', u'', u''', \ldots) = 0, \tag{4}$$

where u' denotes $\frac{du}{d\xi}$.

Step 2. Suppose that the solution of Eq. (4) is the solution of the Riccati-Bernoulli equation

$$u' = au^{2-m} + bu + cu^{m}, \tag{5}$$

where a, b, c, and m are constants to be determined later.

From Eq. (5) and by directly calculating, we get

$$u'' = ab(3-m)u^{2-m} + a^2(2-m)u^{3-2m} + mc^2u^{2m-1} + bc(m+1)u^{m} + \left(2ac+b^2\right)u, \tag{6}$$

$$u''' = \big(ab(3-m)(2-m)u^{1-m} + a^2(2-m)(3-2m)u^{2-2m} + m(2m-1)c^2u^{2m-2} + bcm(m+1)u^{m-1} + \left(2ac+b^2\right)\big)u', \tag{7}$$

$$\ldots$$

Remark When $ac \neq 0$ and $m = 0$, Eq. (5) is a Riccati equation. When $a \neq 0$, $c = 0$, and $m \neq 1$, Eq. (5) is a Bernoulli equation. Obviously, the Riccati equation and Bernoulli equation are special cases of Eq. (5). Because Eq. (5) is firstly proposed, we call Eq. (5) the Riccati-Bernoulli equation in order to avoid introducing new terminology.

Equation (5) has solutions as follows:

Case 1. When $m = 1$, the solution of Eq. (5) is

$$u(\xi) = Ce^{(a+b+c)\xi}. \tag{8}$$

Case 2. When $m \neq 1$, $b = 0$, and $c = 0$, the solution of Eq. (5) is

$$u(\xi) = \left(a(m-1)(\xi + C)\right)^{\frac{1}{m-1}}. \tag{9}$$

Case 3. When $m \neq 1$, $b \neq 0$, and $c = 0$, the solution of Eq. (5) is

$$u(\xi) = \left(-\frac{a}{b} + Ce^{b(m-1)\xi}\right)^{\frac{1}{m-1}}. \tag{10}$$

Case 4. When $m \neq 1$, $a \neq 0$, and $b^2 - 4ac < 0$, the solutions of Eq. (5) are

$$u(\xi) = \left(-\frac{b}{2a} + \frac{\sqrt{4ac-b^2}}{2a}\tan\left(\frac{(1-m)\sqrt{4ac-b^2}}{2}(\xi + C)\right)\right)^{\frac{1}{1-m}} \tag{11}$$

and

$$u(\xi) = \left(-\frac{b}{2a} - \frac{\sqrt{4ac-b^2}}{2a}\cot\left(\frac{(1-m)\sqrt{4ac-b^2}}{2}(\xi + C)\right)\right)^{\frac{1}{1-m}}. \tag{12}$$

Case 5. When $m \neq 1$, $a \neq 0$, and $b^2 - 4ac > 0$, the solutions of Eq. (5) are

$$u(\xi) = \left(-\frac{b}{2a} - \frac{\sqrt{b^2-4ac}}{2a}\coth\left(\frac{(1-m)\sqrt{b^2-4ac}}{2}(\xi + C)\right)\right)^{\frac{1}{1-m}} \tag{13}$$

and

$$u(\xi)=\left(-\frac{b}{2a}-\frac{\sqrt{b^2-4ac}}{2a}\tanh\left(\frac{(1-m)\sqrt{b^2-4ac}}{2}(\xi+C)\right)\right)^{\frac{1}{1-m}}. \tag{14}$$

Case 6. When $m \neq 1$, $a \neq 0$, and $b^2-4ac=0$, the solution of Eq. (5) is

$$u(\xi)=\left(\frac{1}{a(m-1)(\xi+C)}-\frac{b}{2a}\right)^{\frac{1}{1-m}}, \tag{15}$$

where C is an arbitrary constant.

Step 3. Substituting the derivatives of u into Eq. (4) yields an algebraic equation of u. Noticing the symmetry of the right-hand item of Eq. (5) and setting the highest power exponents of u to equivalence in Eq. (4), m can be determined. Comparing the coefficients of u^i yields a set of algebraic equations for a, b, c, and V. Solving the set of algebraic equations and substituting m, a, b, c, V, and $\xi=k(x+Vt)$ into Eq. (8)-(15), we can get traveling wave solutions of Eq. (1).

In the subsequent section, we will give a Bäcklund transformation of the Riccati-Bernoulli equation and some applications to illustrate the validity of the Riccati-Bernoulli sub-ODE method.

3 Bäcklund transformation of the Riccati-Bernoulli equation

When $u_{n-1}(\xi)$ and $u_n(\xi)$ ($u_n(\xi)=u_n(u_{n-1}(\xi))$) are the solutions of Eq. (5), we get

$$\frac{du_n(\xi)}{d\xi}=\frac{du_n(\xi)}{du_{n-1}(\xi)}\frac{du_{n-1}(\xi)}{d\xi}=\frac{du_n(\xi)}{du_{n-1}(\xi)}\left(au_{n-1}^{2-m}+bu_{n-1}+cu_{n-1}^{m}\right),$$

namely

$$\frac{du_n(\xi)}{au_n^{2-m}+bu_n+cu_n^m}=\frac{du_{n-1}(\xi)}{au_{n-1}^{2-m}+bu_{n-1}+cu_{n-1}^m}.$$

Integrating above equation once with respect to ξ and simplifying it, we get

$$u_n(\xi)=\left(\frac{-cA_1+aA_2(u_{n-1}(\xi))^{1-m}}{bA_1+aA_2+aA_1(u_{n-1}(\xi))^{1-m}}\right)^{\frac{1}{1-m}}, \tag{16}$$

where A_1 and A_2 are arbitrary constants.

Equation (16) is a Bäcklund transformation of Eq. (5). If we get a solution of Eq. (5), we can search for new infinite sequence of solutions of Eq. (5) by using Eq. (16). Then an infinite sequence of solutions of Eq. (1) is obtained.

4 Application to the Eckhaus equation

The Eckhaus equation reads

$$i\psi_t+\psi_{xx}+2\left(|\psi|^2\right)_x\psi+|\psi|^4\psi=0, \tag{17}$$

where $\psi=\psi(x,t)$ is a complex-valued function of two real variables x, t.

The Eckhaus equation was found [26] as an asymptotic multiscale reduction of certain classes of nonlinear Schrödinger type equations. A lot of the properties of the Eckhaus equation were obtained [27]. The Eckhaus equation can be linearized by making some transformations of dependent variables [28]. An exact traveling wave solution of the Eckhaus equation was obtained by the $(\frac{G'}{G})$-expansion method [8] and the first integral method [5].

In this section, new type of exact traveling wave solutions of the Eckhaus equation are obtained by using the Riccati-Bernoulli sub-ODE method.

Using the traveling wave transformation

$$\psi(x,t) = u(\xi)e^{i(\alpha x+\beta t)}, \tag{18}$$

Eq. (17) is reduced to

$$k^2u'' - (\alpha^2 + \beta)u + 4ku^2u' + u^5 = 0, \tag{19}$$

where

$$\xi = k(x - 2\alpha t), \tag{20}$$

and k, α, β are real constants to be determined later.

Suppose that the solution of Eq. (19) is the solution of Eq. (5). Substituting Eqs. (5) and (6) into Eq. (19), we get

$$\begin{aligned} &k^2\big(ab(3-m)u^{2-m} + a^2(2-m)u^{3-2m} + mc^2u^{2m-1} + bc(m+1)u^m \\ &\quad + (2ac + b^2)u\big) - (\alpha^2+\beta)u + 4ku^2\big(au^{2-m} + bu + cu^m\big) + u^5 = 0. \end{aligned} \tag{21}$$

Setting $m = -1$ and $c = 0$, Eq. (21) becomes

$$\big(k^2b^2 - (\alpha^2 + \beta)\big)u + \big(4k^2ab + 4kb\big)u^3 + \big(3k^2a^2 + 4ka + 1\big)u^5 = 0. \tag{22}$$

Setting each coefficient of u^j $(j = 1, 3, 5)$ to zero, we get

$$k^2b^2 - (\alpha^2 + \beta) = 0, \tag{23a}$$

$$4k^2ab + 4kb = 0, \tag{23b}$$

$$3k^2a^2 + 4ka + 1 = 0. \tag{23c}$$

Notice that $k \neq 0$, otherwise we can only get trivial solution.

Case A. If $b = 0$, from Eqs. (23a)-(23c) and (5), we get

$$u(x,t) = \pm\frac{1}{\sqrt{-2ka(x - 2\alpha t) + C}}, \tag{24a}$$

$$\alpha^2 + \beta = 0, \tag{24b}$$

$$ka = -1\left(ka = -\frac{1}{3}\right), \tag{24c}$$

where C is an arbitrary real constant.

Case A-1. When $ka = -1$, we get exact traveling wave solutions of Eq. (17),

$$\psi_1(x,t) = \pm\frac{1}{\sqrt{2(x-2\alpha t)+C}}e^{i(\alpha x-\alpha^2 t)}, \tag{25}$$

where C and α are arbitrary real constants.

Case A-2. When $ka = -\frac{1}{3}$, we get exact traveling wave solutions of Eq. (17),

$$\psi_2(x,t) = \pm\frac{1}{\sqrt{\frac{2}{3}(x-2\alpha t)+C}}e^{i(\alpha x-\alpha^2 t)}, \tag{26}$$

where C and α are arbitrary real constants.

Case B. If $b \neq 0$, from Eqs. (23a)-(23c), we get

$$b = \pm\frac{\sqrt{\alpha^2+\beta}}{k}, \tag{27a}$$

$$a = -\frac{1}{k}. \tag{27b}$$

Case B-1. When $b = -\frac{\sqrt{\alpha^2+\beta}}{k}$ and $a = -\frac{1}{k}$, from Eqs. (10) and (18), we get an exact traveling wave solution of Eq. (17),

$$\psi_3(x,t) = \left(-\frac{1}{\sqrt{\alpha^2+\beta}} + Ce^{2\sqrt{\alpha^2+\beta}(x-2\alpha t)}\right)^{-\frac{1}{2}} e^{i(\alpha x+\beta t)}, \tag{28}$$

where C, α, and β are arbitrary real constants.

Especially, if we choose $C = C_1 = \frac{1}{\sqrt{\alpha^2+\beta}}$, Eq. (28) becomes

$$\psi_4(x,t) = \left(\frac{\sqrt{\alpha^2+\beta}}{2}\left(-1+\coth\left(\sqrt{\alpha^2+\beta}(x-2\alpha t)\right)\right)\right)^{\frac{1}{2}} e^{i(\alpha x+\beta t)}, \tag{29}$$

where α and β are arbitrary real constants.

If we choose $C = C_2 = -\frac{1}{\sqrt{\alpha^2+\beta}}$, Eq. (28) becomes

$$\psi_5(x,t) = \left(\frac{\sqrt{\alpha^2+\beta}}{2}\left(-1+\tanh\left(\sqrt{\alpha^2+\beta}(x-2\alpha t)\right)\right)\right)^{\frac{1}{2}} e^{i(\alpha x+\beta t)}, \tag{30}$$

where α and β are arbitrary real constants.

Case B-2. When $b = \frac{\sqrt{\alpha^2+\beta}}{k}$ and $a = -\frac{1}{k}$, from Eqs. (10) and (18), we get an exact traveling wave solution of Eq. (17),

$$\psi_6(x,t) = \left(\frac{1}{\sqrt{\alpha^2+\beta}} + Ce^{-2\sqrt{\alpha^2+\beta}(x-2\alpha t)}\right)^{-\frac{1}{2}} e^{i(\alpha x+\beta t)}, \tag{31}$$

where C, α, and β are arbitrary real constants.

Especially, if we choose $C = C_3 = \frac{1}{\sqrt{\alpha^2+\beta}}$, Eq. (31) becomes

$$\psi_7(x,t) = \left(\frac{\sqrt{\alpha^2+\beta}}{2}\left(1 + \tanh\left(\sqrt{\alpha^2+\beta}(x-2\alpha t)\right)\right)\right)^{\frac{1}{2}} e^{i(\alpha x+\beta t)}, \tag{32}$$

where α and β are arbitrary real constants.

If we choose $C = C_4 = -\frac{1}{\sqrt{\alpha^2+\beta}}$, Eq. (31) becomes

$$\psi_8(x,t) = \left(\frac{\sqrt{\alpha^2+\beta}}{2}\left(1 + \coth\left(\sqrt{\alpha^2+\beta}(x-2\alpha t)\right)\right)\right)^{\frac{1}{2}} e^{i(\alpha x+\beta t)}, \tag{33}$$

where α and β are arbitrary real constants.

Applying Eq. (16) to $\psi_j(x,t)$ $(j = 1,2,\ldots,8)$, we can get an infinite sequence of solutions of Eq. (17). For example, by applying Eq. (16) to Eq. (32), we get a new solution of Eq. (17),

$$\psi_7^*(x,t) = \left(\frac{A_2\sqrt{\alpha^2+\beta}(1+\tanh(\sqrt{\alpha^2+\beta}(x-2\alpha t)))}{2A_2 + A_1\sqrt{\alpha^2+\beta}(-1+\tanh(\sqrt{\alpha^2+\beta}(x-2\alpha t)))}\right)^{\frac{1}{2}} e^{i(\alpha x+\beta t)}, \tag{34}$$

where A_1, A_2, α, and β are arbitrary real constants.

5 Application to the nonlinear fractional Klein-Gordon equation

The nonlinear fractional Klein-Gordon equation [23] reads

$$\frac{\partial^{2\alpha} u(x,t)}{\partial t^{2\alpha}} = \frac{\partial^2 u(x,t)}{\partial x^2} + \beta u(x,t) + \gamma u^3(x,t), \quad t > 0, 0 < \alpha \le 1, \tag{35}$$

where β and γ are known constants.

As is well known, linear and nonlinear Klein-Gordon equations model many problems in classical and quantum mechanics, solitons and condensed matter physics. For example, the nonlinear sine Klein-Gordon equation models a Josephson junction, the motion of rigid pendula attached to a stretched wire, and dislocations in crystals [17, 29–31]. A non-local version of these equations are properly described by the fractional version of them. Exact traveling wave solutions of the nonlinear fractional Klein-Gordon equation were obtained by the homotopy perturbation method [29] and the first integral method [6].

In this section, exact traveling wave solutions of the nonlinear fractional Klein-Gordon equation are obtained by using the Riccati-Bernoulli sub-ODE method.

Using the transformation

$$u(x,t) = u(\xi), \tag{36}$$

with

$$\xi = lx - \frac{\lambda t^\alpha}{\Gamma(1+\alpha)}, \tag{37}$$

where l and λ are constants to be determined later, Eq. (35) becomes

$$u'' - \frac{\beta}{\lambda^2 - l^2}u - \frac{\gamma}{\lambda^2 - l^2}u^3 = 0. \tag{38}$$

Suppose that the solution of Eq. (38) is the solution of Eq. (5). Substituting Eq. (6) into Eq. (38), we get

$$ab(3-m)u^{2-m}+a^2(2-m)u^{3-2m}+mc^2u^{2m-1}$$
$$+bc(m+1)u^m+\left(2ac+b^2\right)u-\frac{\beta}{\lambda^2-l^2}u-\frac{\gamma}{\lambda^2-l^2}u^3=0. \quad (39)$$

Setting $m=0$, Eq. (39) is reduced to

$$3abu^2+2a^2u^3+bc+\left(2ac+b^2\right)u-\frac{\beta}{\lambda^2-l^2}u-\frac{\gamma}{\lambda^2-l^2}u^3=0. \quad (40)$$

Setting each coefficient of u^i $(i=0,1,2,3)$ to zero, we get

$$bc=0, \quad (41a)$$

$$2ac+b^2-\frac{\beta}{\lambda^2-l^2}=0, \quad (41b)$$

$$3ab=0, \quad (41c)$$

$$2a^2-\frac{\gamma}{\lambda^2-l^2}=0. \quad (41d)$$

Solving Eqs. (41a)-(41d), we get

$$b=0, \quad (42a)$$

$$ac=\frac{\beta}{2(\lambda^2-l^2)}, \quad (42b)$$

$$a=\pm\sqrt{\frac{\gamma}{2(\lambda^2-l^2)}}. \quad (42c)$$

Case A. When $\frac{\beta}{\lambda^2-l^2}>0$, substituting Eqs. (42a)-(42c) and (37) into Eqs. (11) and (12), we get exact traveling wave solutions of Eq. (35),

$$u_{1,2}(x,t)=\pm\sqrt{\frac{\beta}{\gamma}}\tan\left(\sqrt{\frac{\beta}{2(\lambda^2-l^2)}}\left(lx-\frac{\lambda t^\alpha}{\Gamma(1+\alpha)}\right)+C\right), \quad (43a)$$

and

$$u_{3,4}(x,t)=\pm\sqrt{\frac{\beta}{\gamma}}\cot\left(\sqrt{\frac{\beta}{2(\lambda^2-l^2)}}\left(lx-\frac{\lambda t^\alpha}{\Gamma(1+\alpha)}\right)+C\right), \quad (43b)$$

where C, l, and λ are arbitrary constants.

Case B. When $\frac{\beta}{\lambda^2-l^2}<0$, substituting Eqs. (42a)-(42c) and (37) into Eqs. (13) and (14), we get exact traveling wave solutions of Eq. (35),

$$u_{5,6}(x,t)=\pm\sqrt{-\frac{\beta}{\gamma}}\tanh\left(\sqrt{-\frac{\beta}{2(\lambda^2-l^2)}}\left(lx-\frac{\lambda t^\alpha}{\Gamma(1+\alpha)}\right)+C\right), \quad (44a)$$

and

$$u_{7,8}(x,t) = \pm\sqrt{-\frac{\beta}{\gamma}}\coth\left(\sqrt{-\frac{\beta}{2(\lambda^2 - l^2)}}\left(lx - \frac{\lambda t^\alpha}{\Gamma(1+\alpha)}\right) + C\right), \tag{44b}$$

where C, l, and λ are arbitrary constants.

Applying Eq. (16) to $u_j(x,t)$ $(j = 1,2,\ldots,8)$, we can get an infinite sequence of solutions of Eq. (35). For example, by applying Eq. (16) to $u_j(x,t)$ $(j = 1,2,\ldots,8)$ once, we get new solutions of Eq. (35),

$$u^*_{1,2}(x,t) = \frac{-\frac{\beta}{\gamma} \pm A_3\sqrt{\frac{\beta}{\gamma}}\tan(\sqrt{\frac{\beta}{2(\lambda^2-l^2)}}(lx - \frac{\lambda t^\alpha}{\Gamma(1+\alpha)}) + C)}{A_3 \pm \sqrt{\frac{\beta}{\gamma}}\tan(\sqrt{\frac{\beta}{2(\lambda^2-l^2)}}(lx - \frac{\lambda t^\alpha}{\Gamma(1+\alpha)}) + C)},$$

$$u^*_{3,4}(x,t) = \frac{-\frac{\beta}{\gamma} \pm A_3\sqrt{\frac{\beta}{\gamma}}\cot(\sqrt{\frac{\beta}{2(\lambda^2-l^2)}}(lx - \frac{\lambda t^\alpha}{\Gamma(1+\alpha)}) + C)}{A_3 \pm \sqrt{\frac{\beta}{\gamma}}\cot(\sqrt{\frac{\beta}{2(\lambda^2-l^2)}}(lx - \frac{\lambda t^\alpha}{\Gamma(1+\alpha)}) + C)},$$

$$u^*_{5,6}(x,t) = \frac{-\frac{\beta}{\gamma} \pm A_3\sqrt{-\frac{\beta}{\gamma}}\tanh(\sqrt{-\frac{\beta}{2(\lambda^2-l^2)}}(lx - \frac{\lambda t^\alpha}{\Gamma(1+\alpha)}) + C)}{A_3 \pm \sqrt{-\frac{\beta}{\gamma}}\tanh(\sqrt{-\frac{\beta}{2(\lambda^2-l^2)}}(lx - \frac{\lambda t^\alpha}{\Gamma(1+\alpha)}) + C)},$$

$$u^*_{7,8}(x,t) = \frac{-\frac{\beta}{\gamma} \pm A_3\sqrt{-\frac{\beta}{\gamma}}\coth(\sqrt{-\frac{\beta}{2(\lambda^2-l^2)}}(lx - \frac{\lambda t^\alpha}{\Gamma(1+\alpha)}) + C)}{A_3 \pm \sqrt{-\frac{\beta}{\gamma}}\coth(\sqrt{-\frac{\beta}{2(\lambda^2-l^2)}}(lx - \frac{\lambda t^\alpha}{\Gamma(1+\alpha)}) + C)},$$

where A_1, A_2, C, l, and λ are arbitrary real constants.

6 Application to the generalized Ostrovsky equation

The generalized Ostrovsky equation reads

$$\left(u_t + 3uu_x - \frac{\beta}{4}u_{xxx}\right)_x - \frac{\varepsilon^2}{2}\left(u + \delta u^2\right) = 0, \tag{45}$$

where β, ε, and δ are known constants.

The generalized Ostrovsky equation is a model for the weakly nonlinear surface and internal waves in a rotating ocean. Exact peaked wave solutions were obtained by the undetermined coefficient method [32].

In this section, exact peaked wave solutions of the generalized Ostrovsky equation are obtained by using the Riccati-Bernoulli sub-ODE method.

Using the transformation

$$u(x,t) = u(\xi), \tag{46}$$

with

$$\xi = k(x + Vt), \tag{47}$$

where k and V are the wave number and wave speed, respectively, Eq. (45) becomes

$$k^2Vu'' + 3k^2\left(u'\right)^2 + 3k^2uu'' - \frac{\beta}{4}k^4u'''' - \frac{\varepsilon^2}{2}\left(u + \delta u^2\right) = 0. \tag{48}$$

Suppose that the solution of Eq. (48) is the solution of Eq. (5). From Eqs. (5) and (6), we get

$$\begin{aligned} u'''' &= \left(ab(3-m)(2-m)(1-m)u^{-m} + a^2(2-m)(3-2m)(2-2m)\right. \\ &\times u^{1-2m} + m(2m-1)(2m-2)c^2u^{2m-3} + bcm(m+1)(m-1)u^{m-2}\big) \\ &\times (u')^2 + \left(ab(3-m)(2-m)u^{1-m} + a^2(2-m)(3-2m)u^{2-2m}\right. \\ &+ m(2m-1)c^2u^{2m-2} + bcm(m+1)u^{m-1} + (2ac+b^2)\big)u''. \end{aligned} \tag{49}$$

Substituting Eqs. (5), (6), and (49) into Eq. (48), we get

$$\begin{aligned} &(k^2V + 3k^2u)\left(ab(3-m)u^{2-m} + a^2(2-m)u^{3-2m}\right. \\ &\quad + mc^2u^{2m-1} + bc(m+1)u^m + (2ac+b^2)u\big) + \frac{\beta}{4}k^4\Sigma = 0, \end{aligned} \tag{50}$$

where

$$\begin{aligned} \Sigma &= (au^{2-m} + bu + cu^m)^2\left(ab(3-m)(2-m)(1-m)u^{-m} + a^2(2-m)(3-2m)\right. \\ &\times (2-2m)u^{1-2m} + m(2m-1)(2m-2)c^2u^{2m-3} + bcm(m+1)(m-1)u^{m-2}\big) \\ &+ \left(ab(3-m)u^{2-m} + a^2(2-m)u^{3-2m} + mc^2u^{2m-1} + bc(m+1)u^m + (2ac+b^2)u\right) \\ &\times \left(ab(3-m)(2-m)u^{1-m} + a^2(2-m)(3-2m)u^{2-2m} + m(2m-1)c^2u^{2m-2}\right. \\ &+ bcm(m+1)u^{m-1} + (2ac+b^2)\big). \end{aligned}$$

Setting $m = 2$ and $c = 0$, Eq. (50) is reduced to

$$\begin{aligned} &\left(k^2Vab + 3k^2a^2 - \frac{\beta}{4}k^4ab^3\right) + \left(k^2Vb^2 - \frac{\beta}{4}k^4b^4 + 9k^2ab - \frac{1}{2}\varepsilon^2\right)u \\ &\quad + \left(6k^2b^2 - \frac{1}{2}\varepsilon^2\delta\right)u^2 = 0. \end{aligned} \tag{51}$$

Setting each coefficient of u^j $(j = 0, 1, 2)$ to zero, we get

$$6k^2b^2 - \frac{1}{2}\varepsilon^2\delta = 0, \tag{52a}$$

$$k^2Vb^2 - \frac{\beta}{4}k^4b^4 + 9k^2ab - \frac{1}{2}\varepsilon^2 = 0, \tag{52b}$$

$$k^2Vab + 3k^2a^2 - \frac{\beta}{4}k^4ab^3 = 0. \tag{52c}$$

Solving Eqs. (52a)-(52c), we get

$$-\frac{a}{b} = -\frac{1}{\delta}, \tag{53a}$$

$$kb = \pm\sqrt{\frac{\delta\varepsilon^2}{12}}, \tag{53b}$$

$$V = \frac{\beta\delta^2\varepsilon^2 - 144}{48\delta}. \tag{53c}$$

Substituting Eqs. (53a)-(53c) and (47) into Eq. (10), we get exact peaked wave solutions of Eq. (45),

$$u_{1,2}(x,t) = -\frac{1}{\delta} + Ce^{\pm\sqrt{\frac{\delta\varepsilon^2}{12}}|x+(\frac{\beta\delta^2\varepsilon^2-144}{48\delta})t|}, \tag{54}$$

where C is an arbitrary constant.

Similar to Sections 4 and 5, by using a Bäcklund transformation, we can get an infinite sequence of solutions of the generalized Ostrovsky equation. It being a similar process, we omit it.

7 Application to the generalized ZK-Burgers equation

The generalized ZK-Burgers equation [33] reads

$$u_t + \alpha u^\lambda u_x + \beta u_{xxx} + \gamma(u_{yy} + u_{zz}) + \sigma u_{xx} = 0, \tag{55}$$

where α, β, γ, σ, and λ are known constants.

The generalized ZK-Burgers equation retains the strong nonlinear aspects of the governing equation in many practical transport problems such as nonlinear waves in a medium with low-frequency pumping or absorption, transport and dispersion of pollutants in rivers, and sediment transport. Wang *et al.* obtained a solitary wave of the generalized ZK-Burgers equation with a positive fractional power term by using the HB method and with the aid of sub-ODEs [33].

In this section, exact traveling wave solutions of the generalized ZK-Burgers equation are obtained by using the Riccati-Bernoulli sub-ODE method.

Using the transformation

$$u(x,y,z,t) = u(\xi), \tag{56}$$

with

$$\xi = k(x + ly + nz + Vt), \tag{57}$$

where k, l, n, and V are constants to be determined later, Eq. (55) becomes

$$kVu' + k\alpha u^\lambda u' + k^3\beta u''' + k^3\gamma(l^2 + n^2)u''' + k^2\sigma u'' = 0. \tag{58}$$

Suppose that the solution of Eq. (55) is the solution of Eq. (5). Noticing $u' \neq 0$ and $k \neq 0$, otherwise we can only get trivial solution. Substituting Eqs. (5), (6), and (7) into Eq. (58), we get

$$\begin{aligned}
&\rho k^2\big(ab(3-m)(2-m)u^{1-m} + a^2(2-m)(3-2m)u^{2-2m} \\
&\quad + m(2m-1)c^2u^{2m-2} + bcm(m+1)u^{m-1} + (2ac+b^2)\big) \\
&\quad + \sigma k\big(a(2-m)u^{1-m} + mcu^{m-1} + b\big) + V + \alpha u^\lambda = 0,
\end{aligned} \tag{59}$$

where

$$\rho = \beta + \gamma\left(l^2 + n^2\right). \tag{60}$$

Setting $m = 1 - \frac{\lambda}{2}$ and $c = 0$, Eq. (59) is reduced to

$$\left(V + \rho k^2 b^2 + \sigma k b\right) + \left(\frac{(\lambda+2)(\lambda+4)\rho k^2 ab}{4} + \frac{(\lambda+2)\sigma ka}{2}\right)u^{\frac{\lambda}{2}} + \left(\alpha + \frac{(\lambda+2)(\lambda+1)\rho k^2 a^2}{2}\right)u^{\lambda} = 0. \tag{61}$$

Setting each coefficient of u^j $(j = 0, \frac{\lambda}{2}, \lambda)$ to zero, we get

$$V + \rho k^2 b^2 + \sigma k b = 0, \tag{62a}$$

$$\frac{(\lambda+2)ka}{2}\left(\frac{(\lambda+4)\rho kb}{2} + \sigma\right) = 0, \tag{62b}$$

$$\alpha + \frac{(\lambda+2)(\lambda+1)\rho k^2 a^2}{2} = 0. \tag{62c}$$

Solving Eqs. (62a)-(62c), we get

$$b = \frac{-2\sigma}{k\rho(\lambda+4)}, \tag{63a}$$

$$a = \pm\frac{1}{k}\sqrt{\frac{-2\alpha}{\rho(\lambda^2+3\lambda+2)}}, \tag{63b}$$

$$V = \frac{2\sigma^2(\lambda+2)}{\rho(\lambda+4)^2}. \tag{63c}$$

Substituting Eqs. (63a)-(63c) and (57) into Eq. (10), we get exact traveling wave solutions of Eq. (55),

$$u_{1,2}(x,t) = \left(\pm\frac{\rho(\lambda+4)}{\sigma}\sqrt{\frac{-\alpha}{2\rho(\lambda^2+3\lambda+2)}} + Ce^{\frac{\lambda\sigma}{\rho(\lambda+4)}\left(x+ly+nz+\left(\frac{2\sigma^2(\lambda+2)}{\rho(\lambda+4)^2}\right)t\right)}\right)^{\frac{-2}{\lambda}}, \tag{64}$$

where C, l, and n are arbitrary constants.

Equation (64) is new type of traveling wave solution of the generalized ZK-Burgers equation. Especially, if we choose $C = C_1 = \frac{\rho(\lambda+4)}{\sigma}\sqrt{\frac{-\alpha}{2\rho(\lambda^2+3\lambda+2)}}$, we get the solitary wave solutions of Eq. (55),

$$u_3(x,t) = \left(\frac{1}{2C_1}\left(1 - \tanh\frac{\eta}{2}\right)\right)^{\frac{2}{\lambda}}, \tag{65}$$

$$u_4(x,t) = \left(\frac{1}{2C_1}\left(-1 + \coth\frac{\eta}{2}\right)\right)^{\frac{2}{\lambda}}, \tag{66}$$

where l, n are arbitrary constants and

$$\eta = \frac{\lambda\sigma}{\rho(\lambda+4)}\left(x + ly + nz + \left(\frac{2\sigma^2(\lambda+2)}{\rho(\lambda+4)^2}\right)t\right). \tag{67}$$

If we choose $C = C_2 = -\frac{\rho(\lambda+4)}{2\sigma}\sqrt{\frac{-2\alpha}{\rho(\lambda^2+3\lambda+2)}}$, we get the solitary wave solutions of Eq. (55),

$$u_5(x,t) = \left(\frac{1}{2C_2}\left(-1 + \coth\frac{\eta}{2}\right)\right)^{\frac{2}{\lambda}}, \tag{68}$$

$$u_6(x,t) = \left(\frac{1}{2C_2}\left(1 - \tanh\frac{\eta}{2}\right)\right)^{\frac{2}{\lambda}}, \tag{69}$$

where l and n are arbitrary constants.

Similar to Sections 4 and 5, by using a Bäcklund transformation, we can get an infinite sequence of solutions of the generalized ZK-Burgers equation. It being a similar process, we omit it.

8 Comparisons and explanations of the solutions

In this section, the physical interpretation of the results of Sections 4-7 are given, respectively. We will compare the Riccati-Bernoulli sub-ODE method with the $(\frac{G'}{G})$-expansion method, the first integral method, and so on. Some of our obtained exact solutions are in the figures represented with the aid of Maple software.

(1) The Eckhaus equation:

Applying the Riccati-Bernoulli sub-ODE method, Eqs. (25), (26), (28), (31), and (34) are new types of exact traveling wave solutions of the Eckhaus equation. Equations (29), (30), (32), and (33), which are expressed by the hyperbolic functions, are a kind of kink-type envelope solitary solutions. They could not be obtained by the method presented in Ref. [27]. Equation (26), which is expressed by the rational functions, could not be obtained by the $(\frac{G'}{G})$-expansion method [8] and the first integral method [5].

(2) The nonlinear fractional Klein-Gordon equation:

Applying the Riccati-Bernoulli sub-ODE method and comparing our results with Golmankhaneh's results [29], it is easy to find that $u_j(x,t)$ $(j = 1,\dots,8)$ are new and identical to results by the first integral method [6]. $u_j(x,t)$ $(j = 1,2,3,4)$, which are expressed by the trigonometric functions, are periodic wave solutions. $u_j(x,t)$ $(j = 5,6,7,8)$, which are expressed by the hyperbolic functions, are a kind of kink-type envelope solitary solutions. The shape of $u = u_1(x,t)$ is represented in Figure 1 with $\alpha = \frac{1}{90}$, $\beta = 1$, $\gamma = 1$, $\lambda = \frac{9}{5}$, $C = 0$ and $l = \frac{3}{2}$ within the interval $-5 \le x \le 5$ and $0 \le t \le \frac{1}{2}$. The shape of $u = u_5(x,t)$ is represented in Figure 2 with $\alpha = \frac{1}{5}$, $\beta = -1$, $\gamma = 1$, $\lambda = 2$, $C = 0$, and $l = 1$ within the interval $-6 \le x \le 6$ and $0 \le t \le 6$.

(3) The generalized Ostrovsky equation:

Applying the Riccati-Bernoulli sub-ODE method, it is easy to find that our results are identical to results presented in Ref. [32]. $u = u_{1,2}(x,t)$ are peaked wave solutions of the generalized Ostrovsky equation. The shape of $u = u_1(x,t)$ is represented in Figure 3 with $\delta = 6$, $\beta = 6$, $\varepsilon = 1$, $\lambda = 2$, and $C = \frac{1}{10}$ within the interval $-5 \le x,t \le 5$.

(4) The generalized ZK-Burgers equation:

By applying the Riccati-Bernoulli sub-ODE method to the generalized ZK-Burgers equation, we find that if λ is a positive fraction, our results degenerate to the results of Ref. [33]. Moreover, we enlarge the value range of parameters λ of the generalized ZK-Burgers equation so that the parameter λ can be an arbitrary constant $(\lambda \neq -1,-2,-4)$. $u_j(x,t)$ $(j = 1,\dots,6)$ are exact traveling wave solutions of the generalized ZK-Burgers equation.

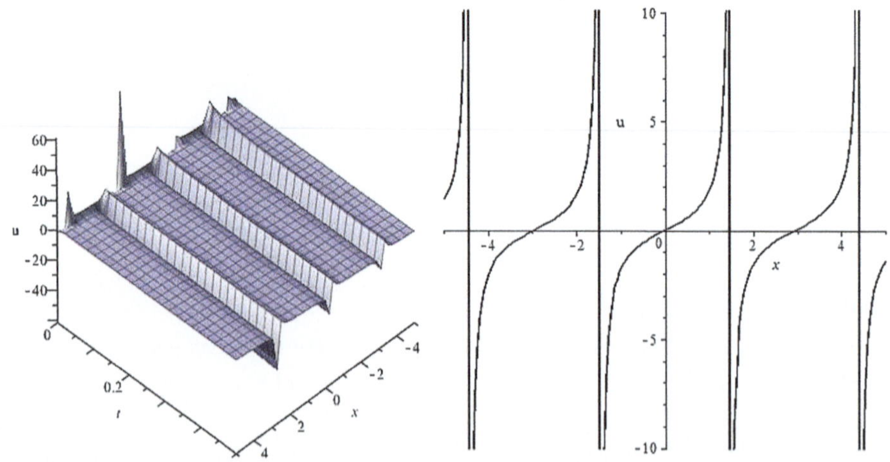

Figure 1 Graph of solution $u = u_1(x,t)$ of the nonlinear fractional Klein-Gordon equation for $\alpha = \frac{1}{90}$, $\beta = 1, \gamma = 1, \lambda = \frac{9}{5}, C = 0$, and $l = \frac{3}{2}$. The left figure shows the 3-D plot and the right figure shows the 2-D plot for $t = 0$.

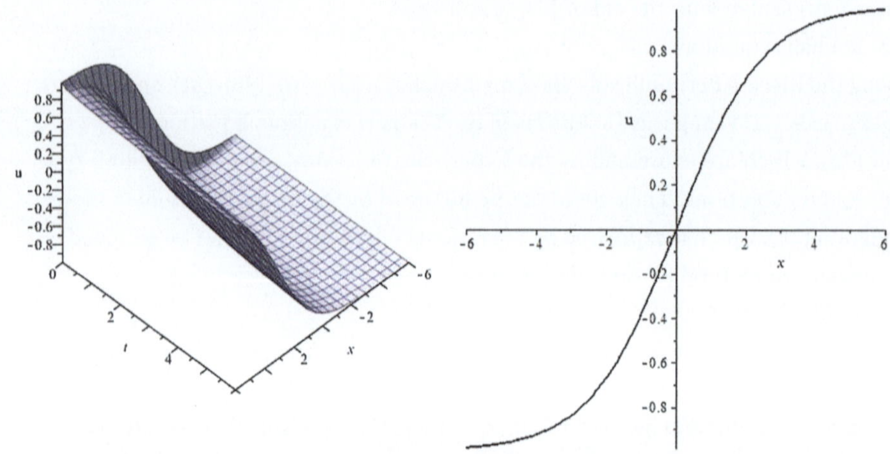

Figure 2 Graph of solution $u = u_5(x,t)$ of the nonlinear fractional Klein-Gordon equation for $\alpha = \frac{1}{5}$, $\beta = -1, \gamma = 1, \lambda = 2, C = 0$, and $l = 1$. The left figure shows the 3-D plot and the right figure shows the 2-D plot for $t = 0$.

$u_j(x,t)$ ($j = 3,4,5,6$), which are expressed by the hyperbolic functions, are a kind of kink-type envelope solitary solutions. The shape of $u = u_1(x,t)$ is represented in Figure 4 with $\alpha = \beta = \gamma = l = n = y = z = 1$, $\lambda = -\sqrt{2}$ and $\sigma = 2$ within the interval $-5 \leq x, t \leq 5$.

Moreover, by using a Bäcklund transformation, we can get an infinite sequence of solutions of these NLPDEs which cannot be obtained by the $(\frac{G'}{G})$-expansion method and the first integral method. The graphical demonstrations of some obtained solutions are shown in Figures 1-4.

9 Conclusions

The Riccati-Bernoulli sub-ODE method is successfully used to establish exact traveling wave solutions, solitary wave solutions and peaked wave solutions of NLPDEs. A Bäcklund transformation of the Riccati-Bernoulli equation is given. By applying a Bäcklund transfor-

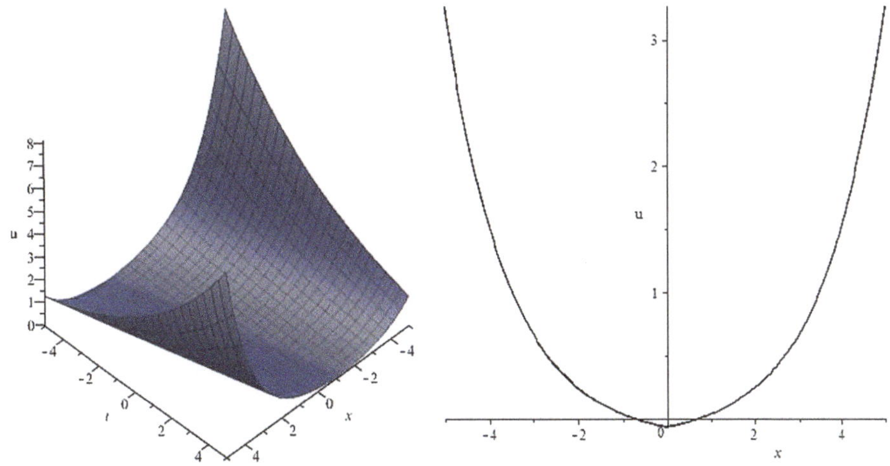

Figure 3 Graph of solution $u = u_5(x,t)$ of the generalized Ostrovsky equation for $\delta = 6$, $\beta = 6$, $\varepsilon = 1$, $\lambda = 2$, and $C = \frac{1}{10}$. The left figure shows the 3-D plot and the right figure shows the 2-D plot for $t = 0$.

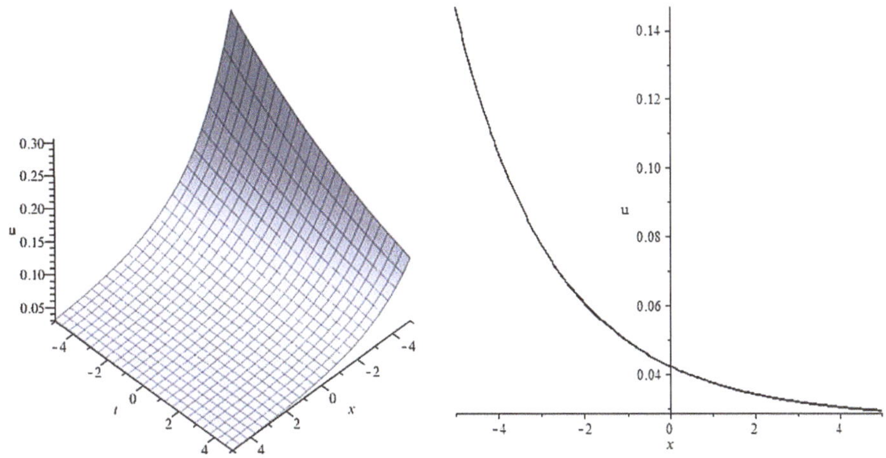

Figure 4 Graph of solution $u = u_5(x,t)$ of the generalized ZK-Burgers equation for $\alpha = \beta = \gamma = l = n = y = z = 1$, $\lambda = -\sqrt{2}$, and $\sigma = 2$. The left figure shows the 3-D plot and the right figure shows the 2-D plot for $t = 3$.

mation of the Riccati-Bernoulli equation to the NLPDEs, an infinite sequence of solutions of the NLPDEs is obtained. The Eckhaus equation, the nonlinear fractional Klein-Gordon equation, the generalized Ostrovsky equation, and the generalized ZK-Burgers equation are chosen to illustrate the validity of the Riccati-Bernoulli sub-ODE method. Many well-known NLPDEs can be handled by this method. The performance of this method is found to be simple and efficient. The availability of computer systems like Maple facilitates the tedious algebraic calculations. The Riccati-Bernoulli sub-ODE method is also a standard and computable method, which allows us to perform complicated and tedious algebraic calculations.

It is well known that it is difficult to propose an uniform analytical method for all types of the NLPDEs, and the Riccati-Bernoulli sub-ODE method is no exception. Similar to the first integral method, the $(\frac{G'}{G})$-expansion method and the homogeneous balance method, the Riccati-Bernoulli sub-ODE method is used to obtain exact solutions of the form of Eq. (1). Constructing more powerful sub-ODE and Bäcklund transformations is future work and aims to search for exact solutions of NLPDEs

Competing interests

The authors declare that they have no competing interests.

Authors' contributions

All authors completed the paper together. All authors read and approved the final manuscript.

Author details

[1]Department of Applied Mathematics, Northwestern Polytechnical University, Xi'an, 710072, P.R. China. [2]School of Mechanics, Civil Engineering and Architecture, Northwestern Polytechnical University, Xi'an, 710072, P.R. China. [3]State Key Laboratory of Structural Analysis of Industrial Equipment, Dalian University of Technology, Dalian, 116023, P.R. China.

Acknowledgements

The research is supported by the National Natural Science Foundation of China (11372252, 11372253 and 11432010) and the Fundamental Research Funds for the Central Universities (3102014JCQ01035).

References

1. Ablowitz, MJ, Clarkson, PA: Solitons, Nonlinear Evolution Equations and Inverse Scattering. Cambridge University Press, New York (1991)
2. Rogers, C, Schief, WK: Bäcklund and Darboux Transformation Geometry and Modern Applications in Solitons Theory. Cambridge University Press, Cambridge (2002)
3. He, JH: An approximate solution technique depending on an artificial parameter: a special example. Commun. Nonlinear Sci. Numer. Simul. **3**, 92-97 (1998)
4. Feng, ZS: The first-integral method to study the Burgers-Korteweg-de Vries equation. J. Phys. A, Math. Gen. **35**, 343-349 (2002)
5. Taghizadeh, N, Mirzazadeh, M, Filiz, T: The first-integral method applied to the Eckhaus equation. Appl. Math. Lett. **25**, 798-802 (2012)
6. Lu, B: The first integral method for some time fractional differential equations. J. Math. Anal. Appl. **395**, 684-693 (2012)
7. Wang, ML, Li, XZ, Zhang, JL: The $(\frac{G'}{G})$-expansion method and travelling wave solutions of nonlinear evolution equations in mathematical physics. Phys. Lett. A **372**, 417-423 (2008)
8. Zhang, H: New application of the $(\frac{G'}{G})$-expansion method. Commun. Nonlinear Sci. Numer. Simul. **14**, 3220-3225 (2009)
9. Khan, K, Akbar, MA: Study of analytical method to seek for exact solutions of variant Boussinesq equations. SpringerPlus **3**, 324-340 (2014)
10. Xu, SL, Liang, JC: Exact soliton solutions to a generalized nonlinear Schrödinger equation. Commun. Theor. Phys. **53**, 159-165 (2010)
11. Wang, ML: Applications of *F*-expansion to periodic wave solutions for a new Hamiltonian amplitude equation. Chaos Solitons Fractals **24**, 1257-1268 (2005)
12. Hirota, R: Exact solution of the Korteweg-de Vries equation for multiple collisions of solitons. Phys. Rev. Lett. **27**, 1192-1194 (1971)
13. Wang, ML: Solitary wave solutions for variant Boussinesq equations. Phys. Lett. A **199**, 169-172 (1995)
14. Wang, ML, Zhou, YB: Application of a homogeneous balance method to exact solutions of nonlinear equations in mathematical physics. Phys. Lett. A **216**, 67-75 (1996)
15. Bai, CL: Extended homogeneous balance method and Lax pairs, Bäcklund transformation. Commun. Theor. Phys. **37**, 645-648 (2002)
16. He, JH: An new approach to nonlinear partial differential equations. Commun. Nonlinear Sci. Numer. Simul. **2**, 230-235 (1997)
17. Yusufoglu, E: The variational iteration method for studying the Klein-Gordon equation. Appl. Math. Lett. **21**, 669-674 (2008)
18. Wazwaz, AM: The tanh method: exact solutions of the sine-Gordon and the sinh-Gordon equations. Appl. Math. Comput. **167**, 1196-1210 (2005)
19. Yan, ZL: Abunbant families of Jacobi elliptic function solutions of the-dimensional integrable Davey-Stewartson-type equation via a new method. Chaos Solitons Fractals **18**, 299-309 (2003)
20. Khan, K, Akbar, MA, Rayhanul Islam, SM: Exact solutions for (1 + 1)-dimensional nonlinear dispersive modified Benjamin-Bona-Mahony equation and coupled Klein-Gordon equations. SpringerPlus **3**, 724-731 (2014)
21. Khan, K, Akbar, MA: Solitary wave solutions of some coupled nonlinear evolution equations. J. Sci. Res. **6**, 273-284 (2014)
22. Khan, K, Akbar, MA: Traveling wave solutions of the (2 + 1)-dimensional Zoomeron equation and the Burgers equations via the MSE method and the Exp-function method. Ain Shams Eng. J. **5**, 247-256 (2014)
23. Ahmed, MT, Khan, K, Akbar, MA: Study of nonlinear evolution equations to construct traveling wave solutions via modified simple equation method. Phys. Rev. Res. Int. **3**, 490-503 (2013)
24. Khan, K, Akbar, MA: The $\exp(-\Phi(\xi))$-expansion method for finding travelling wave solutions of Vakhnenko-Parkes equation. Int. J. Dyn. Syst. Differ. Equ. **5**, 72-83 (2014)

25. Zerarka, A, Ouamane, S, Attaf, A: Construction of exact solutions to a family of wave equations by the functional variable method. Waves Random Complex Media **21**, 44-56 (2011)
26. Calogero, F, Eckhaus, W: Nonlinear evolution equations, rescalings, model PDEs and their integrability: I. Inverse Probl. **3**, 229-262 (1987)
27. Calogero, F, Lillo, SD: The Eckhaus PDE $i\psi_t + \psi_{xx} + 2(|\psi|^2)_x\psi + |\psi|^4\psi = 0$. Inverse Probl. **4**, 633-682 (1987)
28. Calogero, F: The evolution partial differential equation $u_t = u_{xxx} + 3(u_{xx}u^2 + 3u_x^2 u) + 3u_x u^4$. J. Math. Phys. **28**, 538-555 (1987)
29. Golmankhaneh, AK, Baleanu, D: On nonlinear fractional Klein-Gordon equation. Signal Process. **91**, 446-451 (2011)
30. El-Sayed, S: The decomposition method for studying the Klein-Gordon equation. Chaos Solitons Fractals **18**, 1025-1030 (2003)
31. Odibat, Z, Momani, SA: Numerical solution of sine-Gordon equation by variational iteration method. Phys. Lett. A **370**, 437-440 (2007)
32. Lu, Y: A simple method for solving nonlinear wave equations for their peaked soliton solutions and its applications. Acta Phys. Sin. **58**, 7452-7456 (2009)
33. Wang, ML, Li, LX, Li, EQ: Exact solitary wave solutions of nonlinear evolutions with a positive fractional power term. Commun. Theor. Phys. **61**, 7-14 (2014)

2

On properties of meromorphic solutions for difference Painlevé equations

Chang-Wen Peng[1] and Zong-Xuan Chen[2*]

*Correspondence:
chzx@vip.sina.com
[2]School of Mathematical Sciences, South China Normal University, Guangzhou, Guangdong 510631, China
Full list of author information is available at the end of the article

Abstract

In this paper, we mainly investigate properties of finite order transcendental meromorphic solutions of difference Painlevé equations. If f is a finite order transcendental meromorphic solution of difference Painlevé equations, then we get some estimates of the order and the exponent of convergence of poles of $\Delta f(z)$, where $\Delta f(z) = f(z+1) - f(z)$.

MSC: 30D35; 39B12

Keywords: difference Painlevé equation; transcendental meromorphic solution; fixed points

1 Introduction and main results

Let f be a function transcendental and meromorphic in the plane. The forward difference is defined in the standard way by $\Delta f(z) = f(z+1) - f(z)$. In what follows, we assume that the reader is familiar with the basic notions of Nevanlinna value distribution theory (see [1–3]). In addition, we use the notations $\sigma(f)$ to denote the order of growth of the meromorphic function $f(z)$, and $\lambda(f)$ and $\lambda(\frac{1}{f})$ to denote, respectively, the exponents of convergence of zeros and poles of $f(z)$.

Painlevé and his colleagues [4] classified all equations of the Painlevé type of the form

$$\frac{d^2y}{dz^2} = F\left(z, y, \frac{dy}{dz}\right),$$

where F is rational in y and $\frac{dy}{dz}$ and (locally) analytic in z. The first two of these are P_{I} and P_{II}:

$$\frac{d^2y}{dz^2} = 6y^2 + z, \qquad \frac{d^2y}{dz^2} = 2y^2 + zy + \alpha,$$

where α is a constant. The differential Painlevé equations, discovered at the beginning of the last century, have been an important research subject in the field of mathematics and physics.

In the past 18 years, the discrete Painlevé equations became important research problems (see [5]). For example, discrete Painlevé I equations

$$y_{n+1} + y_{n-1} = \frac{\alpha n + \beta}{y_n} + \gamma,$$

$$y_{n+1} + y_{n-1} = \frac{\alpha n + \beta}{y_n} + \frac{\gamma}{y_n^2},$$

and discrete Painlevé II equation

$$y_{n+1} + y_{n-1} = \frac{(\alpha n + \beta)y_n + \gamma}{1 - y_n^2},$$

where α, β and γ are constants, $n \in N$.

Some results on the existence of meromorphic solutions for certain difference equations were obtained by Shimomura [6] and Yanagihara [7] 30 years ago.

Recently, a number of papers (see [8–20]) focused on complex difference equations and difference analogues of Nevanlinna theory. As the difference analogues of Nevanlinna theory are being investigated, many results on the complex difference equations are rapidly obtained.

Ablowitz *et al.* [8] looked at a difference equation of the type

$$y(z+1) + y(z-1) = R(z, y), \tag{1}$$

where R is rational in both of its arguments, and showed the following theorem.

Theorem A (see [8]) *If the second-order difference equation*

$$y(z+1) + y(z-1) = \frac{a_0(z) + a_1(z)y + \cdots + a_p(z)y^p}{b_0(z) + b_1(z)y + \cdots + b_q(z)y^q},$$

where a_i and b_i are polynomials, admits a non-rational meromorphic solution of finite order, then $\max\{p, q\} \leq 2$.

Halburd and Korhonen [15–17] used value distribution theory and a reasoning related to the singularity confinement to single out the difference Painlevé I and II equations from difference equation (1). They obtained that if (1) has a finite order admissible meromorphic solution $f(z)$, then either f satisfies a difference Riccati equation, or (1) can be transformed by a linear change in f to some classical difference equations, which include difference Painlevé I equations

$$f(z+1) + f(z-1) = \frac{az + b}{f(z)} + c, \tag{2}$$

$$f(z+1) + f(z-1) = \frac{az + b}{f(z)} + \frac{c}{f(z)^2}, \tag{3}$$

$$f(z+1) + f(z) + f(z-1) = \frac{az + b}{f(z)} + c, \tag{4}$$

and difference Painlevé II equation

$$f(z+1) + f(z-1) = \frac{(az + b)f(z) + c}{1 - f(z)^2}, \tag{5}$$

where a, b and c are constants.

From above, we see that difference Painlevé I and II equations are the development of the differential and discrete Painlevé I and II equations. So they are an important class of difference equations.

Chen and Chen [10] investigated some properties of meromorphic solutions of difference Painlevé I equation and proved the following Theorem B.

Theorem B (see [10]) *Let a, b, c be constants such that* $|a| + |b| \neq 0$. *If* $f(z)$ *is a finite order transcendental meromorphic solution of the difference Painlevé* I *equation* (4), *then*

(i) $\lambda(\frac{1}{f}) = \lambda(f) = \sigma(f)$;

(ii) *if* $p(z)$ *is a non-constant polynomial, then* $f(z) - p(z)$ *has infinitely many zeros and* $\lambda(f - p) = \sigma(f)$;

(iii) *if* $a \neq 0$, *then* $f(z)$ *has no Borel exceptional value;*
if $a = 0$, *then Borel exceptional values of* $f(z)$ *can only come from the set* $E = \{z \mid 3z^2 - cz - b = 0\}$.

The main aims of this paper are to consider the properties of finite order transcendental meromorphic solutions of difference Painlevé I and II equations (2)-(5), and we obtain the following results.

Theorem 1.1 *Let a, b, c be constants with* $|a| + |b| \neq 0$. *Suppose that* $f(z)$ *is a finite order transcendental meromorphic solution of the difference Painlevé* I *equation* (2). *Then*

(i) *if* $a \neq 0$ *and* $p(z)$ *is a polynomial, then* $f(z) - p(z)$ *has infinitely many zeros and* $\lambda(f - p) = \sigma(f)$;
if $a = 0$, *then Borel exceptional values of* $f(z)$ *can only come from the set* $E = \{z \mid 2z^2 - cz - b = 0\}$;

(ii) $\lambda(\frac{1}{f}) = \lambda(\frac{1}{\Delta f}) = \sigma(\Delta f) = \sigma(f)$.

Theorem 1.2 *Let a, b, c be constants with* $|a| + |b| + |c| \neq 0$. *Suppose that* $f(z)$ *is a finite order transcendental meromorphic solution of the difference Painlevé* II *equation* (5). *Then*

(i) *if* $a \neq 0$ *and* $p(z)$ *is a nonzero polynomial, then* $f(z) - p(z)$ *has infinitely many zeros and* $\lambda(f - p) = \sigma(f)$;
if $a = 0$, *then Borel exceptional values of* $f(z)$ *can only come from the set* $E = \{z \mid 2z^3 + (b - 2)z + c = 0\}$;
if $c \neq 0$, *then* $\lambda(f) = \sigma(f)$;

(ii) $\lambda(\frac{1}{f}) = \lambda(\frac{1}{\Delta f}) = \sigma(\Delta f) = \sigma(f)$.

Remark 1.1 By Theorem 1.2, we conclude that if $ac \neq 0$ and $f(z)$ is a finite order transcendental meromorphic solution of the difference Painlevé II equation (5), then $f(z)$ has no Borel exceptional value.

Theorem 1.3 *Let a, b, c be constants with* $|a| + |b| + |c| \neq 0$. *Suppose that* $f(z)$ *is a finite order transcendental meromorphic solution of the difference Painlevé* I *equation* (3). *Then*

(i) $\lambda(\frac{1}{f}) = \lambda(\frac{1}{\Delta f}) = \sigma(\Delta f) = \sigma(f)$;

(ii) *if* $a = 0$, *then Borel exceptional values of* $f(z)$ *can only come from the set* $E = \{z \mid 2z^3 - bz - c = 0\}$.

Theorem 1.4 *Let a, b, c be constants with $|a| + |b| \neq 0$. Suppose that $f(z)$ is a finite order transcendental meromorphic solution of the difference Painlevé I equation (4). Then $\lambda(\frac{1}{f}) = \lambda(\frac{1}{\Delta f}) = \sigma(\Delta f) = \sigma(f)$.*

Remark 1.2 The following examples show that $\lambda(\Delta f) = \sigma(f)$ may not hold in above several theorems.

Example 1.1 The meromorphic function $f(z) = \tan(\frac{\pi}{2}z)$ satisfies the equation

$$f(z+1) + f(z-1) = \frac{-2}{f(z)},$$

where $a = c = 0$, $b = -2$ satisfying $|a| + |b| = 2 \neq 0$. We obtain that

$$\Delta f = -\cot\left(\frac{\pi}{2}z\right) - \tan\left(\frac{\pi}{2}z\right) = -\frac{2}{\sin(\pi z)}.$$

Thus, $\lambda(\frac{1}{f}) = \lambda(\frac{1}{\Delta f}) = \sigma(\Delta f) = \sigma(f) = 1$ and $\lambda(\Delta f) = 0 \neq \sigma(f)$. The solution $f(z) = \tan(\frac{\pi}{2}z)$ has two Borel exceptional values i and $-i$ satisfying the equation $2z^2 - cz - b = 2z^2 + 2 = 0$.

Example 1.2 The meromorphic function $f(z) = \tan(\frac{\pi}{4}z)$ satisfies the equation

$$f(z+1) + f(z-1) = \frac{4f(z)}{1 - f(z)^2},$$

where $a = c = 0$, $b = 4$ and $|a| + |b| + |c| = 4$. We have

$$\Delta f = \frac{1 + \tan(\frac{\pi}{4}z)}{1 - \tan(\frac{\pi}{4}z)} - \tan\left(\frac{\pi}{4}z\right) = \frac{1 + \tan^2(\frac{\pi}{4}z)}{1 - \tan(\frac{\pi}{4}z)}.$$

Thus, $\lambda(\frac{1}{f}) = \lambda(\frac{1}{\Delta f}) = \sigma(\Delta f) = \sigma(f) = 1$ and $\lambda(\Delta f) = 0 \neq \sigma(f)$. The solution $f(z) = \tan(\frac{\pi}{4}z)$ has two Borel exceptional values $\pm i$ satisfying $2z^3 + (b-2)z + c = 2z^3 + 2z = 0$.

Example 1.3 The meromorphic function $f(z) = \frac{1}{e^{2\pi iz}+z+1}$ satisfies the equation

$$f(z+1) + f(z-1) = \frac{2f(z)}{1 - f(z)^2},$$

where $a = c = 0$, $b = 2$ and $|a| + |b| + |c| = 2$. We have

$$\Delta f = \frac{1}{e^{2\pi iz} + z + 2} - \frac{1}{e^{2\pi iz} + z + 1} = \frac{-1}{(e^{2\pi iz} + z + 2)(e^{2\pi iz} + z + 1)}.$$

Thus, $\lambda(\frac{1}{f}) = \lambda(\frac{1}{\Delta f}) = \sigma(\Delta f) = \sigma(f) = 1$ and $\lambda(\Delta f) = 0$. The solution $f(z) = \frac{1}{e^{2\pi iz}+z+1}$ has a Borel exceptional value 0 satisfying the equation $2z^3 + (b-2)z + c = 2z^3 = 0$.

2 The proof of Theorem 1.1

We need the following lemmas to prove Theorem 1.1.

Lemma 2.1 (see [13, 18]) *Let $w(z)$ be a non-constant finite order meromorphic solution of*

$$P(z,w)=0,$$

where $P(z,w)$ is a difference polynomial in $w(z)$. If $P(z,a)\not\equiv 0$ for a meromorphic function a satisfying $\lim_{r\to\infty}\frac{T(r,a)}{T(r,w)}=0$, then

$$m\left(r,\frac{1}{w-a}\right)=S(r,w)$$

outside of a possible exceptional set of finite logarithmic measure.

Lemma 2.2 (see [11, 16, 17]) *Let f be a non-constant finite order meromorphic function. Then*

$$N(r+1,f)=N(r,f)+S(r,f)$$

outside of a possible exceptional set of finite logarithmic measure.

Remark 2.1 In [12], Chiang and Feng proved that let f be a meromorphic function with exponent of convergence of poles $\lambda(\frac{1}{f})=\lambda<\infty$, $\eta\neq 0$ be fixed, then for each $\varepsilon>0$,

$$N(r,f(z+\eta))=N(r,f)+O(r^{\lambda-1+\varepsilon})+O(\log r).$$

Lemma 2.3 (see [18]) *Let $f(z)$ be a transcendental meromorphic solution of finite order σ of a difference equation of the form*

$$H(z,f)P(z,f)=Q(z,f),$$

where $H(z,f)$ is a difference product of total degree n in $f(z)$ and its shifts, and where $P(z,f)$, $Q(z,f)$ are difference polynomials such that the total degree of $Q(z,f)$ is $\leq n$. Then, for each $\varepsilon>0$,

$$m(r,P(z,f))=O(r^{\sigma-1+\varepsilon})+S(r,f)$$

possibly outside of an exceptional set of finite logarithmic measure.

Lemma 2.4 (Valiron-Mohon'ko, see [2]) *Let $f(z)$ be a meromorphic function. Then, for all irreducible rational functions in f,*

$$R(z,f(z))=\frac{a_0(z)+a_1(z)f(z)+\cdots+a_m(z)f(z)^m}{b_0(z)+b_1(z)f(z)+\cdots+b_n(z)f(z)^n}$$

with meromorphic coefficients $a_i(z)$ $(i=0,1,\ldots,m)$, $b_j(z)$ $(j=0,1,\ldots,n)$, the characteristic function of $R(z,f(z))$ satisfies

$$T(r,R(z,f(z)))=dT(r,f)+O(\Psi(r)),$$

where $d=\deg_f R=\max\{m,n\}$ and $\Psi(r)=\max_{i,j}\{T(r,a_i),T(r,b_j)\}$.

Lemma 2.5 (see [12]) *Let $f(z)$ be a meromorphic function with order $\sigma = \sigma(f) < \infty$, and let η be a fixed nonzero complex number, then for each $\varepsilon > 0$, we have*

$$T\big(r, f(z+\eta)\big) = T(r,f) + O\big(r^{\sigma-1+\varepsilon}\big) + O(\log r).$$

Lemma 2.6 (see [21]) *Let $g : (0,+\infty) \to R$, $h : (0,+\infty) \to R$ be non-decreasing functions. If* (i) *$g(r) \leq h(r)$ outside of an exceptional set of finite linear measure, or* (ii) *$g(r) \leq h(r)$, $r \notin H \cup (0,1]$, where $H \subset (1,\infty)$ is a set of finite logarithmic measure, then for any $\alpha > 1$, there exists $r_0 > 0$ such that $g(r) \leq h(\alpha r)$ for all $r > r_0$.*

Proof of Theorem 1.1 Suppose that $f(z)$ is a transcendental meromorphic solution of finite order $\sigma(f)$ of equation (2).

(i) Let $a \neq 0$. Suppose that $p(z)$ is a polynomial. Set $g(z) = f(z) - p(z)$. Substituting $f(z) = g(z) + p(z)$ into (2), we obtain that

$$g(z+1) + p(z+1) + g(z-1) + p(z-1) = \frac{az+b}{g(z)+p(z)} + c. \tag{6}$$

It follows from (6) that

$$\begin{aligned} P(z,g) := &\big(g(z+1) + p(z+1) + g(z-1) + p(z-1)\big) \\ &\times \big(g(z) + p(z)\big) - (az+b) - c\big(g(z)+p(z)\big) = 0. \end{aligned} \tag{7}$$

By (7), we have

$$P(z,0) = \big(p(z+1) + p(z-1)\big)p(z) - (az+b) - cp(z).$$

If $p(z) \equiv 0$, then $P(z,0) = -(az+b) \not\equiv 0$. If $p(z) \equiv \alpha \in C\backslash\{0\}$, then

$$P(z,0) = 2\alpha^2 - (az+b) - c\alpha \not\equiv 0$$

since $a \neq 0$. Suppose that $p(z)$ is a non-constant polynomial. Set $p(z) = d_k z^k + \cdots + d_0$, where $d_k, \ldots, d_0$ are constants, $d_k \neq 0$ and $k \geq 1$. Then we obtain that

$$\begin{aligned} P(z,0) &= \big(p(z+1) + p(z-1)\big)p(z) - (az+b) - cp(z) \\ &= 2d_k^2 z^{2k} + \cdots \not\equiv 0. \end{aligned}$$

Thus, by Lemma 2.1, we see that

$$m\left(r, \frac{1}{g}\right) = S(r,g)$$

outside of a possible exceptional set of finite logarithmic measure. Thus,

$$N\left(r, \frac{1}{f(z)-p(z)}\right) = N\left(r, \frac{1}{g}\right) = T(r,g) + S(r,g) = T(r,f) + S(r,f) \tag{8}$$

outside of a possible exceptional set of finite logarithmic measure. Hence, by (8) and Lemma 2.6, we have $\lambda(f - p) = \sigma(f)$.

If $a = 0$ and $p(z) = \beta \notin E$, then we have

$$P(z,0) = 2\beta^2 - c\beta - b \not\equiv 0.$$

Using a similar method as above, we obtain $\lambda(f - \beta) = \sigma(f)$. Hence, the Borel exceptional values of $f(z)$ can only come from the set $E = \{z \mid 2z^2 - cz - b = 0\}$.

(ii) Set $z = w + 1$. Substituting $z = w + 1$ into (2), we obtain that

$$f(w+2) + f(w) = \frac{aw + a + b}{f(w+1)} + c.$$

That is,

$$f(w+1)\big(f(w+2) + f(w)\big) = (aw + a + b) + cf(w+1). \tag{9}$$

Substituting $f(w+1) = \Delta f(w) + f(w)$ and $f(w+2) = \Delta f(w+1) + \Delta f(w) + f(w)$ into (9), we have

$$\big(\Delta f(w) + f(w)\big)\big(\Delta f(w+1) + \Delta f(w) + 2f(w)\big) = (aw + a + b) + c\big(\Delta f(w) + f(w)\big).$$

That is,

$$\begin{aligned} -2f(w)^2 = {} & \big(\Delta f(w+1) + 3\Delta f(w) - c\big)f(w) \\ & + \big(\Delta f(w+1) + \Delta f(w) - c\big)\Delta f(w) - (aw + a + b). \end{aligned} \tag{10}$$

Since $N(R, \Delta f(w+1)) \le N(R+1, \Delta f(w)) + o(N(R+1, \Delta f(w)))$, by Lemma 2.2 we see that there is a subset $E_1 \subset (1,\infty)$ of finite logarithmic measure such that for $|w| = R \notin E_1 \cup [0,1]$,

$$N\big(R, \Delta f(w+1)\big) \le N\big(R, \Delta f(w)\big) + S(R,f). \tag{11}$$

By (10) and (11), when $|w| = R \notin E_1 \cup [0,1]$, we obtain that

$$\begin{aligned} 2N\big(R,f(w)\big) &= N\big(R, \big(\Delta f(w+1) + 3\Delta f(w) - c\big)f(w) \\ &\qquad + \big(\Delta f(w+1) + \Delta f(w) - c\big)\Delta f(w) - (aw + a + b)\big) \\ &\le N\big(R,f(w)\big) + 7N\big(R, \Delta f(w)\big) + S(R,f) + O(\log R). \end{aligned}$$

That is,

$$N\big(R,f(w)\big) \le 7N\big(R, \Delta f(w)\big) + S(R,f) \tag{12}$$

for all $|w| = R \notin E_1 \cup [0,1]$. By Lemma 2.6 and (12), for any $\beta_1 > 1$, there exists $R_0 > 0$ such that

$$N\big(R,f(w)\big) \le 7N\big(\beta_1 R, \Delta f(w)\big) + S(R,f) \tag{13}$$

for all $R > R_0$. By (13), we have

$$\lambda\left(\frac{1}{\Delta f(w)}\right) \geq \lambda\left(\frac{1}{f(w)}\right). \tag{14}$$

By Remark 2.1 and (14), we have

$$\lambda\left(\frac{1}{\Delta f(w)}\right) = \lambda\left(\frac{1}{\Delta f(z-1)}\right) = \lambda\left(\frac{1}{\Delta f(z)}\right)$$

and

$$\lambda\left(\frac{1}{f(w)}\right) = \lambda\left(\frac{1}{f(z-1)}\right) = \lambda\left(\frac{1}{f(z)}\right).$$

Hence, we get

$$\lambda\left(\frac{1}{\Delta f(z)}\right) \geq \lambda\left(\frac{1}{f(z)}\right). \tag{15}$$

By (2), we obtain that

$$f(z)\big(f(z+1) + f(z-1)\big) = az + b + cf(z). \tag{16}$$

By (16) and Lemma 2.3, we see that for any given $\varepsilon > 0$, there is a subset $E_2 \subset (1, \infty)$ of finite logarithmic measure such that for $|z| = r \notin E_2 \cup [0, 1]$,

$$m\big(r, f(z+1) + f(z-1)\big) = O\big(r^{\sigma(f)-1+\varepsilon}\big) + S(r, f). \tag{17}$$

By Lemma 2.2, we see that there is a subset $E_3 \subset (1, \infty)$ of finite logarithmic measure such that for $|z| = r \notin E_3 \cup [0, 1]$,

$$N\big(r, f(z+1) + f(z-1)\big) \leq 2N(r+1, f) + S(r, f) = 2N(r, f) + S(r, f). \tag{18}$$

By Lemma 2.4 and (2), we see that

$$T\big(r, f(z+1) + f(z-1)\big) = T(r, f) + S(r, f) \tag{19}$$

since $|a| + |b| \neq 0$. By (17)-(19), when $|z| = r \notin E_3 \cup E_2 \cup [0, 1]$, we have

$$\frac{1}{2}T\big(r, f(z)\big) \leq N\big(r, f(z)\big) + O\big(r^{\sigma(f)-1+\varepsilon}\big) + S(r, f). \tag{20}$$

By Lemma 2.6 and (20), for any $\beta_2 > 1$, there exists $r_0 > 0$ such that

$$\frac{1}{2}T\big(r, f(z)\big) \leq N\big(\beta_2 r, f(z)\big) + O\big(r^{\sigma(f)-1+\varepsilon}\big) + S(r, f)$$

for all $r > r_0$. Thus, we get

$$\lambda\left(\frac{1}{f(z)}\right) \geq \sigma\big(f(z)\big). \tag{21}$$

By (15) and (21), we have $\lambda(\frac{1}{\Delta f}) \geq \lambda(\frac{1}{f}) \geq \sigma(f)$. And we have $\sigma(\Delta f) \leq \sigma(f)$ from Lemma 2.5. Hence,

$$\lambda\left(\frac{1}{f}\right) = \lambda\left(\frac{1}{\Delta f}\right) = \sigma(\Delta f) = \sigma(f). \qquad \square$$

3 The proof of Theorem 1.2

Suppose that $f(z)$ is a transcendental meromorphic solution of finite order $\sigma(f)$ of equation (5).

(i) Suppose that $a \neq 0$ and $p(z)$ is a nonzero polynomial. Set $g_1(z) = f(z) - p(z)$. Substituting $f(z) = g_1(z) + p(z)$ into (5), we obtain that

$$g_1(z+1) + p(z+1) + g_1(z-1) + p(z-1) = \frac{(az+b)(g_1(z)+p(z)) + c}{1 - (g_1(z)+p(z))^2}. \tag{22}$$

It follows from (22) that

$$\begin{aligned} P(z,g_1) := &\big(g_1(z+1) + p(z+1) + g_1(z-1) + p(z-1)\big) \\ &\times \big(g_1(z) + p(z)\big)^2 + (az+b)\big(g_1(z)+p(z)\big) + c \\ &- \big(g_1(z+1) + p(z+1) + g_1(z-1) + p(z-1)\big) = 0. \end{aligned} \tag{23}$$

By (23), we have

$$\begin{aligned} P(z,0) = &\big(p(z+1) + p(z-1)\big)p(z)^2 + (az+b)p(z) \\ &+ c - \big(p(z+1) + p(z-1)\big). \end{aligned}$$

If $p(z) \equiv \alpha \in C \backslash \{0\}$, then

$$P(z,0) = 2\alpha^3 + \alpha(az+b) + c - 2\alpha \not\equiv 0$$

since $a \neq 0$. Suppose that $p(z)$ is a non-constant polynomial. Set $p(z) = d_k z^k + \cdots + d_0$, where $d_k, \ldots, d_0$ are constants, $d_k \neq 0$ and $k \geq 1$. Then we obtain that

$$\begin{aligned} P(z,0) &= \big(p(z+1) + p(z-1)\big)p(z)^2 + (az+b)p(z) \\ &\quad + c - \big(p(z+1) + p(z-1)\big) \\ &= 2d_k^3 z^{3k} + \cdots \not\equiv 0. \end{aligned}$$

Thus, by Lemma 2.1, we see that

$$m\left(r, \frac{1}{g_1}\right) = S(r,g_1)$$

outside of a possible exceptional set of finite logarithmic measure. Thus,

$$N\left(r, \frac{1}{f(z) - p(z)}\right) = N\left(r, \frac{1}{g_1}\right) = T(r,g_1) + S(r,g_1) = T(r,f) + S(r,f) \tag{24}$$

outside of a possible exceptional set of finite logarithmic measure. Hence, by (24) and Lemma 2.6, we have $\lambda(f-p)=\sigma(f)$.

If $a=0$ and $p(z)=\beta\notin E$, then we have

$$P(z,0)=2\beta^3+(b-2)\beta+c\not\equiv 0.$$

Using a similar method as above, we obtain $\lambda(f-\beta)=\sigma(f)$. Hence, the Borel exceptional values of $f(z)$ can only come from the set $E=\{z \mid 2z^3+(b-2)z+c=0\}$.

If $c\neq 0$, then by (5) we have

$$P_1(z,f):=f^2(z)\big(f(z+1)+f(z-1)\big)+(az+b)f(z)+c-f(z+1)-f(z-1).$$

Hence, we have

$$P_1(z,0)\equiv c\not\equiv 0.$$

Using a similar method as above, we obtain $\lambda(f)=\sigma(f)$.

(ii) Set $z=w+1$. Substituting $z=w+1$ into (5), we obtain that

$$f(w+2)+f(w)=\frac{(aw+a+b)f(w+1)+c}{1-f(w+1)^2}.$$

That is,

$$f(w+1)^2\big(f(w+2)+f(w)\big)=f(w+2)+f(w)-c-(aw+a+b)f(w+1). \tag{25}$$

Substituting $f(w+1)=\Delta f(w)+f(w)$ and $f(w+2)=\Delta f(w+1)+\Delta f(w)+f(w)$ into (25), we have

$$\begin{aligned}&\big(\Delta f(w)+f(w)\big)^2\big(\Delta f(w+1)+\Delta f(w)+2f(w)\big)\\&\quad=\Delta f(w+1)+\Delta f(w)+2f(w)-c-(aw+a+b)\big(\Delta f(w)+f(w)\big).\end{aligned}$$

Thus, we obtain that

$$-2f(w)^3=A(w)f(w)+B(w)\Delta f(w)+C(w), \tag{26}$$

where

$$\begin{aligned}A(w)&=\big(\Delta f(w+1)+5\Delta f(w)\big)f(w)\\&\qquad+4\big(\Delta f(w)\big)^2+2\Delta f(w)\Delta f(w+1)+aw+a+b-2,\\B(w)&=\Delta f(w+1)\Delta f(w)+\big(\Delta f(w)\big)^2+aw+a+b-1,\\C(w)&=c-\Delta f(w+1).\end{aligned}$$

Since $N(R,\Delta f(w+1))\leq N(R+1,\Delta f(w))+o(N(R+1,\Delta f(w)))$, by Lemma 2.2 we see that there is a subset $H_1\subset(1,\infty)$ having finite logarithmic measure such that for $|w|=R\notin$

$H_1 \cup [0,1]$,

$$N(R, \Delta f(w+1)) \leq N(R, \Delta f(w)) + S(R,f). \tag{27}$$

By (26) and (27), when $|w| = R \notin H_1 \cup [0,1]$, we obtain that

$$\begin{aligned} 3N(R,f(w)) &= N(R, A(w)f(w) + B(w)\Delta f(w) + C(w)) \\ &\leq 2N(R,f(w)) + 12N(R, \Delta f(w)) + S(R,f) + O(\log R). \end{aligned}$$

That is,

$$N(R,f(w)) \leq 12N(R, \Delta f(w)) + S(R,f) \tag{28}$$

for $|w| = R \notin H_1 \cup [0,1]$. By Lemma 2.6 and (28), for any $\beta_1 > 1$, there exists $R_0 > 0$ such that

$$N(R,f(w)) \leq 12N(\beta_1 R, \Delta f(w)) + S(R,f)$$

for all $R > R_0$. Thus, we have

$$\lambda\left(\frac{1}{\Delta f(w)}\right) \geq \lambda\left(\frac{1}{f(w)}\right). \tag{29}$$

By the same reasoning as in Theorem 1.1(ii), we get

$$\lambda\left(\frac{1}{\Delta f(z)}\right) \geq \lambda\left(\frac{1}{f(z)}\right). \tag{30}$$

By (5), we obtain that

$$f(z)^2(f(z+1) + f(z-1)) = f(z+1) + f(z-1) - (az+b)f(z) - c. \tag{31}$$

By (31) and Lemma 2.3, we see that for any given $\varepsilon > 0$, there is a subset $H_2 \subset (1,\infty)$ having finite logarithmic measure such that for $|z| = r \notin H_2 \cup [0,1]$,

$$m(r, f(z+1) + f(z-1)) = O(r^{\sigma(f)-1+\varepsilon}) + S(r,f). \tag{32}$$

By Lemma 2.2, we see that there is a subset $H_3 \subset (1,\infty)$ of finite logarithmic measure such that for $|z| = r \notin H_3 \cup [0,1]$,

$$N(r, f(z+1) + f(z-1)) \leq 2N(r+1,f) + S(r,f) = 2N(r,f) + S(r,f). \tag{33}$$

By Lemma 2.4 and (5), we see that

$$T(r, f(z+1) + f(z-1)) = 2T(r, f(z)) + S(r,f) \tag{34}$$

since $|a| + |b| + |c| \neq 0$. By (32)-(34), when $|z| = r \notin H_3 \cup H_2 \cup [0,1]$, we have

$$T(r,f(z)) \leq N(r,f(z)) + O(r^{\sigma(f)-1+\varepsilon}) + S(r,f). \tag{35}$$

By Lemma 2.6 and (35), for any $\beta_2 > 1$, there exists $r_1 > 0$ such that

$$T(r,f(z)) \leq N(\beta_2 r,f(z)) + O(r^{\sigma(f)-1+\varepsilon}) + S(r,f)$$

for all $r > r_1$. Thus, we get

$$\lambda\left(\frac{1}{f(z)}\right) \geq \sigma(f(z)). \tag{36}$$

By (30) and (36), we see that $\lambda(\frac{1}{\Delta f}) \geq \lambda(\frac{1}{f}) \geq \sigma(f)$. And we have $\sigma(\Delta f) \leq \sigma(f)$ from Lemma 2.5. Hence,

$$\lambda\left(\frac{1}{f}\right) = \lambda\left(\frac{1}{\Delta f}\right) = \sigma(\Delta f) = \sigma(f).$$

4 Proofs of Theorems 1.3 and 1.4

Proof of Theorem 1.3 Suppose that $f(z)$ is a transcendental meromorphic solution of finite order $\sigma(f)$ of equation (3).

(i) Using the same methods as in the proofs of Theorems 1.1(ii) and 1.2(ii), we have

$$-2f(w)^3 = A(w)f(w) + B(w)\Delta f(w) + C(w), \tag{37}$$

where

$$\begin{aligned} A(w) &= (\Delta f(w+1) + 5\Delta f(w))f(w) \\ &\quad + 4(\Delta f(w))^2 + 2\Delta f(w)\Delta f(w+1) - (aw + a + b), \\ B(w) &= (\Delta f(w+1) + \Delta f(w))\Delta f(w) - (aw + a + b), \\ C(w) &= -c. \end{aligned}$$

Set $|w| = R$. Since $N(R, \Delta f(w+1)) \leq N(R+1, \Delta f(w)) + o(N(R+1, \Delta f(w)))$, by Lemma 2.2 we see that

$$N(R, \Delta f(w+1)) \leq N(R, \Delta f(w)) + S(R,f) \tag{38}$$

outside of a possible exceptional set of finite logarithmic measure. By (37) and (38), we obtain that

$$\begin{aligned} 3N(R,f(w)) &= N(R, A(w)f(w) + B(w)\Delta f(w) + C(w)) \\ &\leq 2N(R,f(w)) + 10N(R, \Delta f(w)) \\ &\quad + S(R,f) + O(\log R) \end{aligned} \tag{39}$$

outside of a possible exceptional set of finite logarithmic measure. By (39) and Lemma 2.6, we have

$$\lambda\left(\frac{1}{\Delta f(w)}\right) \geq \lambda\left(\frac{1}{f(w)}\right). \tag{40}$$

By Remark 2.1 and (40), we obtain that

$$\lambda\left(\frac{1}{\Delta f(w)}\right) = \lambda\left(\frac{1}{\Delta f(z-1)}\right) = \lambda\left(\frac{1}{\Delta f(z)}\right)$$

and

$$\lambda\left(\frac{1}{f(w)}\right) = \lambda\left(\frac{1}{f(z-1)}\right) = \lambda\left(\frac{1}{f(z)}\right).$$

Hence, we get

$$\lambda\left(\frac{1}{\Delta f(z)}\right) \geq \lambda\left(\frac{1}{f(z)}\right). \tag{41}$$

By (3), we obtain that

$$f(z)^2\left(f(z+1)+f(z-1)\right) = (az+b)f(z)+c. \tag{42}$$

By (42) and Lemma 2.3, we see that

$$m\left(r,f(z+1)+f(z-1)\right) = O\left(r^{\sigma(f)-1+\varepsilon}\right)+S(r,f) \tag{43}$$

outside of a possible exceptional set of finite logarithmic measure. By Lemma 2.2, we get

$$N\left(r,f(z+1)+f(z-1)\right) \leq 2N\left(r+1,f(z)\right)+S(r,f) = 2N(r,f)+S(r,f) \tag{44}$$

outside of a possible exceptional set of finite logarithmic measure.

If $c \neq 0$, by Lemma 2.4 and (3), we see that

$$T\left(r,f(z+1)+f(z-1)\right) = 2T\left(r,f(z)\right)+S(r,f). \tag{45}$$

If $c = 0$, by the same reason, we have

$$T\left(r,f(z+1)+f(z-1)\right) = T\left(r,f(z)\right)+S(r,f) \tag{46}$$

since $|a|+|b|+|c| = |a|+|b| \neq 0$. By (43)-(46), we have

$$T\left(r,f(z)\right) \leq N\left(r,f(z)\right)+O\left(r^{\sigma(f)-1+\varepsilon}\right)+S(r,f) \tag{47}$$

outside of a possible exceptional set of finite logarithmic measure. By Lemma 2.6 and (47), we get

$$\lambda\left(\frac{1}{f(z)}\right) \geq \sigma\left(f(z)\right). \tag{48}$$

By (41) and (48), we have $\lambda(\frac{1}{\Delta f}) \geq \lambda(\frac{1}{f}) \geq \sigma(f)$. And we have $\sigma(\Delta f) \leq \sigma(f)$ from Lemma 2.5. Hence,

$$\lambda\left(\frac{1}{f}\right) = \lambda\left(\frac{1}{\Delta f}\right) = \sigma(\Delta f) = \sigma(f).$$

(ii) Let $a = 0$ and $p(z) = \beta \notin E$. Using the same methods as in the proof of Theorem 1.1(i), we have

$$P(z,0) = 2\beta^3 - b\beta - c \not\equiv 0.$$

Thus, we obtain $\lambda(f - \beta) = \sigma(f)$. Hence, the Borel exceptional values of $f(z)$ can only come from the set $E = \{z \mid 2z^3 - bz - c = 0\}$. □

Proof of Theorem 1.4 Suppose that $f(z)$ is a transcendental meromorphic solution of finite order $\sigma(f)$ of equation (4). Set $z = w + 1$. Using the same method as in the proof of Theorem 1.3(i), we have

$$\begin{aligned}-3f(w)^2 &= \big(\Delta f(w+1) + 5\Delta f(w) - c\big)f(w)\\ &\quad + \big(\Delta f(w+1) + 2\Delta f(w) - c\big)\Delta f(w) - (aw + a + b).\end{aligned}$$

By the same reason as that in Theorem 1.3(i), when $|w| = R$, we have

$$2N\big(R,f(w)\big) \leq N\big(R,f(w)\big) + 5N\big(R,\Delta f(w)\big) + S(R,f) + O(\log R) \tag{49}$$

possibly outside of an exceptional set of finite logarithmic measure.

By Lemma 2.6 and (49), we obtain that $\lambda(\frac{1}{\Delta f(w)}) \geq \lambda(\frac{1}{f(w)})$. By the same method as above, we have $\lambda(\frac{1}{\Delta f(z)}) \geq \lambda(\frac{1}{f(z)})$. And we see that $\lambda(\frac{1}{f(z)}) = \sigma(f(z))$ from Theorem B and $\sigma(\Delta f) \leq \sigma(f)$ from Lemma 2.5. Hence, we have

$$\lambda\left(\frac{1}{f}\right) = \lambda\left(\frac{1}{\Delta f}\right) = \sigma(\Delta f) = \sigma(f).$$ □

Competing interests
The authors declare that they have no competing interests.

Authors' contributions
C-WP completed the main part of this article, C-WP and Z-XC corrected the main theorems. All authors read and approved the final manuscript.

Author details
[1]School of Mathematics and Computer Sciences, Guizhou Normal College, Guiyang, Guizhou 550018, China. [2]School of Mathematical Sciences, South China Normal University, Guangzhou, Guangdong 510631, China.

Acknowledgements
The authors thank the referee for his/her valuable suggestions. This work is supported by the State Natural Science Foundation of China (No. 61462016), the Science and Technology Foundation of Guizhou Province (Nos. [2014]2125; [2014]2142) and the Doctoral Foundation of Guizhou Normal University (No. 11904-0514021).

References

1. Hayman, WK: Meromorphic Functions. Clarendon, Oxford (1964)
2. Laine, I: Nevanlinna Theory and Complex Differential Equations. de Gruyter, Berlin (1993)
3. Yang, L: Value Distribution Theory. Science Press, Beijing (1993)
4. Painlevé, P: Mémoire sur les équations différentielles dont l'integrale générale est uniforme. Bull. Soc. Math. Fr. **28**, 201-261 (1900)
5. Fokas, AS: From continuous to discrete Painlevé equations. J. Math. Anal. Appl. **180**, 342-360 (1993)
6. Shimomura, S: Entire solutions of a polynomial difference equation. J. Fac. Sci., Univ. Tokyo, Sect. IA Math. **28**, 253-266 (1981)
7. Yanagihara, N: Meromorphic solutions of some difference equations. Funkc. Ekvacioj **23**, 309-326 (1980)
8. Ablowitz, M, Halburd, RG, Herbst, B: On the extension of Painlevé property to difference equations. Nonlinearity **13**, 889-905 (2000)
9. Bergweiler, W, Langley, JK: Zeros of differences of meromorphic functions. Math. Proc. Camb. Philos. Soc. **142**, 133-147 (2007)
10. Chen, MR, Chen, ZX: On properties of meromorphic solution of certain difference Painlevé I equation. Bull. Aust. Math. Soc. **85**, 463-475 (2012)
11. Chen, ZX: Value distribution of meromorphic solutions of certain difference Painlevé equations. J. Math. Anal. Appl. **364**, 556-566 (2010)
12. Chiang, YM, Feng, JF: On the Nevanlinna characteristic of $f(z+\eta)$ and difference equations in the complex plane. Ramanujan J. **16**, 105-129 (2008)
13. Halburd, RG, Korhonen, RJ: Difference analogue of the lemma on the logarithmic derivative with applications to difference equations. J. Math. Anal. Appl. **314**, 477-487 (2006)
14. Halburd, RG, Korhonen, RJ: Nevanlinna theory for the difference operator. Ann. Acad. Sci. Fenn., Math. **31**, 463-478 (2006)
15. Halburd, RG, Korhonen, RJ: Existence of finite-order meromorphic solutions as a detector of integrability in difference equations. Physica D **218**, 191-203 (2006)
16. Halburd, RG, Korhonen, RJ: Meromorphic solution of difference equation, integrability and the discrete Painlevé equations. J. Phys. A **40**, 1-38 (2007)
17. Halburd, RG, Korhonen, RJ: Finite-order meromorphic solutions and the discrete Painlevé equations. Proc. Lond. Math. Soc. **94**, 443-474 (2007)
18. Laine, I, Yang, CC: Clunie theorems for difference and q-difference polynomials. J. Lond. Math. Soc. **76**, 556-566 (2007)
19. Peng, CW, Chen, ZX: On a conjecture concerning some nonlinear difference equations. Bull. Malays. Math. Soc. **36**, 221-227 (2013)
20. Peng, CW, Chen, ZX: Properties of meromorphic solutions of some certain difference equations. Kodai Math. J. **37**, 97-119 (2014)
21. Gundersen, G: Finite order solutions of second order linear differential equations. Trans. Am. Math. Soc. **305**, 415-429 (1988)

Homoclinic solutions for a class of nonlinear difference systems with classical (ϕ_1, ϕ_2)-Laplacian

Xingyong Zhang* and Yun Wang

*Correspondence: zhangxingyong1@163.com
Department of Mathematics, Faculty of Science, Kunming University of Science and Technology, Kunming, Yunnan 650500, P.R. China

Abstract

In this paper, we consider the existence of homoclinic solutions for a class of nonlinear difference systems involving classical (ϕ_1, ϕ_2)-Laplacian. First, we improve some inequalities in known literature. Then, by using the variational method, some new existence results are obtained. Finally, some examples are given to verify our results.

MSC: 37J45; 58E50; 34C25

Keywords: difference systems; classical (ϕ_1, ϕ_2)-Laplacian; homoclinic solutions; variational method

1 Introduction and main results

Let $\mathbb{R}$ denote the real numbers and $\mathbb{Z}$ the integers. Given $a < b$ in $\mathbb{Z}$. Let $\mathbb{Z}[a,b] = \{a, a+1, \dots, b\}$. Let $T > 1$ and N be fixed positive integers.

In this paper, we investigate the existence of homoclinic solutions for the following nonlinear difference systems involving classical (ϕ_1, ϕ_2)-Laplacian:

$$\begin{cases} \Delta\phi_1(\Delta u_1(t-1)) + \nabla_{u_1} V(t, u_1(t), u_2(t)) = f_1(t), \\ \Delta\phi_2(\Delta u_2(t-1)) + \nabla_{u_2} V(t, u_1(t), u_2(t)) = f_2(t), \end{cases} \tag{1.1}$$

where $t \in \mathbb{Z}$, $u_m(t) \in \mathbb{R}^N$, $m = 1, 2$, $V(t, x_1, x_2) = -K(t, x_1, x_2) + W(t, x_1, x_2)$, $K, W : \mathbb{Z} \times \mathbb{R}^N \times \mathbb{R}^N \to \mathbb{R}$ and ϕ_m, $m = 1, 2$, satisfy the following condition:

($\mathcal{A}0$) ϕ_m is a homeomorphism from $\mathbb{R}^N$ onto $\mathbb{R}^N$ such that $\phi_m(0) = 0$, $\phi_m = \nabla\Phi_m$, with $\Phi_m \in C^1(\mathbb{R}^N, [0, +\infty])$ strictly convex and $\Phi_m(0) = 0$, $m = 1, 2$.

Remark 1.1 Assumption ($\mathcal{A}0$) is given in [1], which is used to characterize the classical homeomorphism. If, furthermore, $\Phi_m : \mathbb{R}^N \to \mathbb{R}$ is coercive (*i.e.*, $\Phi_m(x) \to +\infty$ as $|x| \to \infty$), there exists $\delta_m > 0$ such that

$$\Phi_m(x) \geq \delta_m\big(|x| - 1\big), \quad x \in \mathbb{R}^N, \tag{1.2}$$

where $\delta_m = \min_{|x|=1} \Phi_m(x)$, $m = 1, 2$ (see [1]).

We call $u = (u_1, u_2)$ a nontrivial homoclinic solution of system (1.1) if u satisfies system (1.1), $u \neq 0$ and $u(t) \to 0$ as $t \to \infty$.

It is well known that the variational method has become an important tool to study the existence and multiplicity of solutions for various difference systems. Lots of contributions have been obtained (for example, see [1–20]). It is remarkable that, to the best of our knowledge, few people investigated system (1.1). Recently, in [1] and [2], by using the variational approach, J Mawhin investigated the following second order nonlinear difference systems with ϕ-Laplacian:

$$\Delta\phi\left[\Delta u(n-1)\right] = \nabla_u F\left[n, u(n)\right] + h(n) \quad (n \in \mathbb{Z}), \tag{1.3}$$

where $\phi = \nabla\Phi$, Φ strictly convex, is a homeomorphism of $\mathbb{R}^N$ onto the ball $B_a \subset \mathbb{R}^N$ or of B_a onto $\mathbb{R}^N$. By using the variational approach, under different conditions, the author obtained that system (1.3) has at least one or $N+1$ geometrically distinct T-periodic solutions. It is interesting that J Mawhin considered three kinds of ϕ: (1) $\phi : \mathbb{R}^N \to \mathbb{R}^N$ is a classical homeomorphism, for example, $\phi(x) = |x|^{p-1}x$ for some $p > 1$ and all $x \in \mathbb{R}^N$; (2) $\phi : \mathbb{R}^N \to B_a$ $(a < +\infty)$ is a bounded homeomorphism, for example, $\phi(x) = \frac{x}{\sqrt{1+|x|^2}} \in B_1$ for all $x \in \mathbb{R}^N$; (3) $\phi : B_a \subset \mathbb{R}^N \to \mathbb{R}^N$ is a singular homeomorphism, for example, $\phi(x) = \frac{x}{\sqrt{1-|x|^2}}$ for all $x \in B_1$. Recently, in [17], we generalized some results in [2] for classical homeomorphism and bounded homeomorphism to system (1.1), which seem to be the first results for system (1.1).

In 2011, He and Chen [16] investigated the existence of homoclinic solutions for the following discrete p-Laplacian systems:

$$\Delta\left(\left|\Delta u(t-1)\right|^{p-2}\Delta u(t-1)\right) = \nabla F\left(t, u(t)\right) + f(t), \quad t \in \mathbb{Z}, u \in \mathbb{R}^N, \tag{1.4}$$

where $p > 1$. They obtained homoclinic orbits as the limit of the subharmonics for system (1.4).

In this paper, motivated by [1, 2, 15, 16] and [17], we first improve some inequalities in [16] and then investigate the existence of homoclinic solutions for system (1.1) with classical homeomorphism. Next we make the following assumption:

($\mathcal{A}1$) *Let $p > 1$. Assume that there exist positive constants d_1, d_2, d_3, d_4 such that*

$$d_1|x|^p \le \Phi_1(x) \le d_3|x|^p, \qquad d_2|y|^p \le \Phi_2(y) \le d_4|y|^p, \quad \forall x, y \in \mathbb{R}^N$$

and

$$\left(\phi_1(x), x\right) \le p\Phi_1(x), \qquad \left(\phi_2(y), y\right) \le p\Phi_2(y), \quad \forall x, y \in \mathbb{R}^N.$$

For every $s \in \mathbb{N}$, define

$$l^s = \left\{ g : \mathbb{Z} \to \mathbb{R}^N, \sum_{t=-\infty}^{+\infty} |g(t)|^s < \infty \right\}$$

with the norm

$$\|g\|_{l^s} = \left(\sum_{t=-\infty}^{+\infty} |g(t)|^s \right)^{1/s}.$$

Let $p' > 1$ be such that $\frac{1}{p} + \frac{1}{p'} = 1$ and

$$C_* = \left((2T)^{-p'/p} + \min\left\{ \frac{(T+1)^{p'+1} + T^{p'+1} - 2}{(2T)^{p'}(p'+1)}, \frac{T}{2^{p'/p}} \right\} \right)^{1/p'}.$$

Next, we present our main results.

Theorem 1.1 *Assume that* $(\mathcal{A}1)$ *holds,* $f_i \not\equiv 0$, $i = 1, 2$, W *and* K *satisfy the following conditions:*

($\mathcal{V}$) $V(t, x_1, x_2) = -K(t, x_1, x_2) + W(t, x_1, x_2)$, *where* $K, W : \mathbb{Z} \times \mathbb{R}^N \times \mathbb{R}^N \to \mathbb{R}$, $K(t, x_1, x_2)$ *and* $W(t, x_1, x_2)$ *are* T*-periodic and for every* $t \in \mathbb{Z}$, $K, W \in C^1(\mathbb{Z} \times \mathbb{R}^N \times \mathbb{R}^N, \mathbb{R})$;

(H1) *there exist* $\gamma \in (1, p)$ *and* $a_1, a_2 > 0$ *such that*

$$K(t, x_1, x_2) \geq a_1 |x_1|^\gamma + a_2 |x_2|^\gamma \quad \text{for all } (t, x_1, x_2) \in \mathbb{Z}[0, T-1] \times \mathbb{R}^N \times \mathbb{R}^N;$$

(H2) $K(t, 0, 0) \equiv 0$ *and*

$$\begin{aligned}&\left(x_1, \nabla_{x_1} K(t, x_1, x_2)\right) + \left(x_2, \nabla_{x_2} K(t, x_1, x_2)\right)\\ &\quad \leq pK(t, x_1, x_2) \quad \text{for all } (t, x_1, x_2) \in \mathbb{Z}[0, T-1] \times \mathbb{R}^N \times \mathbb{R}^N;\end{aligned}$$

(H3) (i) *there exist* $r \in (0, 1]$, $0 < b_1 < a_1 C_*^{\gamma - p}$, *and* $0 < b_2 < a_2 C_*^{\gamma - p}$ *such that*

$$\begin{aligned}&W(t, x_1, x_2) \leq b_1 |x_1|^p + b_2 |x_2|^p,\\ &\quad \forall t \in \mathbb{Z}[0, T-1], |x_1| \leq rC_*, |x_2| \leq rC_*;\end{aligned} \tag{1.5}$$

(ii) *there exist* $r > 1$, $0 < b_1 < a_1 (C_* r)^{\gamma - p}$, *and* $0 < b_2 < a_2 (C_* r)^{\gamma - p}$ *such that* (1.5) *holds;*

(H4)

$$\lim_{|x_1| + |x_2| \to +\infty} \frac{W(t, x_1, x_2)}{|x_1|^p + |x_2|^p} > d_3 + d_4 + 2^{p-1} A_0 \quad \text{for all } t \in \mathbb{Z}[0, T-1],$$

where

$$A_0 = \max_{|x_1| \leq 1, |x_2| \leq 1, t \in \mathbb{Z}[0, T-1]} K(t, x_1, x_2);$$

(H5) *there exist positive constants* ξ, η_1, η_2 *and* $\nu \in [0, \gamma - 1)$ *such that*

$$\begin{aligned}0 &\leq \left(p + \frac{1}{\xi + \eta_1 |x_1|^\nu + \eta_2 |x_2|^\nu}\right) W(t, x_1, x_2)\\ &\leq \left(\nabla_{x_1} W(t, x_1, x_2), x_1\right) + \left(\nabla_{x_2} W(t, x_1, x_2), x_2\right)\end{aligned}$$

for all $(t, x_1, x_2) \in \mathbb{Z}[0, T-1] \times \mathbb{R}^N \times \mathbb{R}^N$;

(H6) $f_1, f_2 \in l^{p'} \cap l^{\frac{p-\nu}{p-\nu-1}}$ *and*

(i) *when* $r \in (0,1]$,

$$\max\{\|f_1\|_{l^{p'}}, \|f_2\|_{l^{p'}}\} < \frac{1}{2^{p-1}} \min\{d_1, d_2, a_1 C_*^{\gamma-p} - b_1, a_2 C_*^{\gamma-p} - b_2\} r^{p-1};$$

(ii) *when* $r \in (1, +\infty)$,

$$\max\{\|f_1\|_{l^{p'}}, \|f_2\|_{l^{p'}}\} < \frac{1}{2^{p-1}} \min\{d_1, d_2, a_1 (C_* r)^{\gamma-p} - b_1, a_2 (C_* r)^{\gamma-p} - b_2\} r^{p-1}.$$

Then system (1.1) *possesses a nontrivial homoclinic solution.*

Theorem 1.2 *Assume that* ($\mathcal{A}$1) *holds,* $f_i \neq 0$, $i = 1, 2$, W *and* K *satisfy* ($\mathcal{V}$), (H1)-(H5) *and the following conditions:*

(H6)′ $f_1, f_2 \in l^1$ *and*

(i) *when* $r \in (0,1]$,

$$\max\{\|f_1\|_{l^1}, \|f_2\|_{l^1}\} < \frac{1}{2^{p-1} C_*} \min\{d_1, d_2, a_1 C_*^{\gamma-p} - b_1, a_2 C_*^{\gamma-p} - b_2\} r^{p-1};$$

(ii) *when* $r \in (1, +\infty)$,

$$\max\{\|f_1\|_{l^1}, \|f_2\|_{l^1}\} < \frac{1}{2^{p-1} C_*} \min\{d_1, d_2, a_1 (C_* r)^{\gamma-p} - b_1, a_2 (C_* r)^{\gamma-p} - b_2\} r^{p-1}.$$

Then system (1.1) *possesses a nontrivial homoclinic solution.*

Theorem 1.3 *Assume that* ($\mathcal{A}$1) *holds,* $f_i \neq 0$, $i = 1, 2$, W *and* K *satisfy* ($\mathcal{V}$), (H2), (H4), (H5) *and the following conditions:*

(H1)′ *there exist* $a_1, a_2 > 0$ *such that*

$$K(t, x_1, x_2) \geq a_1 |x_1|^p + a_2 |x_2|^p \quad \textit{for all } (t, x_1, x_2) \in \mathbb{Z}[0, T-1] \times \mathbb{R}^N \times \mathbb{R}^N;$$

(H3)′ *there exist* $r > 0$ *and* $0 < b_1 < a_1$, $0 < b_2 < a_2$ *such that*

$$W(t, x_1, x_2) \leq b_1 |x_1|^p + b_2 |x_2|^p, \quad \forall |x_1| \leq r C_*, |x_2| \leq r C_*;$$

(H6)″ $f_1, f_2 \in l^{p'} \cap l^{\frac{p-\nu}{p-\nu-1}}$ *and*

$$\max\{\|f_1\|_{l^{p'}}, \|f_2\|_{l^{p'}}\} < \frac{1}{2^{p-1}} \min\{d_1, d_2, a_1 - b_1, a_2 - b_2\} r^{p-1}.$$

Then system (1.1) *possesses a nontrivial homoclinic solution.*

Theorem 1.4 *Assume that* $(\mathcal{A}1)$ *holds,* $f_i \neq 0$, $i = 1, 2$, W *and* K *satisfy* $(\mathcal{V})$, (H1)′, (H2), (H3)′, (H4), (H5) *and the following condition*:

(H6)‴ $f_1, f_2 \in l^1$ *and*

$$\max\left\{ \|f_1\|_{l^1}, \|f_2\|_{l^1} \right\} < \frac{1}{2^{p-1}C_*} \min\{d_1, d_2, a_1 - b_1, a_2 - b_2\} r^{p-1}.$$

Then system (1.1) *possesses a nontrivial homoclinic solution.*

Remark 1.2 Theorem 1.3 and Theorem 1.4 show that f_1, f_2 can be large when r is large.

2 Preliminaries

Similar to [15] and [16], we will obtain the homoclinic orbit of system (1.1) as a limit of solutions of a sequence of difference systems:

$$\begin{cases} \Delta\phi_1(\Delta u_1(t-1)) + \nabla_{u_1} V(t, u_1(t), u_2(t)) = f_{1,k}(t), \\ \Delta\phi_2(\Delta u_2(t-1)) + \nabla_{u_2} V(t, u_1(t), u_2(t)) = f_{2,k}(t), \end{cases} \tag{2.1}$$

where $f_{m,k} : \mathbb{Z} \to \mathbb{R}^N$ is a $2kT$-periodic extension of restriction of f_m to the interval $\mathbb{Z}[-kT, kT-1]$, $k \in \mathbb{N}$, $m = 1, 2$.

Next, we present some basic notations. We use $|\cdot|$ to denote the usual Euclidean norm in $\mathbb{R}^N$. Define

$$\begin{aligned} \mathcal{V} = \big\{ u = (u_1, u_2)^\tau = \{u(t)\} \,|\, u(t) = (u_1(t), u_2(t))^\tau \in \mathbb{R}^{2N}, \\ u_m = \{u_m(t)\}, u_m(t) \in \mathbb{R}^N, m = 1, 2, t \in \mathbb{Z} \big\}. \end{aligned}$$

$\mathcal{H}$ is defined as a subspace of $\mathcal{V}$ by

$$\mathcal{H}_k = \big\{ u = \{u(t)\} \in \mathcal{V} \,|\, u(t + 2kT) = u(t), t \in \mathbb{Z} \big\}.$$

Define

$$\mathcal{H}_{m,k} = \big\{ u_m = \{u_m(t)\} \,|\, u_m(t + 2kT) = u_m(t), u_m(t) \in \mathbb{R}^N, t \in \mathbb{Z} \big\}, \quad m = 1, 2.$$

Then $\mathcal{H}_k = \mathcal{H}_{1,k} \times \mathcal{H}_{2,k}$. For $u_m \in \mathcal{H}_{m,k}$, set

$$\|u_m\|_{s,k} = \left(\sum_{t=-kT}^{kT-1} |u_m(t)|^s \right)^{1/s}, \quad m = 1, 2, s > 1.$$

Moreover, l^∞_{2kT} denote the space of all bounded real functions on $\mathbb{Z}[-kT, kT-1]$ endowed with the norm

$$\|u_m\|_{l^\infty_{2kT}} = \max_{t \in \mathbb{Z}[-kT, kT-1]} |u_m(t)|, \quad m = 1, 2.$$

For $1 < p < +\infty$, on $\mathcal{H}_{m,k}$, we define

$$\|u_m\|_{\mathcal{H}_{m,k}} = \left(\sum_{t=-kT}^{kT-1} |\Delta u_m(t)|^p + \sum_{t=-kT}^{kT-1} |u_m(t)|^p \right)^{1/p}, \quad m = 1, 2.$$

For $u = (u_1, u_2)^\tau \in \mathcal{H}_k$, define

$$\|u\|_{\mathcal{H}_k} = \|u_1\|_{\mathcal{H}_{1,k}} + \|u_2\|_{\mathcal{H}_{2,k}}.$$

Then $(\mathcal{H}_k, \|u\|_{\mathcal{H}_k})$, $(\mathcal{H}_{1,k}, \|u\|_{\mathcal{H}_{1,k}})$ and $(\mathcal{H}_{2,k}, \|u\|_{\mathcal{H}_{2,k}})$ are reflexive Banach spaces.

Lemma 2.1 *Let $a, b \in \mathbb{Z}$, $a \geq 1$, $b \geq 0$, $\varrho > 1$, $u_m \in \mathcal{H}_{m,k}$, $m = 1, 2$. Then, for every $t \in \mathbb{Z}$,*

$$\begin{aligned} |u_m(t)| &\leq (a+b+1)^{-1/\varrho}\left(\sum_{s=t-a}^{t+b}|u_m(s)|^\varrho\right)^{1/\varrho} \\ &\quad + \min\left\{\frac{[(a+1)^{p'+1}+(b+1)^{p'+1}-2]^{1/p'}}{(a+b+1)^{p'}(p'+1)^{1/p'}}, \frac{\max\{a,b\}}{(a+b+1)^{1/p}}\right\} \\ &\quad \cdot \left(\sum_{s=t-a}^{t+b}|\Delta u_m(s)|^p\right)^{1/p}, \end{aligned} \tag{2.2}$$

where $m = 1, 2$.

Proof Fix $t \in \mathbb{Z}$. For every $\tau \in \mathbb{Z}[t-a, t-1]$, we have

$$u_m(t) = u_m(\tau) + \sum_{s=\tau}^{t-1}\Delta u_m(s) \tag{2.3}$$

and for every $\tau \in \mathbb{Z}[t, t+b]$,

$$u_m(t) = u_m(\tau) - \sum_{s=t}^{\tau-1}\Delta u_m(s). \tag{2.4}$$

Summing (2.3) over $\mathbb{Z}[t-a, t-1]$ and (2.4) over $\mathbb{Z}[t, t+b]$, we have

$$\begin{aligned} au_m(t) &= \sum_{\tau=t-a}^{t-1}u_m(\tau) + \sum_{\tau=t-a}^{t-1}\sum_{s=\tau}^{t-1}\Delta u_m(s) \\ &= \sum_{\tau=t-a}^{t-1}u_m(\tau) + \sum_{s=t-a}^{t-1}(s-t+a+1)\Delta u_m(s) \end{aligned} \tag{2.5}$$

and

$$\begin{aligned} (b+1)u_m(t) &= \sum_{\tau=t}^{t+b}u_m(\tau) - \sum_{\tau=t}^{t+b}\sum_{s=t}^{\tau-1}\Delta u_m(s) \\ &= \sum_{\tau=t}^{t+b}u_m(\tau) - \sum_{s=t}^{t+b-1}(t+b-s)\Delta u_m(s). \end{aligned} \tag{2.6}$$

Set

$$\phi(s) = \begin{cases} s-t+a+1, & t-a \leq s \leq t-1, \\ t+b-s, & t \leq s \leq t+b. \end{cases}$$

Combining (2.5) with (2.6) and using Hölder's inequality, we obtain

$$
\begin{aligned}
&(a+b+1)|u_m(t)| \\
&= \left| \sum_{\tau=t-a}^{t+b} u_m(\tau) + \sum_{s=t-a}^{t-1} (s-t+a+1)\Delta u_m(s) - \sum_{s=t}^{t+b-1} (t+b-s)\Delta u_m(s) \right| \\
&\le \sum_{\tau=t-a}^{t+b} |u_m(\tau)| + \sum_{s=t-a}^{t-1} (s-t+a+1)|\Delta u_m(s)| + \sum_{s=t}^{t+b-1} (t+b-s)|\Delta u_m(s)| \\
&= \sum_{\tau=t-a}^{t+b} |u_m(\tau)| + \sum_{s=t-a}^{t+b-1} \phi(s)|\Delta u_m(s)| = \sum_{\tau=t-a}^{t+b} |u_m(\tau)| + \sum_{s=t-a}^{t+b} \phi(s)|\Delta u_m(s)| \\
&\le (a+b+1)^{(\varrho-1)/\varrho} \left(\sum_{\tau=t-a}^{t+b} |u_m(\tau)|^{\varrho} \right)^{1/\varrho} + \left(\sum_{s=t-a}^{t+b} [\phi(s)]^{p'} \right)^{1/p'} \left(\sum_{s=t-a}^{t+b} |\Delta u_m(s)|^{p} \right)^{1/p} \\
&= (a+b+1)^{(\varrho-1)/\varrho} \left(\sum_{\tau=t-a}^{t+b} |u_m(\tau)|^{\varrho} \right)^{1/\varrho} \\
&\quad + \left(\sum_{s=t-a}^{t-1} (s-t+a+1)^{p'} + \sum_{s=t}^{t+b} (t+b-s)^{p'} \right)^{1/p'} \left(\sum_{s=t-a}^{t+b} |\Delta u_m(s)|^{q} \right)^{1/q}.
\end{aligned}
\tag{2.7}
$$

Since

$$
\begin{aligned}
&\sum_{s=t-a}^{t-1} (s-t+a+1)^{p'} = \sum_{s=1}^{a} s^{p'} < \frac{(a+1)^{p'+1}-1}{p'+1}, \\
&\sum_{s=t}^{t+b} (t+b-s)^{p'} = \sum_{k=1}^{b} k^{p'} < \frac{(b+1)^{p'+1}-1}{p'+1}
\end{aligned}
\tag{2.8}
$$

and

$$
\begin{aligned}
&\sum_{s=t-a}^{t-1} (s-t+a+1)^{p'} + \sum_{s=t}^{t+b} (t+b-s)^{p'} \\
&\le \sum_{s=t-a}^{t-1} a^{p'} + \sum_{s=t}^{t+b} b^{p'} \le \sum_{s=t-a}^{t+b} \max\{a^{p'}, b^{p'}\} = \max\{a^{p'}, b^{p'}\}(b+a+1).
\end{aligned}
\tag{2.9}
$$

Equation (2.7) implies that

$$
\begin{aligned}
&(a+b+1)|u_m(t)| \\
&\le (a+b+1)^{(\varrho-1)/\varrho} \left(\sum_{\tau=t-a}^{t+b} |u_m(\tau)|^{\varrho} \right)^{1/\varrho} \\
&\quad + \left(\min\left\{ \frac{(a+1)^{p'+1} + (b+1)^{p'+1} - 2}{p'+1}, \max\{a^{p'}, b^{p'}\}(b+a+1) \right\} \right)^{1/p'} \\
&\quad \cdot \left(\sum_{s=t-a}^{t+b} |\Delta u_m(s)|^{p} \right)^{1/p},
\end{aligned}
$$

which implies that (2.2) holds. Thus the proof is complete. □

Corollary 2.1 *Let $u_m \in \mathcal{H}_{m,k}$, $m = 1, 2$. Then*

$$\|u_m\|_{l^\infty_{2kT}} \le (2T)^{-1/\varrho}\left(\sum_{s=-kT}^{kT-1} |u_m(s)|^\varrho\right)^{1/\varrho} + \min\left\{\frac{[(T+1)^{p'+1} + T^{p'+1} - 2]^{1/p'}}{(2T)^{p'}(p'+1)^{1/p'}}, \frac{T^{1/p'}}{2^{1/p}}\right\}\left(\sum_{s=-kT}^{kT-1} |\Delta u_m(s)|^p\right)^{1/p}, \tag{2.10}$$

where $m = 1, 2$.

Proof Obviously, there exists $t^* \in \mathbb{Z}[-kT, kT-1]$ such that

$$|u_m(t^*)| = \|u_m\|_{l^\infty_{2kT}} = \max_{t\in\mathbb{Z}[-kT,kT-1]} |u_m(s)|.$$

In Lemma 2.1, let $a = T$ and $b = T - 1$,

$$\begin{aligned}
|u_m(t^*)| &\le (2T)^{-1/\varrho}\left(\sum_{s=t^*-T}^{t^*+T-1} |u_m(s)|^\varrho\right)^{1/\varrho} \\
&\quad + \min\left\{\frac{[(T+1)^{p'+1} + T^{p'+1} - 2]^{1/p'}}{(2T)^{p'}(p'+1)^{1/p'}}, \frac{T}{(2T)^{1/p}}\right\}\left(\sum_{s=t^*-T}^{t^*+T-1} |\Delta u_m(s)|^p\right)^{1/p} \\
&\le (2T)^{-1/\varrho}\left(\sum_{s=t^*-kT}^{t^*+kT-1} |u_m(s)|^\varrho\right)^{1/\varrho} \\
&\quad + \min\left\{\frac{[(T+1)^{p'+1} + T^{p'+1} - 2]^{1/p'}}{(2T)^{p'}(p'+1)^{1/p'}}, \frac{T}{(2T)^{1/p}}\right\}\left(\sum_{s=t^*-kT}^{t^*+kT-1} |\Delta u_m(s)|^p\right)^{1/p} \\
&= (2T)^{-1/\varrho}\left(\sum_{s=-kT}^{kT-1} |u_m(s)|^\varrho\right)^{1/\varrho} \\
&\quad + \min\left\{\frac{[(T+1)^{p'+1} + T^{p'+1} - 2]^{1/p'}}{(2T)^{p'}(p'+1)^{1/p'}}, \frac{T^{1/p'}}{2^{1/p}}\right\}\left(\sum_{s=-kT}^{kT-1} |\Delta u_m(s)|^p\right)^{1/p}.
\end{aligned}$$

The proof is complete. □

Corollary 2.2 *Let $u_m \in \mathcal{H}_{m,k}$, $m = 1, 2$. Then*

$$\|u_m\|_{l^\infty_{2kT}} \le \left((2T)^{-p'/p} + \min\left\{\frac{(T+1)^{p'+1} + T^{p'+1} - 2}{(2T)^{p'^2}(p'+1)^{1/p'}}, \frac{T}{2^{p'/p}}\right\}\right)^{1/p'} \cdot \left(\sum_{s=-kT}^{kT-1} |u_m(s)|^p + \sum_{s=-kT}^{kT-1} |\Delta u_m(s)|^p\right)^{1/p}, \tag{2.11}$$

where $m = 1, 2$.

Proof In Corollary 2.1, let $\varrho = p$ and then use Hölder's inequality. Then the proof is completed easily. □

Remark 2.1 As $a \geq 1$, Lemma 2.1, Corollary 2.1, and Corollary 2.2 improve Lemma 3.1, Corollary 3.1, and Corollary 3.2 in [16], respectively.

By Lemma 2.3 and Lemma 2.4 in [17], we have the following two lemmas.

Lemma 2.2 (see [17]) *For any $u = (u_1, u_2), v = (v_1, v_2) \in \mathcal{H}_k$, the following two equalities hold*:

$$-\sum_{t=-kT}^{kT-1} (\Delta\phi_1(\Delta u_1(t-1)), v_1(t)) = \sum_{t=-kT}^{kT-1} (\Delta\phi_1(\Delta u_1(t)), \Delta v_1(t)), \tag{2.12}$$

$$-\sum_{t=-kT}^{kT-1} (\Delta\phi_2(\Delta u_2(t-1)), v_2(t)) = \sum_{t=-kT}^{kT-1} (\Delta\phi_2(\Delta u_2(t)), \Delta v_2(t)). \tag{2.13}$$

Lemma 2.3 (see [17]) *Let $L : \mathbb{Z}[-kT, kT-1] \times \mathbb{R}^N \times \mathbb{R}^N \times \mathbb{R}^N \times \mathbb{R}^N \to \mathbb{R}$, $(t, x_1, x_2, y_1, y_2) \to L(t, x_1, x_2, y_1, y_2)$ and assume that L is continuously differential in (x_1, x_2, y_1, y_2) for all $t \in \mathbb{Z}[-kT, kT-1]$. Then the function $\varphi_k : \mathcal{H}_k \to \mathbb{R}$ defined by*

$$\varphi_k(u) = \varphi_k(u_1, u_2) = \sum_{t=-kT}^{kT-1} L(t, u_1(t), u_2(t), \Delta u_1(t), \Delta u_2(t))$$

is continuously differentiable on $\mathcal{H}_k$ and

$$\begin{aligned}
\langle \varphi_k'(u), v\rangle &= \langle \varphi_k'(u_1, u_2), (v_1, v_2)\rangle \\
&= \sum_{t=-kT}^{kT-1} \big[(D_{x_1} L(t, u_1(t), u_2(t), \Delta u_1(t), \Delta u_2(t)), v_1(t)) \\
&\quad + (D_{y_1} L(t, u_1(t), u_2(t), \Delta u_1(t), \Delta u_2(t)), \Delta v_1(t)) \\
&\quad + (D_{x_2} L(t, u_1(t), u_2(t), \Delta u_1(t), \Delta u_2(t)), v_2(t)) \\
&\quad + (D_{y_2} L(t, u_1(t), u_2(t), \Delta u_1(t), \Delta u_2(t)), \Delta v_2(t)) \big],
\end{aligned}$$

where $u, v \in \mathcal{H}_k$.

Let

$$L(t, x_1, x_2, y_1, y_2) = \Phi_1(y_1) + \Phi_2(y_2) + K(t, x_1, x_2) - W(t, x_1, x_2) + (f_{1,k}(t), x_1) + (f_{2,k}(t), x_2)$$

and define $\eta_k : \mathcal{H}_k \to [0, +\infty)$ by

$$\eta_k(u) = \eta_k(u_1, u_2) = \sum_{t=-kT}^{kT-1} \big[\Phi_1(\Delta u_1(t)) + \Phi_2(\Delta u_2(t)) + K(t, u_1(t), u_2(t))\big].$$

Then

$$\begin{aligned}
\varphi_k(u) &= \varphi_k(u_1, u_2) \\
&= \sum_{t=-kT}^{kT-1} \big[\Phi_1(\Delta u_1(t)) + \Phi_2(\Delta u_2(t)) + K(t, u_1(t), u_2(t))
\end{aligned}$$

$$- W\big(t,u_1(t),u_2(t)\big) + \big(f_{1,k}(t),u_1(t)\big) + \big(f_{2,k}(t),u_2(t)\big)\big]$$

$$= \eta_k(u) + \sum_{t=-kT}^{kT-1}\big[-W\big(t,u_1(t),u_2(t)\big) + \big(f_{1,k}(t),u_1(t)\big) + \big(f_{2,k}(t),u_2(t)\big)\big]. \tag{2.14}$$

It follows from $(\mathcal{A}0)$, $(\mathcal{V})$ and Lemma 2.3 that

$$\begin{aligned}\langle\varphi_k'(u),v\rangle &= \langle\varphi_k'(u_1,u_2),(v_1,v_2)\rangle\\ &= \sum_{t=-kT}^{kT-1}\big[\big(\phi_1\big(\Delta u_1(t)\big),\Delta v_1(t)\big) + \big(\phi_2\big(\Delta u_2(t)\big),\Delta v_2(t)\big)\\ &\quad + \big(\nabla_{u_1}K\big(t,u_1(t),u_2(t)\big),v_1(t)\big) + \big(\nabla_{u_2}K\big(t,u_1(t),u_2(t)\big),v_2(t)\big)\\ &\quad - \big(\nabla_{u_1}W\big(t,u_1(t),u_2(t)\big),v_1(t)\big) - \big(\nabla_{u_2}W\big(t,u_1(t),u_2(t)\big),v_2(t)\big)\\ &\quad + \big(f_{1,k}(t),v_1(t)\big) + \big(f_{2,k}(t),v_2(t)\big)\big], \quad \forall u,v\in\mathcal{H}_k.\end{aligned} \tag{2.15}$$

By Lemma 2.2, it is easy to see that critical points of φ_k in $\mathcal{H}_k$ are $2kT$-periodic solutions of system (2.1).

We shall use one linking method in [21] to obtain the critical points of φ (the details can be seen in [21]). Let $(E,\|\cdot\|)$ be a Banach space. Define a continuous map $\Gamma:[0,1]\times E\to E$ by $\Gamma(t,x)=\Gamma(t)x$, where $\Gamma(t)$ satisfies the following conditions:

(1) $\Gamma(0)=I$, the identity map.

(2) For each $t\in[0,1)$, $\Gamma(t)$ is a homeomorphism of E onto E and $\Gamma^{-1}(t)\in C(E\times[0,1),E)$.

(3) $\Gamma(1)E$ is a single point in E and $\Gamma(t)A$ converges uniformly to $\Gamma(1)E$ as $t\to 1$ for each bounded set $A\subset E$.

(4) For each $t_0\in[0,1)$ and each bounded set $A\subset E$,

$$\sup_{\substack{0\le t\le t_0\\ u\in A}}\left\{\|\Gamma(t)u\| + \|\Gamma^{-1}(t)u\|\right\} < \infty.$$

Let Φ be the set of all continuous maps Γ as defined above.

Definition 2.1 (see [21], Definition 3.2) We say that A links B[hm] if A and B are subsets of E such that $A\cap B=\emptyset$, and for each $\Gamma\in\Phi$, there is $t'\in(0,1]$ such that $\Gamma(t')A\cap B\neq\emptyset$.

Example 1 (see [21], p.21) Let B be an open set in E, and let A consist of two points e_1, e_2 with $e_1\in B$ and $e_2\notin\bar{B}$. Then A links ∂B[hm].

We use the following theorem to prove our main results.

Theorem 2.1 (see [21], Theorem 3.4 and Theorem 2.12) *Let E be a Banach space, $\varphi\in C^1(E,\mathbb{R})$ and A and B be two subsets of E such that A links B*[hm]. *Assume that*

$$\sup_A\varphi\le\inf_B\varphi$$

and

$$c:=\inf_{\Gamma\in\Phi}\sup_{\substack{s\in[0,1]\\ u\in A}}\varphi\big(\Gamma(s)u\big)<\infty.$$

Let $\psi(t)$ be a positive, nonincreasing, locally Lipschitz continuous function on $[0,\infty)$ satisfying $\int_0^\infty \psi(r)\,dr = \infty$. Then there exists a sequence $\{u_n\} \subset E$ such that $\varphi(u_n) \to c$ and $\varphi'(u_n)/\psi(\|u_n\|) \to 0$, as $n \to \infty$. Moreover, if $c = \sup_A \varphi$, then there is a sequence $\{u_n\} \subset E$ satisfying $\varphi(u_n) \to c$, $\varphi'(u_n) \to 0$, and $d(u_n, B) \to 0$, as $n \to \infty$.

Remark 2.2 Since A links B, by Definition 2.1, it is easy to know that $c \geq \inf_B \varphi$. By [21], if we let $\psi(r) = \frac{1}{1+r}$, the sequence $\{u_n\}$ is the Cerami sequence that is $\{u_n\}$ satisfying

$$\varphi(u_n) \to c, \qquad (1 + \|u_n\|)\|\varphi'(u_n)\| \to 0, \quad \text{as } n \to \infty.$$

3 Proofs

Lemma 3.1 *Suppose that* (H2) *holds. Then*

$$K(t,x_1,x_2) \leq 2^{p-1} K\left(t, \frac{x_1}{|x|}, \frac{x_2}{|x|}\right)\left(|x_1|^p + |x_2|^p\right) \quad \text{for all } t \in \mathbb{Z}[0, T-1], |x| \geq 1;$$

$$K(t,x_1,x_2) \geq \frac{1}{2} K\left(t, \frac{x_1}{|x|}, \frac{x_2}{|x|}\right)\left(|x_1|^p + |x_2|^p\right) \quad \text{for all } t \in \mathbb{Z}[0, T-1], |x| \leq 1.$$

Proof Define the function $\xi \in (0,+\infty) \to K(t, \xi^{-1}x_1, \xi^{-1}x_2)(\xi^p + \xi^p)$. Then we have

$$\begin{aligned}
&\left(K\left(t,\xi^{-1}x_1,\xi^{-1}x_2\right)\left(\xi^p+\xi^p\right)\right)'_\xi \\
&\quad = -\left(\nabla_{x_1} K\left(t,\xi^{-1}x_1,\xi^{-1}x_2\right), \xi^{-2}x_1\right)\left(\xi^p+\xi^p\right) \\
&\qquad - \left(\nabla_{x_2} K\left(t,\xi^{-1}x_1,\xi^{-1}x_2\right), \xi^{-2}x_2\right)\left(\xi^p+\xi^p\right) + K\left(t,\xi^{-1}x_1,\xi^{-1}x_2\right)\left(p\xi^{p-1}+p\xi^{p-1}\right) \\
&\quad \geq -\left(\nabla_{x_1} K\left(t,\xi^{-1}x_1,\xi^{-1}x_2\right), \xi^{-1}x_1\right)\left(\xi^{p-1}+\xi^{p-1}\right) \\
&\qquad - \left(\nabla_{x_2} K\left(t,\xi^{-1}x_1,\xi^{-1}x_2\right), \xi^{-1}x_2\right)\left(\xi^{p-1}+\xi^{p-1}\right) + pK\left(t,\xi^{-1}x_1,\xi^{-1}x_2\right)\left(\xi^{p-1}+\xi^{p-1}\right) \\
&\quad \geq 0.
\end{aligned}$$

Hence the function $\xi \in (0,+\infty) \to K(t, \xi^{-1}x_1, \xi^{-1}x_2)(\xi^p + \xi^p)$ is nondecreasing. Moreover, note that

$$\frac{|x_1|^p + |x_2|^p}{2} \leq |x|^p \leq \left(|x_1| + |x_2|\right)^p \leq 2^{p-1}\left(|x_1|^p + |x_2|^p\right).$$

Then the proof can be completed easily. □

Lemma 3.2 *Suppose that* (H1) *holds. Then, for any $u \in \mathcal{H}_k$,*

$$\begin{aligned}
\eta_k(u) &\geq \min\left\{d_1\|u_1\|^p_{\mathcal{H}_{1,k}}, a_1 C_*^{\gamma-p}\|u_1\|^\gamma_{\mathcal{H}_{1,k}}\right\} \\
&\quad + \min\left\{d_2\|u_2\|^p_{\mathcal{H}_{1,k}}, a_2 C_*^{\gamma-p}\|u_2\|^\gamma_{\mathcal{H}_{2,k}}\right\}, \quad \forall k \in \mathbb{N}.
\end{aligned}$$

Proof It follows from $(\mathcal{A}1)$, (H1), $\gamma \in (1,p)$ and Corollary 2.2 that

$$\begin{aligned}
\eta_k(u) &= \sum_{t=-kT}^{kT-1} \left[\Phi_1\left(\Delta u_1(t)\right) + \Phi_2\left(\Delta u_2(t)\right) + K\left(t, u_1(t), u_2(t)\right)\right] \\
&\geq \sum_{t=-kT}^{kT-1} \left[d_1\left|\Delta u_1(t)\right|^p + d_2\left|\Delta u_2(t)\right|^p + a_1\left|u_1(t)\right|^\gamma + a_2\left|u_2(t)\right|^\gamma\right]
\end{aligned}$$

$$\geq \sum_{t=-kT}^{kT-1} \left[d_1|\Delta u_1(t)|^p + d_2|\Delta u_2(t)|^p + a_1\|u_1\|_{l^\infty_{2kT}}^{\gamma-p}|u_1(t)|^p + a_2\|u_2\|_{l^\infty_{2kT}}^{\gamma-p}|u_2(t)|^p\right]$$

$$\geq \sum_{t=-kT}^{kT-1} \left[d_1|\Delta u_1(t)|^p + d_2|\Delta u_2(t)|^p + a_1\left(C_*\|u_1\|_{\mathcal{H}_{1,k}}\right)^{\gamma-p}|u_1(t)|^p\right.$$
$$\left. + a_2\left(C_*\|u_2\|_{\mathcal{H}_{2,k}}\right)^{\gamma-p}|u_2(t)|^p\right]$$
$$\geq \min\left\{d_1, a_1\left(C_*\|u_1\|_{\mathcal{H}_{1,k}}\right)^{\gamma-p}\right\}\|u_1\|^p_{\mathcal{H}_{1,k}} + \min\left\{d_2, a_2\left(C_*\|u_2\|_{\mathcal{H}_{1,k}}\right)^{\gamma-p}\right\}\|u_2\|^p_{\mathcal{H}_{2,k}}$$
$$= \min\left\{d_1\|u_1\|^p_{\mathcal{H}_{1,k}}, a_1C_*^{\gamma-p}\|u_1\|^\gamma_{\mathcal{H}_{1,k}}\right\} + \min\left\{d_2\|u_2\|^p_{\mathcal{H}_{2,k}}, a_2C_*^{\gamma-p}\|u_2\|^\gamma_{\mathcal{H}_{2,k}}\right\}. \qquad \square$$

Proof of Theorem 1.1 We divide the proof into the following Lemmas 3.3-3.5.

Lemma 3.3 *Under the assumptions of Theorem* 1.1, *for every $k \in \mathbb{N}$, system* (2.1) *has a nontrivial solution u_k in $\mathcal{H}_k$.*

Proof We first construct A and B which satisfy the assumptions in Theorem 2.1.

(i) When $r \in (0,1]$, by Corollary 2.2, (H1), (H3)(i), Hölder's inequality and $\gamma < p$, for $u \in \mathcal{H}_k$ with $\|u\|_{\mathcal{H}_k} = r$, we have $\|u_1\|_{l^\infty_{2kT}} \leq C_*\|u_1\|_{\mathcal{H}_{1,k}} \leq rC_*$ and $\|u_2\|_{l^\infty_{2kT}} \leq C_*\|u_2\|_{\mathcal{H}_{2,k}} \leq rC_*$,

$$\varphi_k(u) \geq \eta_k(u) - b_1\sum_{t=-kT}^{kT-1}|u_1(t)|^p - b_2\sum_{t=-kT}^{kT-1}|u_2(t)|^p$$
$$- \left(\sum_{t=-kT}^{kT-1}|f_{1,k}(t)|^{p'}\right)^{1/p'}\left(\sum_{t=-kT}^{kT-1}|u_1(t)|^p\right)^{1/p}$$
$$- \left(\sum_{t=-kT}^{kT-1}|f_{2,k}(t)|^{p'}\right)^{1/p'}\left(\sum_{t=-kT}^{kT-1}|u_2(t)|^p\right)^{1/p}$$
$$\geq \sum_{t=-kT}^{kT-1}\left[d_1|\Delta u_1(t)|^p + d_2|\Delta u_2(t)|^p + a_1|u_1(t)|^\gamma + a_2|u_2(t)|^\gamma\right] - b_1\sum_{t=-kT}^{kT-1}|u_1(t)|^p$$
$$- b_2\sum_{t=-kT}^{kT-1}|u_2(t)|^p - \left(\sum_{t=-kT}^{kT-1}|f_{1,k}(t)|^{p'}\right)^{1/p'}\left(\sum_{t=-kT}^{kT-1}|u_1(t)|^p\right)^{1/p}$$
$$- \left(\sum_{t=-kT}^{kT-1}|f_{2,k}(t)|^{p'}\right)^{1/p'}\left(\sum_{t=-kT}^{kT-1}|u_2(t)|^p\right)^{1/p}$$
$$\geq \sum_{t=-kT}^{kT-1} d_1|\Delta u_1(t)|^p + \sum_{t=-kT}^{kT-1} d_2|\Delta u_2(t)|^p$$
$$+ a_1\left(C_*\|u_1\|_{\mathcal{H}_{1,k}}\right)^{\gamma-p}\sum_{t=-kT}^{kT-1}|u_1(t)|^p - b_1\sum_{t=-kT}^{kT-1}|u_1(t)|^p$$
$$- b_2\sum_{t=-kT}^{kT-1}|u_2(t)|^p + a_2\left(C_*\|u_2\|_{\mathcal{H}_{2,k}}\right)^{\gamma-p}\sum_{t=-kT}^{kT-1}|u_2(t)|^p$$
$$- \|f_1\|_{l^{p'}}\|u_1\|_{\mathcal{H}_{1,k}} - \|f_2\|_{l^{p'}}\|u_2\|_{\mathcal{H}_{2,k}}$$
$$\geq \min\left\{d_1, a_1(C_*r)^{\gamma-p} - b_1\right\}\|u_1\|^p_{\mathcal{H}_{1,k}} + \min\left\{d_2, a_2(C_*r)^{\gamma-p} - b_2\right\}\|u_2\|^p_{\mathcal{H}_{2,k}}$$

$$
\begin{aligned}
&- \|f_1\|_{l^{p'}} \|u_1\|_{\mathcal{H}_{1,k}} - \|f_2\|_{l^{p'}} \|u_2\|_{\mathcal{H}_{2,k}} \\
&\geq \min\{d_1, a_1 C_*^{\gamma-p} - b_1\} \|u_1\|_{\mathcal{H}_{1,k}}^p + \min\{d_2, a_2 C_*^{\gamma-p} - b_2\} \|u_2\|_{\mathcal{H}_{2,k}}^p \\
&- \|f_1\|_{l^{p'}} \|u_1\|_{\mathcal{H}_{1,k}} - \|f_2\|_{l^{p'}} \|u_2\|_{\mathcal{H}_{2,k}} \\
&\geq \min\{d_1, d_2, a_1 C_*^{\gamma-p} - b_1, a_2 C_*^{\gamma-p} - b_2\} \frac{1}{2^{p-1}} \left(\|u_1\|_{\mathcal{H}_{1,k}} + \|u_2\|_{\mathcal{H}_{2,k}}\right)^p \\
&- \max\{\|f_1\|_{l^{p'}}, \|f_2\|_{l^{p'}}\} \left(\|u_1\|_{\mathcal{H}_{1,k}} + \|u_2\|_{\mathcal{H}_{2,k}}\right).
\end{aligned} \tag{3.1}
$$

(H6)(i) implies that there exists $\alpha > 0$ such that

$$\varphi_k(u) \geq \alpha > 0 \quad \text{for all } u \in \mathcal{H}_k \text{ with } \|u\|_{\mathcal{H}_k} = r, \forall k \in \mathbb{N}.$$

(ii) When $r \in (1, +\infty)$, by Corollary 2.2, (H1), (H3)(ii), Hölder's inequality and $\gamma < p$, for $u \in \mathcal{H}_k$ with $\|u\|_{\mathcal{H}_k} = r$, we have

$$
\begin{aligned}
\varphi_k(u) &\geq \min\{d_1, a_1 (C_* r)^{\gamma-p} - b_1\} \|u_1\|_{\mathcal{H}_{1,k}}^p + \min\{d_2, a_2 (C_* r)^{\gamma-p} - b_2\} \|u_2\|_{\mathcal{H}_{2,k}}^p \\
&- \|f_1\|_{l^{p'}} \|u_1\|_{\mathcal{H}_{1,k}} - \|f_2\|_{l^{q'}} \|u_2\|_{\mathcal{H}_{2,k}} \\
&\geq \min\{d_1, d_2, a_1 (C_* r)^{\gamma-p} - b_1, a_2 (C_* r)^{\gamma-p} - b_2\} \frac{1}{2^{p-1}} \left(\|u_1\|_{\mathcal{H}_{1,k}} + \|u_2\|_{\mathcal{H}_{2,k}}\right)^p \\
&- \max\{\|f_1\|_{l^{p'}}, \|f_2\|_{l^{p'}}\} \left(\|u_1\|_{\mathcal{H}_{1,k}} + \|u_2\|_{\mathcal{H}_{2,k}}\right).
\end{aligned} \tag{3.2}
$$

(H6)(ii) implies that there exists $\alpha > 0$ such that

$$\varphi_k(u) \geq \alpha > 0 \quad \text{for all } u \in \mathcal{H}_k \text{ with } \|u\|_{\mathcal{H}_k} = r, \forall k \in \mathbb{N}.$$

By Lemma 3.1 and the T-periodicity of K, there exists a constant $B_0 > 0$ such that

$$K(t, x_1, x_2) \leq 2^{p-1} A_0 \left(|x_1|^p + |x_2|^p\right) + B_0 \quad \text{for all } (t, x_1, x_2) \in \mathbb{Z} \times \mathbb{R}^N \times \mathbb{R}^N, \tag{3.3}$$

where

$$A_0 = \max_{|x_1| \leq 1, |x_2| \leq 1, t \in \mathbb{Z}[0, T-1]} K(t, x_1, x_2).$$

By (H4), we know that there exist $\varepsilon_0 > 0$ and $L > 0$ such that

$$
\begin{aligned}
&W(t, x_1, x_2) \geq \left(d_3 + d_4 + 2^{p-1} A_0 + \varepsilon_0\right) \left(|x_1|^p + |x_2|^p\right) \\
&\quad \text{for all } t \in \mathbb{Z}[0, T-1] \text{ and } \forall |x| \geq L.
\end{aligned} \tag{3.4}
$$

By (3.4) and the T-periodicity of W, there exists a constant $B_1 > 0$ such that

$$W(t, x_1, x_2) \geq \left(d_3 + d_4 + 2^{p-1} A_0 + \varepsilon_0\right) \left(|x_1|^p + |x_2|^p\right) - B_1 \tag{3.5}$$

for all $(t, x_1, x_2) \in \mathbb{Z}[0, T-1] \times \mathbb{R}^N \times \mathbb{R}^N$. For any $k \in \mathbb{N}$, define $w^{(k)} \in \mathcal{H}_k$ by

$$w^{(k)}(t) = \left(w_1^{(k)}(t), w_2^{(k)}(t)\right) = \begin{cases} (1, 0, \ldots, 0, 1, 0, \ldots, 0) & \text{if } t = 0, \\ 0 & \text{if } t \in \mathbb{Z}[-kT, kT-1]/\{0\}, \end{cases}$$

where

$$w_i^{(k)}(t) = \begin{cases} (1,0,\ldots,0) & \text{if } t = 0, \\ 0 & \text{if } t \in \mathbb{Z}[-kT, kT-1]/\{0\}, \end{cases} \qquad i = 1, 2.$$

Since $K(t,0,0) \equiv 0$ and $W(t,0,0) \equiv 0$, which are implied by (H2) and (H5), then by (3.3) and (3.5) we have

$$\begin{aligned} \varphi_k(\xi w^{(k)}) &= \sum_{t=-kT}^{kT-1} \Big[\Phi_1(\xi \Delta w_1^{(k)}(t)) + \Phi_2(\xi \Delta w_2^{(k)}(t)) + K(t, \xi w_1^{(k)}(t), \xi w_2^{(k)}(t)) \\ &\quad - W(t, \xi w_1^{(k)}(t), \xi w_2^{(k)}(t)) + \xi (f_{1,k}(t), w_1^{(k)}(t)) + \xi (f_{2,k}(t), w_2^{(k)}(t))\Big] \\ &\le d_3 |\xi|^p \sum_{t=-kT}^{kT-1} |\Delta w_1^{(k)}(t)|^p + d_4 |\xi|^p \sum_{t=-kT}^{kT-1} |\Delta w_2^{(k)}(t)|^p \\ &\quad + K(0, \xi w_1^{(k)}(0), \xi w_2^{(k)}(0)) - W(0, \xi w_1^{(k)}(0), \xi w_2^{(k)}(0)) \\ &\quad + \xi (f_{1,k}(0), w_1^{(k)}(0)) + \xi (f_{2,k}(0), w_2^{(k)}(0)) \\ &= d_3 |\xi|^p (|\Delta w_1^{(k)}(-1)|^p + |\Delta w_1^{(k)}(0)|^p) + d_4 |\xi|^p (|\Delta w_2^{(k)}(-1)|^p + |\Delta w_2^{(k)}(0)|^p) \\ &\quad + K(0, \xi w_1^{(k)}(0), \xi w_2^{(k)}(0)) - W(0, \xi w_1^{(k)}(0), \xi w_2^{(k)}(0)) \\ &\quad + \xi (f_{1,k}(0), w_1^{(k)}(0)) + \xi (f_{2,k}(0), w_2^{(k)}(0)) \\ &\le 2d_3 |\xi|^p + 2d_4 |\xi|^p + 2^{p-1} A_0 |\xi|^p + 2^{p-1} A_0 |\xi|^p \\ &\quad + B_0 - (d_3 + d_4 + 2^{p-1} A_0 + \varepsilon_0)(|\xi|^p + |\xi|^p) \\ &\quad + B_1 + |\xi| |f_{1,k}(0)| + |\xi| |f_{2,k}(0)| \\ &\le -2\varepsilon_0 |\xi|^p + |\xi| |f_{1,k}(0)| + |\xi| |f_{2,k}(0)| + B_0 + B_1. \end{aligned} \tag{3.6}$$

So there exists $\xi_0 \in \mathbb{R}$ such that $\|\xi_0 w^{(k)}\| > r$ and $\varphi_k(\xi_0 w^{(k)}) < 0$. Moreover, it is clear that $\varphi_k(0) = 0$. Let $e_1 = \xi_0 w^{(k)}$ and

$$A = \{0, e_1\}, \qquad B = \{u \in \mathcal{H}_k : \|u\|_{\mathcal{H}_k} < r\}.$$

Then $0 \in B$ and $e_1 \notin \bar{B}$. So by Example 1 in Section 2, we know that A links ∂B[hm]. So by Theorem 2.1 and Remark 2.2, we have

$$c_k = \inf_{\Gamma \in \Phi} \sup_{\substack{s \in [0,1] \\ u \in A}} \varphi_k(\Gamma(s)u) \ge \inf_{\partial B} \varphi_k > \alpha > 0, \tag{3.7}$$

and there exists a sequence $\{u_n = (u_1^{(n)}, u_2^{(n)})\}_{n=1}^{\infty} \subset \mathcal{H}_k$ such that

$$\varphi_k(u_n) \to c_k, \qquad (1 + \|u_n\|_{\mathcal{H}_k}) \|\varphi_k'(u_n)\| \to 0.$$

Then there exists a constant $C_{1k} > 0$ such that

$$|\varphi_k(u_n)| \le C_{1k}, \qquad (1 + \|u_n\|_{\mathcal{H}_k}) \|\varphi_k'(u_n)\| \le C_{1k} \quad \text{for all } n \in \mathbb{N}. \tag{3.8}$$

It follows from (H5) and the T-periodicity and continuity of W, $\nabla_{x_1} W$ and $\nabla_{x_2} W$ that

$$\begin{aligned}&\left[\left(\nabla_{x_1} W(t,x_1,x_2),x_1\right) + \left(\nabla_{x_2} W(t,x_1,x_2),x_2\right) - pW(t,x_1,x_2)\right]\left(\zeta + \eta_1|x_1|^{\nu} + \eta_2|x_2|^{\nu}\right)\\ &\quad \geq W(t,x_1,x_2) \geq 0, \quad \forall (t,x_1,x_2) \in \mathbb{Z} \times \mathbb{R}^N \times \mathbb{R}^N. \end{aligned} \tag{3.9}$$

So by (3.5) and $p - \nu > 1$, there exists $C_2 > 0$ such that

$$\begin{aligned}&\left[\left(\nabla_{x_1} W(t,x_1,x_2),x_1\right) + \left(\nabla_{x_2} W(t,x_1,x_2),x_2\right) - pW(t,x_1,x_2)\right]\\
&\quad \geq \frac{W(t,x_1,x_2)}{\zeta + \eta_1|x_1|^{\nu} + \eta_2|x_2|^{\nu}}\\
&\quad \geq \frac{(d_3 + d_4 + 2^{p-1}A_0 + \varepsilon_0)(|x_1|^p + |x_2|^p) - B_1}{\zeta + \eta_1|x_1|^{\nu} + \eta_2|x_2|^{\nu}}\\
&\quad \geq \frac{(d_3 + d_4 + 2^{p-1}A_0 + \varepsilon_0)\frac{1}{2^{p-1}}(|x_1| + |x_2|)^p - B_1}{\zeta + 2\max\{\eta_1,\eta_2\}(|x_1| + |x_2|)^{\nu}}\\
&\quad \geq \frac{(d_3 + d_4 + 2^{p-1}A_0 + \varepsilon_0)\frac{1}{2^{p-1}}}{4\max\{\eta_1,\eta_2\}}\left(|x_1| + |x_2|\right)^{p-\nu} - C_2\\
&\quad \geq \frac{(d_3 + d_4 + 2^{p-1}A_0 + \varepsilon_0)\frac{1}{2^{p-1}}}{4\max\{\eta_1,\eta_2\}}\left(|x_1|^{p-\nu} + |x_2|^{p-\nu}\right) - C_2, \quad \forall x \in \mathbb{R}^N. \end{aligned} \tag{3.10}$$

Hence, it follows from (H2), (3.8) and (3.10) that

$$\begin{aligned}&pC_{1k} + C_{1k}\\
&\quad \geq p\varphi_k(u_n) - \langle \varphi_k'(u_n), u_n\rangle\\
&\quad = p\varphi_k(u_1^{(n)}, u_2^{(n)}) - \langle \varphi_k'(u_1^{(n)}, u_2^{(n)}), (u_1^{(n)}, u_2^{(n)})\rangle\\
&\quad \geq \sum_{t=-kT}^{kT-1}\Big[\left(\nabla_{u_1} W(t,u_1^{(n)}(t),u_2^{(n)}(t)),u_1^{(n)}(t)\right) + \left(\nabla_{u_2} W(t,u_1^{(n)}(t),u_2^{(n)}(t)),u_2^{(n)}(t)\right)\\
&\qquad - pW(t,u_1^{(n)}(t),u_2^{(n)}(t))\Big]\\
&\qquad + (p-1)\sum_{t=-kT}^{kT-1}(f_{1,k}(t),u_1^{(n)}(t)) + (p-1)\sum_{t=-kT}^{kT-1}(f_{2,k}(t),u_2^{(n)}(t))\\
&\quad \geq \frac{(d_3 + d_4 + 2^{p-1}A_0 + \varepsilon_0)\frac{1}{2^{p-1}}}{4\max\{\eta_1,\eta_2\}}\sum_{t=-kT}^{kT-1}\left(|u_1^{(n)}(t)|^{p-\nu} + |u_2^{(n)}(t)|^{p-\nu}\right) - 2kTC_2\\
&\qquad - (p-1)\sum_{t=-kT}^{kT-1}|f_{1,k}(t)||u_1^{(n)}(t)| - (p-1)\sum_{t=-kT}^{kT-1}|f_{2,k}(t)||u_2^{(n)}(t)|\\
&\quad \geq \frac{(d_3 + d_4 + 2^{p-1}A_0 + \varepsilon_0)\frac{1}{2^{p-1}}}{4\max\{\eta_1,\eta_2\}}\sum_{t=-kT}^{kT-1}\left(|u_1^{(n)}(t)|^{p-\nu} + |u_2^{(n)}(t)|^{p-\nu}\right) - 2kTC_2\\
&\qquad - (p-1)\left(\sum_{t=-kT}^{kT-1}|f_{1,k}(t)|^{\frac{p-\nu}{p-\nu-1}}\right)^{\frac{p-\nu-1}{p-\nu}}\left(\sum_{t=-kT}^{kT-1}|u_1^{(n)}(t)|^{p-\nu}\right)^{1/(p-\nu)}\\
&\qquad - (p-1)\left(\sum_{t=-kT}^{kT-1}|f_{2,k}(t)|^{\frac{p-\nu}{p-\nu-1}}\right)^{\frac{p-\nu-1}{p-\nu}}\left(\sum_{t=-kT}^{kT-1}|u_2^{(n)}(t)|^{p-\nu}\right)^{1/(p-\nu)}. \end{aligned} \tag{3.11}$$

The fact $p - \nu > 1$ and the above inequality show that $\sum_{t=-kT}^{kT-1} |u_1^{(n)}(t)|^{p-\nu}$ and $\sum_{t=-kT}^{kT-1} |u_2^{(n)}(t)|^{p-\nu}$ are bounded. By ($\mathcal{A}1$), (H1), (H6), (3.8), (3.9), (3.11), Hölder's inequality and Corollary 2.2, we have

$$
\begin{aligned}
& d_1 \| u_1^{(n)} \|_{\mathcal{H}_{1,k}}^p + d_2 \| u_2^{(n)} \|_{\mathcal{H}_{2,k}}^p \\
& = d_1 \sum_{t=-kT}^{kT-1} |\Delta u_1^{(n)}(t)|^p + d_1 \sum_{t=-kT}^{kT-1} |u_1^{(n)}(t)|^p + d_2 \sum_{t=-kT}^{kT-1} |\Delta u_2^{(n)}(t)|^p + d_2 \sum_{t=-kT}^{kT-1} |u_2^{(n)}(t)|^p \\
& \leq \varphi_k(u^{(n)}) - \sum_{t=-kT}^{kT-1} K(t, u_1^{(n)}(t), u_2^{(n)}(t)) \\
& \quad + \sum_{t=-kT}^{kT-1} W(t, u_1^{(n)}(t), u_2^{(n)}(t)) + d_1 \sum_{t=-kT}^{kT-1} |u_1^{(n)}(t)|^p \\
& \quad + d_2 \sum_{t=-kT}^{kT-1} |u_2^{(n)}(t)|^p - \sum_{t=-kT}^{kT-1} (f_{1k}(t), u_1^{(n)}(t)) - \sum_{t=-kT}^{kT-1} (f_{2k}(t), u_2^{(n)}(t)) \\
& \leq \varphi_k(u^{(n)}) \\
& \quad + \sum_{t=-kT}^{kT-1} [(\nabla_{u_1} W(t, u_1^{(n)}(t), u_2^{(n)}(t)), u_1^{(n)}(t)) + (\nabla_{u_2} W(t, u_1^{(n)}(t), u_2^{(n)}(t)), u_2^{(n)}(t)) \\
& \quad - pW(t, u_1^{(n)}(t), u_2^{(n)}(t))](\zeta + \eta_1 |u_1^{(n)}(t)|^\nu + \eta_2 |u_2^{(n)}(t)|^\nu) \\
& \quad + d_1 \sum_{t=-kT}^{kT-1} |u_1^{(n)}(t)|^p + d_2 \sum_{t=-kT}^{kT-1} |u_2^{(n)}(t)|^p + \left(\sum_{t=-kT}^{kT-1} |u_1^{(n)}(t)|^p \right)^{\frac{1}{p}} \left(\sum_{t\in\mathbb{Z}} |f_1(t)|^{p'} \right)^{\frac{1}{p'}} \\
& \quad + \left(\sum_{t=-kT}^{kT-1} |u_2^{(n)}(t)|^p \right)^{\frac{1}{p}} \left(\sum_{t\in\mathbb{Z}} |f_2(t)|^{p'} \right)^{\frac{1}{p'}} \\
& \leq C_{1k} + d_1 \sum_{t=-kT}^{kT-1} |u_1^{(n)}(t)|^p + d_2 \sum_{t=-kT}^{kT-1} |u_2^{(n)}(t)|^p + \| u_1^{(n)} \|_{\mathcal{H}_{1,k}} \left(\sum_{t\in\mathbb{Z}} |f_1(t)|^{p'} \right)^{\frac{1}{p'}} \\
& \quad + \| u_2^{(n)} \|_{\mathcal{H}_{2,k}} \left(\sum_{t\in\mathbb{Z}} |f_2(t)|^{p'} \right)^{\frac{1}{p'}} \\
& \quad + (\zeta + \eta_1 \| u_1^{(n)} \|_{l_{2kT}^\infty}^\nu + \eta_2 \| u_2^{(n)} \|_{l_{2kT}^\infty}^\nu) \\
& \quad \cdot \sum_{t=-kT}^{kT-1} [(\nabla_{u_1} W(t, u_1^{(n)}(t), u_2^{(n)}(t)), u_1^{(n)}(t)) + (\nabla_{u_2} W(t, u_1^{(n)}(t), u_2^{(n)}(t)), u_2^{(n)}(t)) \\
& \quad - pW(t, u_1^{(n)}(t), u_2^{(n)}(t))] \\
& \leq C_{1k} + d_1 \| u_1^{(n)} \|_{l_{2kT}^\infty}^\nu \sum_{t=-kT}^{kT-1} |u_1^{(n)}(t)|^{p-\nu} + d_2 \| u_2^{(n)} \|_{l_{2kT}^\infty}^\nu \sum_{t=-kT}^{kT-1} |u_2^{(n)}(t)|^{p-\nu} \\
& \quad + \| u_1^{(n)} \|_{\mathcal{H}_{1,k}} \left(\sum_{t\in\mathbb{Z}} |f_1(t)|^{p'} \right)^{\frac{1}{p'}} + \| u_2^{(n)} \|_{\mathcal{H}_{2,k}} \left(\sum_{t\in\mathbb{Z}} |f_2(t)|^{p'} \right)^{\frac{1}{p'}} \\
& \quad + (\zeta + \eta_1 \| u_1^{(n)} \|_{l_{2kT}^\infty}^\nu + \eta_2 \| u_2^{(n)} \|_{l_{2kT}^\infty}^\nu)
\end{aligned}
$$

$$\cdot\Big[(p+1)C_{1k}+(p-1)\|u_1^{(n)}\|_{\mathcal{H}_{1,k}}\Big(\sum_{t\in\mathbb{Z}}|f_1(t)|^{p'}\Big)^{\frac{1}{p'}}$$

$$+(p-1)\|u_2^{(n)}\|_{\mathcal{H}_{2,k}}\Big(\sum_{t\in\mathbb{Z}}|f_2(t)|^{p'}\Big)^{\frac{1}{p'}}\Big]$$

$$\leq C_{1k}+d_1C_*^{\nu}\|u_1^{(n)}\|_{\mathcal{H}_{1,k}}^{\nu}\sum_{t=-kT}^{kT-1}|u_1^{(n)}(t)|^{p-\nu}+d_2C_*^{\nu}\|u_2^{(n)}\|_{\mathcal{H}_{2,k}}^{\nu}\sum_{t=-kT}^{kT-1}|u_2^{(n)}(t)|^{p-\nu}$$

$$+\|u_1^{(n)}\|_{\mathcal{H}_{1,k}}\Big(\sum_{t\in\mathbb{Z}}|f_1(t)|^{p'}\Big)^{\frac{1}{p'}}+\|u_2^{(n)}\|_{\mathcal{H}_{2,k}}\Big(\sum_{t\in\mathbb{Z}}|f_2(t)|^{p'}\Big)^{\frac{1}{p'}}$$

$$+\big(\zeta+\eta_1C_*^{\nu}\|u_1^{(n)}\|_{\mathcal{H}_{1,k}}^{\nu}+\eta_2C_*^{\nu}\|u_2^{(n)}\|_{\mathcal{H}_{2,k}}^{\nu}\big)$$

$$\cdot\Big[(p+1)C_{1k}+(p-1)\|u_1^{(n)}\|_{\mathcal{H}_{1,k}}\Big(\sum_{t\in\mathbb{Z}}|f_1(t)|^{p'}\Big)^{\frac{1}{p'}}$$

$$+(p-1)\|u_2^{(n)}\|_{\mathcal{H}_{2,k}}\Big(\sum_{t\in\mathbb{Z}}|f_2(t)|^{p'}\Big)^{\frac{1}{p'}}\Big]. \tag{3.12}$$

Since $\nu < p-1$, (3.12) and the boundedness of $\sum_{t=-kT}^{kT-1}|u_1^{(n)}(t)|^{p-\nu}$ and $\sum_{t=-kT}^{kT-1}|u_2^{(n)}(t)|^{p-\nu}$ imply that $\|u_1^{(n)}\|_{\mathcal{H}_{1,k}}$ and $\|u_2^{(n)}\|_{\mathcal{H}_{2,k}}$ are bounded. Since $\mathcal{H}$ is a finite-dimensional space, $\{u^{(n)}=(u_1^{(n)},u_2^{(n)})\}$ has a convergence subsequence, still denoted by $\{u^{(n)}=(u_1^{(n)},u_2^{(n)})\}$, such that $u^{(n)}=(u_1^{(n)},u_2^{(n)})\to u_k=(u_{1k},u_{2k})$ as $n\to\infty$. Moreover, by the continuity of φ_k and φ_k', we obtain $\varphi_k'(u_k)=0$ and $\varphi_k(u_k)=c_k>0$. It is clear that $u_k\neq 0$ and so u_k is a desired nontrivial solution of system (2.1). The proof is complete. □

Lemma 3.4 *Let $\{u_k=(u_{1k},u_{2k})\}_{k\in\mathbb{N}}$ be the solutions of system (2.1). Then there exists $M_1>0$ such that $\|u_{1k}\|_{l_{2kT}^{\infty}}\leq M_1$ and $\|u_{2k}\|_{l_{2kT}^{\infty}}\leq M_1$.*

Proof First, we prove that the sequence $\{c_k\}_{k\in\mathbb{N}}$ is bounded. For every $k\in\mathbb{N}$, define $\Gamma_k:[0,1]\times\mathcal{H}_k\to\mathcal{H}_k$ by

$$\Gamma_k(s)v=(1-s)v,\quad v\in\mathcal{H}_k.$$

Then $\Gamma\in\Phi$. Note that the set $A=\{0,e_1\}$. So (3.7) and the argument of (3.6) imply that

$$\varphi_k(u_k)=c_k\leq\sup_{s\in[0,1]u\in A}\varphi_k\big((1-s)u\big)$$

$$=\sup_{s\in[0,1]}\varphi_k\big((1-s)e_1\big)$$

$$=\sup_{s\in[0,1]}\varphi_k\big((1-s)e_1\big)$$

$$=\sup_{s\in[0,1]}\Big\{\sum_{t=-kT}^{kT-1}\big[\Phi_1\big((1-s)\Delta w_1^{(k)}(t)\big)+\Phi_2\big((1-s)\Delta w_2^{(k)}(t)\big)$$

$$+K\big(t,(1-s)w_1^{(k)}(t),(1-s)w_2^{(k)}(t)\big)-W\big(t,(1-s)w_1^{(k)}(t),(1-s)w_2^{(k)}(t)\big)$$

$$+(1-s)\big(f_{1,k}(t),w_1^{(k)}(t)\big)+(1-s)\big(f_{2,k}(t),w_2^{(k)}(t)\big)\big]\Big\}$$

$$\begin{aligned}
&\leq \sup_{s\in[0,1]}\left\{-2\varepsilon_0|1-s|^p + |1-s|\left|f_{1,k}(0)\right| + |1-s|\left|f_{2,k}(0)\right| + B_0 + B_1\right\} \\
&\leq \left|f_1(0)\right| + \left|f_2(0)\right| + B_0 + B_1 := M_2,
\end{aligned} \tag{3.13}$$

where M_2 is independent of $k \in \mathbb{N}$, which implies that the sequence $\{c_k\}_{k\in\mathbb{N}}$ is bounded. Moreover, $\varphi_k'(u_k) = 0$. Then it follows from ($\mathcal{A}1$), (H2), and (3.10) that

$$\begin{aligned}
pM_2 \geq pc_k &= p\varphi_k(u_k) - \langle\varphi_k'(u_k), u_k\rangle \\
&= p\varphi_k(u_{1k}, u_{2k}) - \langle\varphi_k'(u_{1k}, u_{2k}), (u_{1k}, u_{2k})\rangle \\
&\geq \sum_{t=-kT}^{kT-1}\big[\big(\nabla_{u_1} W\big(t, u_{1k}(t), u_{2k}(t)\big), u_{1k}(t)\big) + \big(\nabla_{u_2} W\big(t, u_{1k}(t), u_{2k}(t)\big), u_{2k}(t)\big) \\
&\quad - pW\big(t, u_{1k}(t), u_{2k}(t)\big)\big] \\
&\quad + (p-1)\sum_{t=-kT}^{kT-1}\big(f_{1k}(t), u_{1k}(t)\big) + (p-1)\sum_{t=-kT}^{kT-1}\big(f_{2k}(t), u_{2k}(t)\big) \\
&\geq \sum_{t=-kT}^{kT-1}\frac{W(t, u_{1k}(t), u_{2k}(t))}{\xi + \eta_1|u_{1k}(t)|^\nu + \eta_2|u_{2k}(t)|^\nu} + (p-1)\sum_{t=-kT}^{kT-1}\big(f_{1k}(t), u_{1k}(t)\big) \\
&\quad + (p-1)\sum_{t=-kT}^{kT-1}\big(f_{2k}(t), u_{2k}(t)\big).
\end{aligned}$$

So

$$\begin{aligned}
&\sum_{t=-kT}^{kT-1}\frac{W(t, u_{1k}(t), u_{2k}(t))}{\xi + \eta_1|u_{1k}(t)|^\nu + \eta_2|u_{2k}(t)|^\nu} \\
&\quad \leq pM_2 - (p-1)\sum_{t=-kT}^{kT-1}\big(f_{1k}(t), u_{1k}(t)\big) - (p-1)\sum_{t=-kT}^{kT-1}\big(f_{2k}(t), u_{2k}(t)\big).
\end{aligned}$$

Then

$$\begin{aligned}
\eta_k(u_k) &= \varphi_k(u_k) + \sum_{t=-kT}^{kT-1}\frac{W(t, u_{1k}(t), u_{2k}(t))}{\xi + \eta_1|u_{1k}(t)|^\nu + \eta_2|u_{2k}(t)|^\nu}\left(\xi + \eta_1\left|u_{1k}(t)\right|^\nu + \eta_2\left|u_{2k}(t)\right|^\nu\right) \\
&\quad - \sum_{t=-kT}^{kT-1}\big(f_{1,k}(t), u_{1k}(t)\big) - \sum_{t=-kT}^{kT-1}\big(f_{2,k}(t), u_{2k}(t)\big) \\
&\leq \varphi_k(u_k) + \big(\xi + \eta_1\|u_{1k}\|^\nu_{l^\infty_{2kT}} + \eta_2\|u_{2k}\|^\nu_{l^\infty_{2kT}}\big)\sum_{t=-kT}^{kT-1}\frac{W(t, u_{1k}(t), u_{2k}(t))}{\xi + \eta_1|u_{1k}(t)|^\nu + \eta_2|u_{2k}(t)|^\nu} \\
&\quad - \sum_{t=-kT}^{kT-1}\big(f_{1,k}(t), u_{1k}(t)\big) - \sum_{t=-kT}^{kT-1}\big(f_{2,k}(t), u_{2k}(t)\big) \\
&\leq \varphi_k(u_k) + \big(\xi + \eta_1 C_*^\nu\|u_{1k}\|^\nu_{\mathcal{H}_{1,k}} + \eta_2 C_*^\nu\|u_{2k}\|^\nu_{\mathcal{H}_{2,k}}\big) \\
&\quad \cdot\left[pM_2 - (p-1)\sum_{t=-kT}^{kT-1}\big(f_{1,k}(t), u_{1k}(t)\big) - (p-1)\sum_{t=-kT}^{kT-1}\big(f_{2,k}(t), u_{2k}(t)\big)\right]
\end{aligned}$$

$$
\begin{aligned}
&- \sum_{t=-kT}^{kT-1} \big(f_{1,k}(t), u_{1k}(t)\big) - \sum_{t=-kT}^{kT-1} \big(f_{2,k}(t), u_{2k}(t)\big) \\
&= M_2 + p\xi M_2 + p\eta_1 C_*^{\nu} M_2 \|u_{1k}\|^{\nu}_{\mathcal{H}_{1,k}} + p\eta_2 C_*^{\nu} M_2 \|u_{2k}\|^{\nu}_{\mathcal{H}_{2,k}} \\
&\quad - (p-1)\xi \sum_{t=-kT}^{kT-1} \big(f_{1,k}(t), u_{1k}(t)\big) - (p-1)\xi \sum_{t=-kT}^{kT-1} \big(f_{2,k}(t), u_{2k}(t)\big) \\
&\quad - (p-1)\eta_1 C_*^{\nu} \|u_{1k}\|^{\nu}_{\mathcal{H}_{1,k}} \sum_{t=-kT}^{kT-1} \big(f_{1,k}(t), u_{1k}(t)\big) \\
&\quad - (p-1)\eta_2 C_*^{\nu} \|u_{2k}\|^{\nu}_{\mathcal{H}_{2,k}} \sum_{t=-kT}^{kT-1} \big(f_{2,k}(t), u_{2k}(t)\big) \\
&\quad - (p-1)\eta_1 C_*^{\nu} \|u_{1k}\|^{\nu}_{\mathcal{H}_{1,k}} \sum_{t=-kT}^{kT-1} \big(f_{2,k}(t), u_{2k}(t)\big) \\
&\quad - (p-1)\eta_2 C_*^{\nu} \|u_{2k}\|^{\nu}_{\mathcal{H}_{2,k}} \sum_{t=-kT}^{kT-1} \big(f_{1,k}(t), u_{1k}(t)\big) \\
&\quad - \sum_{t=-kT}^{kT-1} \big(f_{1,k}(t), u_{1k}(t)\big) - \sum_{t=-kT}^{kT-1} \big(f_{2,k}(t), u_{2k}(t)\big) \\
&\le (1+p\xi)M_2 + \big[(p-1)\xi + 1\big]\left(\sum_{t\in\mathbb{Z}} |f_1(t)|^{p'}\right)^{1/p'} \left(\sum_{t=-kT}^{kT-1} |u_{1k}(t)|^{p}\right)^{1/p} \\
&\quad + \big[(p-1)\xi + 1\big]\left(\sum_{t\in\mathbb{Z}} |f_2(t)|^{p'}\right)^{1/p'} \left(\sum_{t=-kT}^{kT-1} |u_{2k}(t)|^{p}\right)^{1/p} \\
&\quad + p\eta_1 C_*^{\nu} M_2 \|u_{1k}\|^{\nu}_{\mathcal{H}_{1,k}} + p\eta_2 C_*^{\nu} M_2 \|u_{2k}\|^{\nu}_{\mathcal{H}_{2,k}} \\
&\quad + (p-1)\eta_1 C_*^{\nu} \|u_{1k}\|^{\nu}_{\mathcal{H}_{1,k}} \left(\left(\sum_{t\in\mathbb{Z}} |f_1(t)|^{p'}\right)^{1/p'} \left(\sum_{t=-kT}^{kT-1} |u_{1k}(t)|^{p}\right)^{1/p} \right. \\
&\quad \left. + \left(\sum_{t\in\mathbb{Z}} |f_2(t)|^{p'}\right)^{1/p'} \left(\sum_{t=-kT}^{kT-1} |u_{2k}(t)|^{p}\right)^{1/p} \right) \\
&\quad + (p-1)\eta_2 C_*^{\nu} \|u_{2k}\|^{\nu}_{\mathcal{H}_{2,k}} \left(\left(\sum_{t\in\mathbb{Z}} |f_1(t)|^{p'}\right)^{1/p'} \left(\sum_{t=-kT}^{kT-1} |u_{1k}(t)|^{p}\right)^{1/p} \right. \\
&\quad \left. + \left(\sum_{t\in\mathbb{Z}} |f_2(t)|^{p'}\right)^{1/p'} \left(\sum_{t=-kT}^{kT-1} |u_{2k}(t)|^{p}\right)^{1/p} \right) \\
&\le (1+p\xi)M_2 + \big[(p-1)\xi + 1\big]\left(\sum_{t\in\mathbb{Z}} |f_1(t)|^{p'}\right)^{1/p'} \|u_{1k}\|_{\mathcal{H}_{1,k}} \\
&\quad + \big[(p-1)\xi + 1\big]\left(\sum_{t\in\mathbb{Z}} |f_2(t)|^{p'}\right)^{1/p'} \|u_{2k}\|_{\mathcal{H}_{2,k}} \\
&\quad + p\eta_1 C_*^{\nu} M_2 \|u_{1k}\|^{\nu}_{\mathcal{H}_{1,k}} + p\eta_2 C_*^{\nu} M_2 \|u_{2k}\|^{\nu}_{\mathcal{H}_{2,k}} \\
&\quad + (p-1)\eta_1 C_*^{\nu} \|u_{1k}\|^{\nu}_{\mathcal{H}_{1,k}}
\end{aligned}
$$

$$
\begin{aligned}
&\cdot\left(\left(\sum_{t\in\mathbb{Z}}|f_1(t)|^{p'}\right)^{1/p'}\|u_{1k}\|_{\mathcal{H}_{1,k}}+\left(\sum_{t\in\mathbb{Z}}|f_2(t)|^{p'}\right)^{1/p'}\|u_{2k}\|_{\mathcal{H}_{2,k}}\right)\\
&+(p-1)\eta_2 C_*^{\nu}\|u_{2k}\|_{\mathcal{H}_{2,k}}^{\nu}\\
&\cdot\left(\left(\sum_{t\in\mathbb{Z}}|f_1(t)|^{p'}\right)^{1/p'}\|u_{1k}\|_{\mathcal{H}_{1,k}}+\left(\sum_{t\in\mathbb{Z}}|f_2(t)|^{p'}\right)^{1/p'}\|u_{2k}\|_{\mathcal{H}_{2,k}}\right)\\
\le{}&(1+p\xi)M_2+\left[(p-1)\xi+1\right]\left(\sum_{t\in\mathbb{Z}}|f_1(t)|^{p'}\right)^{1/p'}\|u_{1k}\|_{\mathcal{H}_{1,k}}\\
&+\left[(p-1)\xi+1\right]\left(\sum_{t\in\mathbb{Z}}|f_2(t)|^{p'}\right)^{1/p'}\|u_{2k}\|_{\mathcal{H}_{2,k}}\\
&+p\eta_1 C_*^{\nu}M_2\|u_{1k}\|_{\mathcal{H}_{1,k}}^{\nu}+p\eta_2 C_*^{\nu}M_2\|u_{2k}\|_{\mathcal{H}_{2,k}}^{\nu}\\
&+(p-1)\eta_1 C_*^{\nu}\|u_{1k}\|_{\mathcal{H}_{1,k}}^{\nu+1}\left(\sum_{t\in\mathbb{Z}}|f_1(t)|^{p'}\right)^{1/p'}\\
&+(p-1)\eta_1 C_*^{\nu}\|u_{1k}\|_{\mathcal{H}_{1,k}}^{\nu}\|u_{2k}\|_{\mathcal{H}_{2,k}}\left(\sum_{t\in\mathbb{Z}}|f_2(t)|^{p'}\right)^{1/p'}\\
&+(p-1)\eta_2 C_*^{\nu}\|u_{2k}\|_{\mathcal{H}_{2,k}}^{\nu+1}\left(\sum_{t\in\mathbb{Z}}|f_2(t)|^{p'}\right)^{1/p'}\\
&+(p-1)\eta_2 C_*^{\nu}\|u_{2k}\|_{\mathcal{H}_{2,k}}^{\nu}\|u_{1k}\|_{\mathcal{H}_{1,k}}\left(\sum_{t\in\mathbb{Z}}|f_1(t)|^{p'}\right)^{1/p'}\\
\le{}&(1+p\xi)M_2+\left[(p-1)\xi+1\right]\left(\sum_{t\in\mathbb{Z}}|f_1(t)|^{p'}\right)^{1/p'}\|u_{1k}\|_{\mathcal{H}_{1,k}}\\
&+\left[(p-1)\xi+1\right]\left(\sum_{t\in\mathbb{Z}}|f_2(t)|^{p'}\right)^{1/p'}\|u_{2k}\|_{\mathcal{H}_{2,k}}\\
&+p\eta_1 C_*^{\nu}M_2\|u_{1k}\|_{\mathcal{H}_{1,k}}^{\nu}+p\eta_2 C_*^{\nu}M_2\|u_{2k}\|_{\mathcal{H}_{2,k}}^{\nu}\\
&+(p-1)\eta_1 C_*^{\nu}\|u_{1k}\|_{\mathcal{H}_{1,k}}^{\nu+1}\left(\sum_{t\in\mathbb{Z}}|f_1(t)|^{p'}\right)^{1/p'}\\
&+(p-1)\eta_1 C_*^{\nu}\left(\frac{\nu}{\nu+1}\|u_{1k}\|_{\mathcal{H}_{1,k}}^{\nu+1}+\frac{1}{\nu+1}\|u_{2k}\|_{\mathcal{H}_{2,k}}^{\nu+1}\right)\left(\sum_{t\in\mathbb{Z}}|f_2(t)|^{p'}\right)^{1/p'}\\
&+(p-1)\eta_2 C_*^{\nu}\|u_{2k}\|_{\mathcal{H}_{2,k}}^{\nu+1}\left(\sum_{t\in\mathbb{Z}}|f_2(t)|^{p'}\right)^{1/p'}\\
&+(p-1)\eta_2 C_*^{\nu}\left(\frac{\nu}{\nu+1}\|u_{2k}\|_{\mathcal{H}_{2,k}}^{\nu+1}+\frac{1}{\nu+1}\|u_{1k}\|_{\mathcal{H}_{1,k}}^{\nu+1}\right)\left(\sum_{t\in\mathbb{Z}}|f_1(t)|^{p'}\right)^{1/p'}. \qquad (3.14)
\end{aligned}
$$

Thus (3.14) and Lemma 3.2 imply that

$$
\begin{aligned}
&(1+p\xi)M_2+\left[(p-1)\xi+1\right]\left(\sum_{t\in\mathbb{Z}}|f_1(t)|^{p'}\right)^{1/p'}\|u_{1k}\|_{\mathcal{H}_{1,k}}\\
&+\left[(p-1)\xi+1\right]\left(\sum_{t\in\mathbb{Z}}|f_2(t)|^{p'}\right)^{1/p'}\|u_{2k}\|_{\mathcal{H}_{2,k}}
\end{aligned}
$$

$$
\begin{aligned}
&+ p\eta_1 C_*^\nu M_2 \|u_{1k}\|_{\mathcal{H}_{1,k}}^\nu + p\eta_2 C_*^\nu M_2 \|u_{2k}\|_{\mathcal{H}_{2,k}}^\nu \\
&+ (p-1)\eta_1 C_*^\nu \|u_{1k}\|_{\mathcal{H}_{1,k}}^{\nu+1} \left(\sum_{t\in\mathbb{Z}} |f_1(t)|^{p'} \right)^{1/p'} \\
&+ (p-1)\eta_1 C_*^\nu \left(\frac{\nu}{\nu+1} \|u_{1k}\|_{\mathcal{H}_{1,k}}^{\nu+1} + \frac{1}{\nu+1} \|u_{2k}\|_{\mathcal{H}_{2,k}}^{\nu+1} \right) \left(\sum_{t\in\mathbb{Z}} |f_2(t)|^{p'} \right)^{1/p'} \\
&+ (p-1)\eta_2 C_*^\nu \|u_{2k}\|_{\mathcal{H}_{2,k}}^{\nu+1} \left(\sum_{t\in\mathbb{Z}} |f_2(t)|^{p'} \right)^{1/p'} \\
&+ (p-1)\eta_2 C_*^\nu \left(\frac{\nu}{\nu+1} \|u_{2k}\|_{\mathcal{H}_{2,k}}^{\nu+1} + \frac{1}{\nu+1} \|u_{1k}\|_{\mathcal{H}_{1,k}}^{\nu+1} \right) \left(\sum_{t\in\mathbb{Z}} |f_1(t)|^{p'} \right)^{1/p'} \\
\geq{}& \min\left\{ d_1 \|u_1\|_{\mathcal{H}_{1,k}}^p, a_1 C_*^{\gamma-p} \|u_1\|_{\mathcal{H}_{1,k}}^\gamma \right\} + \min\left\{ d_2 \|u_2\|_{\mathcal{H}_{2,k}}^p, a_2 C_*^{\gamma-p} \|u_2\|_{\mathcal{H}_{2,k}}^\gamma \right\}.
\end{aligned}
$$

Note that $p > \gamma > \nu + 1$. So (H6) implies there exists $M_3 > 0$ (independent of k) such that

$$\|u_{1k}\|_{\mathcal{H}_{1,k}} \leq M_3, \qquad \|u_{2k}\|_{\mathcal{H}_{2,k}} \leq M_3 \quad \text{for every } k \in \mathbb{N}.$$

By Corollary 2.1,

$$\|u_{1k}\|_{l_{2kT}^\infty} \leq C_* M_3, \qquad \|u_{2k}\|_{l_{2kT}^\infty} \leq C_* M_3 \quad \text{for every } k \in \mathbb{N}.$$

Let $M_1 = \max\{C_{1*}M_3, C_{2*}M_3\}$. Thus the proof is complete. □

Lemma 3.5 *Let $\{u_k\}$ be determined by Lemma* 3.4. *Then there exists a subsequence $\{u_{k_j} = (u_{1k_j}, u_{2k_j})\}$ of $\{u_k\}_{k\in\mathbb{N}}$ convergent to a certain function $u_\infty = (u_{1\infty}, u_{2\infty})$ and when $f_1 \neq 0$ and $f_2 \neq 0$, u_∞ is a nontrivial solution of system* (1.1) *such that $u_\infty(t) \to 0$ and $\Delta u_\infty(t-1) \to 0$ as $t \to \pm\infty$.*

Proof Note that

$$\|u_{1k}\|_{\mathcal{H}_{1,k}} \leq M_1, \qquad \|u_{2k}\|_{\mathcal{H}_{2,k}} \leq M_1 \quad \text{for every } k \in \mathbb{N}.$$

Then, similar to the argument in [15] or [16], one can prove that $\{u_{mk}\}_{k\in\mathbb{N}}$ has a convergent subsequence $\{u_{mk_j}\}$ such that $u_{mk_j} \to u_{m\infty}$ and $u_{m\infty}(t) \to 0$ and $\Delta u_{m\infty}(t-1) \to 0$ as $t \to \pm\infty$, where $m = 1, 2$. Let $u_\infty = (u_{1\infty}, u_{2\infty})$. By (3.13) and the continuity of Φ_m, $K(t,\cdot,\cdot)$, $W(t,\cdot,\cdot)$ and φ_k', similar to the argument in [15] or [16], the proof is easy to be completed. □

□

Proof of Theorem 1.2 The proof is easy to be completed by replacing

$$
\begin{aligned}
\sum_{t=-kT}^{kT-1} (f_m(t), u_m(t)) &\leq \left(\sum_{t=-kT}^{kT-1} |f_m(t)|^{p'} \right)^{1/p'} \left(\sum_{t=-kT}^{kT-1} |u_m(t)|^p \right)^{1/p} \\
&\leq \|u_m\|_{\mathcal{H}_{m,k}} \left(\sum_{t\in\mathbb{Z}} |f_m(t)|^{p'} \right)^{1/p'}
\end{aligned}
$$

with

$$\sum_{t=-kT}^{kT-1}(f_m(t),u_m(t)) \le \|u_m\|_{l^\infty_{2kT}} \sum_{t=-kT}^{kT-1}|f_m(t)| \le C_*\|u_m\|_{\mathcal{H}_{m,k}} \sum_{t\in\mathbb{Z}}|f_m(t)|, \quad m=1,2,$$

in the proofs of Lemma 3.3 and Lemma 3.4. □

Proofs of Theorem 1.3 *and Theorem* 1.4 We only note that in the proof of Lemma 3.3, when $\gamma = p$, we do not need to consider the case that $r \in (0,1]$ alone and it is sufficient that $r > 0$. Other proofs are the same as those of Theorem 1.1 and Theorem 1.2, respectively. □

4 Examples

We first give two examples about Φ which satisfy assumption $(\mathcal{A}1)$.

(I) An example with $N = 1$. Define $\Phi_m : \mathbb{R} \to \mathbb{R}^N$, $m = 1, 2$, by

$$\Phi_1(x) = \begin{cases} \alpha_1|x|^p, & x \ge 0, \\ \alpha_2|x|^p, & x < 0, \end{cases} \qquad \Phi_2(y) = \begin{cases} \beta_1|y|^p, & y \ge 0, \\ \beta_2|y|^p, & y < 0, \end{cases}$$

where $\alpha_1, \alpha_2 \in [d_1, d_3]$, $\beta_1, \beta_2 \in [d_2, d_4]$. Then it is easy to verify that Φ_m, $m = 1, 2$, satisfies $(\mathcal{A}1)$.

(II) As described in [1], the following classical case with p-Laplacian also satisfies the assumption $(\mathcal{A}1)$. Define $\Phi_m : \mathbb{R}^N \to \mathbb{R}^N$, $m = 1, 2$, by

$$\Phi_1(x) = \alpha|x|^p, \qquad \Phi_2(y) = \beta|y|^p,$$

where $\alpha \in [d_1, d_3]$, $\beta \in [d_2, d_4]$.

Next, we present some examples of K and W which satisfy those assumptions in Theorem 1.1. There are lots of examples of K. For example, let

$$K(t,x_1,x_2) = a_1(t)|x_1|^\gamma + a_2(t)|x_2|^\gamma, \quad (t,x_1,x_2) \in \mathbb{Z}[0,T-1] \times \mathbb{R}^N \times \mathbb{R}^N,$$

where $\gamma \in (1,p)$, $a_i, i = 1,2 : \mathbb{Z} \to \mathbb{R}^+$ are T-periodic. Let $a_i = \min_{t\in\mathbb{Z}[0,T-1]} a_i(t)$. Then it is easy to see that K satisfies (H1) and (H2).

For W, we assume that

$$W(t,x_1,x_2) = b(t)\big(|x_1|^p + |x_2|^p\big)\ln\big(|x_1|^p + |x_2|^p + 1\big), \quad (t,x_1,x_2) \in \mathbb{Z}[0,T-1] \times \mathbb{R}^N \times \mathbb{R}^N,$$

where $b : \mathbb{Z} \to \mathbb{R}^+$ is T-periodic. Let $b^+ = \max_{t\in\mathbb{Z}[0,T-1]}\{b(t)\}$. Then

$$W(t,x_1,x_2) \le b^+\big(|x_1|^p + |x_2|^p\big)\ln\big(|rC_*|^p + |rC_*|^p + 1\big)$$
$$\text{for all } t \in \mathbb{Z}[0,T-1], |x_1| \le rC_*, |x_2| \le rC_*.$$

Let $b_1 = b_2 = b^+\ln(|rC_*|^p + |rC_*|^p + 1)$. If r is sufficiently small, then (H3)(i) holds. It is easy to see that

$$\lim_{|x_1|+|x_2|\to+\infty} \frac{W(t,x_1,x_2)}{|x_1|^p + |x_2|^p} = +\infty \quad \text{for all } t \in \mathbb{Z}[0,T-1].$$

So (H4) holds. Let $\nu \in (0, \gamma - 1)$. Note that

$$p\xi\left(|x_1|^p + |x_2|^p\right) \geq \ln\left(|x_1|^p + |x_2|^p + 1\right), \qquad p\left(\eta_1|x_1|^\nu + \eta_2|x_2|^\nu\right) \geq \ln\left(|x_1|^p + |x_2|^p + 1\right)$$

for all $(x_1, x_2) \in \mathbb{R}^N \times \mathbb{R}^N$, when we choose sufficiently large ξ, η_1 and η_2. Hence

$$\begin{aligned}
&p\xi\left(|x_1|^p + |x_2|^p\right) + p\left(|x_1|^p + |x_2|^p\right)\left(\eta_1|x_1|^\nu + \eta_2|x_2|^\nu\right) \\
&\quad \geq \ln\left(|x_1|^p + |x_2|^p + 1\right) + \ln\left(|x_1|^p + |x_2|^p + 1\right)\left(|x_1|^p + |x_2|^p\right) \\
&\iff \quad p\left(\xi + \eta_1|x_1|^\nu + \eta_2|x_2|^\nu\right)\left(|x_1|^p + |x_2|^p\right) \\
&\qquad\qquad \geq \ln\left(|x_1|^p + |x_2|^p + 1\right)\left(|x_1|^p + |x_2|^p + 1\right) \\
&\iff \quad p\left(\xi + \eta_1|x_1|^\nu + \eta_2|x_2|^\nu\right)\left(|x_1|^p + |x_2|^p\right)^2 \\
&\qquad\qquad \geq \left(|x_1|^p + |x_2|^p\right)\ln\left(|x_1|^p + |x_2|^p + 1\right)\left(|x_1|^p + |x_2|^p + 1\right) \\
&\iff \quad \frac{p(|x_1|^p + |x_2|^p)^2}{|x_1|^p + |x_2|^p + 1} \geq \frac{(|x_1|^p + |x_2|^p)\ln(|x_1|^p + |x_2|^p + 1)}{\xi + \eta_1|x_1|^\nu + \eta_2|x_2|^\nu} \\
&\iff \quad \left(\nabla_{x_1} W(t, x_1, x_2), x_1\right) + \left(\nabla_{x_2} W(t, x_1, x_2), x_2\right) - pW(t, x_1, x_2) \\
&\qquad\qquad \geq \frac{W(t, x_1, x_2)}{\xi + \eta_1|x_1|^\nu + \eta_2|x_2|^\nu}
\end{aligned}$$

for all $(t, x_1, x_2) \in \mathbb{Z}[0, T-1] \times \mathbb{R}^N \times \mathbb{R}^N$, which implies (H5) holds.

Competing interests
The authors declare that they have no competing interests.

Authors' contributions
The authors read and approved the final manuscript.

Acknowledgements
This project is supported by the National Natural Science Foundation of China (No. 11301235), Tianyuan Fund for Mathematics of the National Natural Science Foundation of China (No. 11226135) and the Fund for Fostering Talents in Kunming University of Science and Technology (No. KKSY201207032).

References
1. Mawhin, J: Periodic solutions of second order nonlinear difference systems with ϕ-Laplacian: a variational approach. Nonlinear Anal. **75**, 4672-4687 (2012)
2. Mawhin, J: Periodic solutions of second order Lagrangian difference systems with bounded or singular ϕ-Laplacian and periodic potential. Discrete Contin. Dyn. Syst., Ser. S **6**, 1065-1076 (2013)
3. Bonanno, G, Candito, P: Nonlinear difference equations investigated via critical points methods. Nonlinear Anal. **70**, 3180-3186 (2009)
4. Candito, P, Giovannelli, N: Multiple solutions for a discrete boundary value problem. Comput. Math. Appl. **56**, 959-964 (2008)
5. Guo, ZM, Yu, JS: The existence of periodic and subharmonic solutions to subquadratic second-order difference equations. J. Lond. Math. Soc. **68**, 419-430 (2003)
6. Guo, ZM, Yu, JS: Existence of periodic and subharmonic solutions for second-order superlinear difference equations. Sci. China Ser. A **46**, 506-515 (2003)
7. Zhou, Z, Yu, JS, Guo, ZM: Periodic solutions of higher-dimensional discrete systems. Proc. R. Soc. Edinb., Sect. A **134**, 1013-1022 (2004)
8. Xue, YF, Tang, CL: Existence of a periodic solution for subquadratic second-order discrete Hamiltonian system. Nonlinear Anal. **67**, 2072-2080 (2007)
9. Lin, XY, Tang, XH: Existence of infinitely many homoclinic orbits in discrete Hamiltonian systems. J. Math. Anal. Appl. **373**, 59-72 (2011)
10. He, T, Chen, W: Periodic solutions of second order convex systems involving the p-Laplacian. Appl. Math. Comput. **206**, 124-132 (2008)
11. Tang, XH, Zhang, X: Periodic solutions for second-order discrete Hamiltonian systems. J. Differ. Equ. Appl. **17**, 1413-1430 (2011)

12. Zhang, X, Tang, X: Existence of solutions for a nonlinear discrete system involving the p-Laplacian. Appl. Math. **57**, 11-30 (2012)
13. Zhang, X: Notes on periodic solutions for a nonlinear discrete system involving the p-Laplacian. Bull. Malays. Math. Soc. **37**, 499-509 (2014)
14. Zhang, Q, Tang, XH, Zhang, QM: Existence of periodic solutions for a class of discrete Hamiltonian systems. Discrete Dyn. Nat. Soc. **2011**, Article ID 463480 (2011)
15. Ma, M, Guo, ZM: Homoclinic orbits and subharmonics for nonlinear second order difference equations. Nonlinear Anal. **67**, 1737-1745 (2007)
16. He, X, Chen, P: Homoclinic solutions for second order discrete p-Laplacian systems. Adv. Differ. Equ. **2011**, 57 (2011)
17. Wang, Y, Zhang, X: Multiple periodic solutions for a class of nonlinear difference systems with classical or bounded (ϕ_1, ϕ_2)-Laplacian. Adv. Differ. Equ. **2014**, 218 (2014)
18. Cabada, A, Iannizzotto, A, Tersian, S: Multiple solutions for discrete boundary value problems. J. Math. Anal. Appl. **356**, 418-428 (2009)
19. Bereanu, C, Jebelean, P, Serban, C: Ground state and mountain pass solutions for discrete $p(\cdot)$-Laplacian. Bound. Value Probl. **2012**, 104 (2012)
20. Zhang, Q: Existence of homoclinic solutions for a class of difference systems involving p-Laplacian. Adv. Differ. Equ. **2014**, 291 (2014)
21. Schechter, M: Minimax Systems and Critical Point Theory. Birkhäuser, Boston (2009)

4

Existence of time periodic solutions for a class of non-resonant discrete wave equations

Guang Zhang[1*], Wenying Feng[2] and Yubin Yan[3]

*Correspondence: lxyzhg@tjcu.edu.cn
[1]School of Science, Tianjin University of Commerce, Tianjin, 300134, P.R. China
Full list of author information is available at the end of the article

Abstract

In this paper, a class of discrete wave equations with Dirichlet boundary conditions are obtained by using the center-difference method. For any positive integers m and T, when the existence of time mT-periodic solutions is considered, a strongly indefinite discrete system needs to be established. By using a variant generalized weak linking theorem, a non-resonant superlinear (or superquadratic) result is obtained and the Ambrosetti-Rabinowitz condition is improved. Such a method cannot be used for the corresponding continuous wave equations or the continuous Hamiltonian systems; however, it is valid for some general discrete Hamiltonian systems.

Keywords: wave equation; Hamiltonian system; strongly indefinite discrete system; time mT-periodic solution; variant generalized weak linking theorem; Ambrosetti-Rabinowitz condition

1 Introduction

The existence of time periodic solutions for a nonlinear wave equation of the form

$$u_{tt} - u_{xx} + g(t,u) = 0, \quad 0 < x < \pi, t \in R, \tag{1}$$

with the Dirichlet boundary conditions

$$u(0,t) = 0 = u(\pi,t) \tag{2}$$

had been established in Vitt [1] when the distributed self-oscillating systems been considered.

We note that the eigenvalues of the operator $\partial_{tt} - \partial_{xx}$ in the space of functions $u(x,t)$, $2\pi/\omega$-periodic in time and such that, say, $u(\cdot,t) \in H_0^1(0,\pi)$ for all t, are $-\omega^2 l^2 + j^2$, $l \in Z$, $j \geq 1$. Therefore, when ω^2 is irrational, the eigenvalues accumulate to 0. In this case, the inverse operator of $\partial_{tt} - \partial_{xx}$ is unbounded and the standard implicit function theorem is not applicable. When ω^2 is rational, the number of 0-spectra is infinite, thus, this will introduce the presence of an infinite-dimensional bifurcation equation. Consequently, when we consider the existence of the time periodic solutions for problem (1)-(2), two main difficulties must be overcome: the 'small denominators' problem and the presence of an infinite-dimensional bifurcation equation. To this end, two main methods are used, that is, the variations viewpoint (see Rabinowitz [2] and Brézis and Coron [3]) and the KAM theory (see Berti and Bolle [4] and Gentile and Mastropietro [5]).

Let Z be a set of all integers, R be a set of all real numbers, and $Z^+ = \{0,1,2,\ldots\}$. For any integers k and l with $k < l$, denote $[k,l] = \{k, k+1, \ldots, l\}$. By using the center-difference method for the space variable x and the time variable t, we can obtain a discrete analog of (1)-(2) of the form

$$\begin{cases} \frac{1}{h^2}\Delta^2 u(\frac{i\pi}{N+1},(n-1)h) - (\frac{N+1}{\pi})^2\nabla^2 u(\frac{(i-1)\pi}{N+1}, nh) \\ \quad + g(nh, u(\frac{i\pi}{N+1}, nh)) = 0, \quad i \in [1,N], n \in Z, \\ u(\frac{0\pi}{N+1}, nh) = u(\frac{(N+1)\pi}{N+1}, nh) = 0, \quad n \in Z, \end{cases} \tag{3}$$

where $h > 0$ is the time step size, N is a positive integer and the space step size is $\pi/(N+1)$. Let

$$u\left(\frac{i\pi}{N+1}, nh\right) = u_n^i,$$

then we have

$$\begin{cases} \Delta^2 u_{n-1}^i - \delta^2\nabla^2 u_n^{i-1} + f(n, u_n^i) = 0, \quad i \in [1,N], n \in Z, \\ u_n^0 = 0 = u_n^{N+1}, \quad n \in Z, \end{cases} \tag{4}$$

where

$$\Delta^2 u_{n-1}^i = u_{n+1}^i - 2u_n^i + u_{n-1}^i,$$

$$\nabla^2 u_n^{i-1} = u_n^{i+1} - 2u_n^i + u_n^{i-1},$$

$$\delta^2 = h^2\left(\frac{N+1}{\pi}\right)^2,$$

and

$$f(n, u_n^i) = h^2 g\left(nh, u\left(\frac{i\pi}{N+1}, nh\right)\right).$$

Problem (4) can be rewritten by the vector and matrix as

$$\Delta^2 U_{n-1} + \delta^2 A U_n + \nabla V(n, U_n) = 0, \quad n \in Z, \tag{5}$$

where

$$U_n = \mathrm{col}(u_n^1, u_n^2, u_n^N),$$

$$A = \begin{pmatrix} 2 & -1 & 0 & \cdots & 0 \\ -1 & 2 & -1 & & 0 \\ & \cdots & & \cdots & \\ 0 & & -1 & 2 & -1 \\ 0 & \cdots & 0 & -1 & 2 \end{pmatrix}_{N\times N},$$

and

$$V(n, U_n) = \sum_{i=1}^{N} \int_0^{u_n^i} f(n,s)\, ds,$$

which implies that

$$\nabla V(n, U_n) = \mathrm{col}\big(f(n, u_n^1), f(n, u_n^2), \ldots, f(n, u_n^N)\big).$$

For a given positive integer T, we assume that the function $f : Z \times R \to R$ is continuous about the two variables and satisfies the periodic condition

$$f(n+T, \cdot) = f(n, \cdot) \quad \text{for } n \in Z. \tag{6}$$

Let $L = -(\Delta^2 + \delta^2 A)$. For any positive integers T and m, $\sigma(L)$ denotes the spectrum of the linear operator L in E_{mT} which will be defined in the next section. We can see that $\sigma(L) \cap (0, \infty) \neq \emptyset$ and $\sigma(L) \cap (-\infty, 0) \neq \emptyset$. In this paper, we will consider the existence of mT-periodic solutions for problem (4) when the condition $0 \notin \sigma(L)$ holds. However, 0 is the spectrum point of the linear operator $\partial_{tt} - \partial_{xx}$ for problem (1)-(2). Thus, our method cannot be used for the corresponding continuous wave equations.

Clearly, system (5) is also a discrete second order Hamiltonian system. Recently, the existence of T-periodic solutions for system (5) has also been extensively studied when A is symmetric matrix and L is positive definite. By using critical point theory, many solvability conditions are given, such as the superquadratic condition and subquadratic condition (see Deng *et al.* [6] and Yan *et al.* [7]), the convex condition (see Jiang [8]), and the asymptotically linear condition (see Guo and Yu [9]). Our method can be extended to a general system, however, the linear operator L is strongly indefinite. On the other hand, in the superquadratic case, all papers required the Ambrosetti-Rabinowitz condition, that is, there exist constants $\mu > 2$ and $M > 0$ such that

$$0 < \mu H(t, U) \leq \big(\nabla H(t, U), U\big), \quad |U| \geq M, t \in [0, T], \tag{7}$$

where

$$|U| = \left(\sum_{i=1}^{N} |u^i|^2\right)^{1/2}, \quad U \in R^N.$$

In this paper, we will give a more general condition.

In fact, system (5) is also a discrete analog of the second order Hamiltonian system of the form

$$U''(t) + BU(t) + \nabla H(t, U) = 0, \quad t \in R. \tag{8}$$

The existence of periodic solutions of (8) have also been extensively discussed since Poincaré [10]. The importance of periodic solutions for finite-dimensional Hamiltonian system was pointed out by Poincaré in [10]. Poincaré stressed their importance formulating a conjecture. This conjecture stimulates the systematic study of periodic solutions by Poincaré himself, Lyapunov [11], Birkhoff and Lewis [12], Moser [13], Weinstein [14], Rabinowitz [15], Ekeland and Lasry [16], *etc.* In Pugh and Robinson [17] a positive answer to the conjecture was given, but only in a generic sense (namely in the C^2-category of Hamiltonian functions): the periodic orbits are dense on every compact and regular energy surface. However, for specific systems, the conjecture is still open (and far from having been proved); see [18].

Since Rabinowitz's pioneering work [15] of 1978, variational methods have been widely used in the study of existence of solutions of Hamiltonian systems; see Long [19] and Luan and Mao [20]. Recently, some authors had improved Ambrosetti-Rabinowitz condition by using the linking theorem; for example, see Schechter and Zou [21] and Chen and Ma [22]. In [21], the authors considered the existence of solutions for a Schrödinger equation and the classical Ambrosetti-Rabinowitz condition is replaced by a general superquadratic condition: there exist constants $a > 0$ and $p > 2$ such that

$$|\nabla V(t,U)| \le a\left(1 + |U|^{p-1}\right) \quad \text{for } U \in R^N. \tag{9}$$

In [22], Chen and Ma established the existence of periodic solutions of (8) by using a similar method. However, the condition (9) will be improved in our result.

In the next section of the present paper, we will give some preliminary results which will be used in the proof of the main results. The exact spectrum of the linear operator L in E_{mT} will be given. In this case, we easily give the conditions $0 \notin \sigma(L)$, $\sigma(L) \cap (0, \infty) \neq \emptyset$, and $\sigma(L) \cap (-\infty, 0) \neq \emptyset$. Thus, in this paper, we only consider the non-resonant strongly indefinite problem. Our approach is based on an application of a variant generalized weak linking strongly indefinite problem developed by Schechter and Zou [21]; also see Chen and Ma [22]. Thus, a variant generalized weak linking theorem is also given in this section. In Section 3, our main result will be obtained by using the variant generalized weak linking theorem, and the Ambrosetti-Rabinowitz condition will be improved.

2 Some preliminary results

In this section, we recall some basic facts which will be used in the proof of the main results.

Let

$$X = \left\{U = \{U_n\}_{n\in Z} : U_n \in R^N, n \in Z\right\}.$$

For any given positive integers T and m, E_{mT} is defined by

$$E_{mT} = \{U \in X : U_{n+mT} = U_n, n \in Z\}.$$

E_{mT} can be equipped with the inner product $\langle\cdot,\cdot\rangle_{mT}$ and norm $\|\cdot\|_{mT}$ as follows:

$$\langle U, V\rangle_{mT} = \sum_{n=1}^{mT}(U_n, V_n) = \sum_{n=1}^{mT}\sum_{i=1}^{N} u_n^i v_n^i, \quad U, V \in E_{mT},$$

$$\|U\|_{mT} = \left(\sum_{n=1}^{mT}\sum_{i=1}^{N}\left(u_n^i\right)^2\right)^{1/2}.$$

Clearly, we can also define one other norm for E_{mT}. Note that the space E_{mT} is finite dimensional, thus, they are equivalent.

It is easy to see that $(E_{mT}, \langle \cdot, \cdot \rangle_{mT})$ is a finite-dimensional Hilbert space and linearly homeomorphic to $R^{mT \times N}$. For convenience, we identify $U \in E_{mT}$ with $U = \mathrm{col}(U_1, U_2, \ldots, U_{mT})$. In this case, we will consider the existence of solutions for the discrete wave equation

$$\Delta^2 u_{n-1}^i - \delta^2 \nabla^2 u_n^{i-1} + f(n, u_n^i) = 0, \quad (i, n) \in [1, N] \times [1, mT], \tag{10}$$

with the space Dirichlet boundary conditions

$$u_n^0 = 0 = u_n^{N+1}, \quad n \in [1, mT], \tag{11}$$

and the time periodic boundary conditions

$$u_0^i = u_{mT}^i \quad \text{and} \quad u_1^i = u_{mT+1}^i, \quad i \in [1, N], \tag{12}$$

or the discrete Hamiltonian system

$$\Delta^2 U_{n-1} + \delta^2 A U_n + \nabla V(n, U_n) = 0, \quad n \in [1, mT], \tag{13}$$

with the time periodic boundary conditions

$$U_0 = U_{mT} \quad \text{and} \quad U_1 = U_{mT+1}. \tag{14}$$

Define the functional H on E_{mT} as follows:

$$\begin{aligned} H(U) &= \sum_{j=1}^{mT} \sum_{i=1}^{N} \left[\frac{1}{2} |u_{j+1}^i - u_j^i|^2 - \frac{\delta^2}{2} |u_j^{i+1} - u_j^i|^2 - \int_0^{u_j^i} f(j, s)\, ds \right] \\ &= \frac{1}{2} \sum_{j=1}^{mT} \sum_{i=1}^{N} \left(|\Delta u_j^i|^2 - \frac{\delta^2}{2} |\nabla u_j^i|^2 \right) - \sum_{j=1}^{mT} \sum_{i=1}^{N} \int_0^{u_j^i} f(j, s)\, ds \\ &= \frac{1}{2} \sum_{j=1}^{mT} (\Delta U_j, \Delta U_j) - \frac{\delta^2}{2} \sum_{j=1}^{mT} (A U_j, U_j) - \sum_{j=1}^{mT} H(j, U_j). \end{aligned}$$

A vector $W \in E_{mT}$ is called a critical point of the functional H if the gradient of H at W is zero, *i.e.*,

$$\left. \frac{\partial H(U)}{u_j^i} \right|_{U=W} = 0 \quad \text{for } i \in [1, N], j \in [1, mT]. \tag{15}$$

At the same time, $c = H(W)$ is called a critical value of H. So we can obtain the following result.

Lemma 1 *A vector $W \in E_{mT}$ is a critical point of the functional $H(U)$ (or $-H(U)$) if, and only if, W is a solution of problem* (10)-(12), *in fact, it is also a solution of* (13)-(14).

Its proof is similar to Lemma 1 in [23], thus, it will be omitted.

In the following, we will consider the eigenvalue problem of the form

$$LU = -\left(\Delta^2 u_{n-1}^i - \delta^2 \nabla^2 u_n^{i-1}\right) = \lambda u_n^i, \quad (i,n) \in [1,N] \times [1,mT], \tag{16}$$

with the boundary conditions (11) and (12).

It is well known that the eigenvalue problem

$$\begin{cases} -\Delta^2 x_{n-1} = \gamma x_n, & n \in [1,N], \\ x_0 = x_{N+1} = 0, \end{cases}$$

has the eigenvalues

$$\gamma_k = 4\sin^2 \frac{k\pi}{2(N+1)}, \quad k \in [1,N],$$

and that the eigenvalue problem

$$\begin{cases} -\Delta^2 x_{n-1} = \eta x_n, & n \in [1,mT], \\ x_0 = x_{mT}, & x_1 = x_{mT+1}, \end{cases}$$

has the eigenvalues

$$\eta_l = 4\sin^2 \frac{(l-1)\pi}{mT}, \quad l \in [1,mT].$$

See Cheng [24]. Thus, we see that all eigenvalues of the linear problem (16)-(11)-(12) are

$$\lambda_{kl} = 4\sin^2 \frac{(l-1)\pi}{mT} - 4\delta^2 \sin^2 \frac{k\pi}{2(N+1)}$$

for $k \in [1,N]$ and $l \in [1,mT]$. That is,

$$\sigma(L) = \left\{4\left[\sin^2 \frac{(l-1)\pi}{mT} - \delta^2 \sin^2 \frac{k\pi}{2(N+1)}\right], (k,l) \in [1,N] \times [1,mT]\right\}. \tag{17}$$

We note that

$$\gamma_{\max} = \max_{k\in[1,N]} \gamma_k = 4\sin^2 \frac{N\pi}{2(N+1)},$$

$$\gamma_{\min} = \min_{k\in[1,N]} \gamma_k = 4\sin^2 \frac{\pi}{2(N+1)},$$

$$\eta_{\max} = \max_{l\in[1,mT]} \eta_l = \begin{cases} 4, & mT \text{ is even}, \\ 4\cos^2 \frac{\pi}{mT}, & mT \text{ is odd}, \end{cases}$$

and

$$\eta_{\min} = \min_{l\in[1,mT]} \eta_l = 0.$$

Thus, we have

$$\begin{aligned}\lambda_{\max} &= \max_{(k,l)\in[1,N]\times[1,mT]} \lambda_{kl} \\ &= \eta_{\max} - \delta^2\gamma_{\min} \\ &= \begin{cases} 4(1-\delta^2\sin^2\frac{\pi}{2(N+1)}), & mT \text{ is even}, \\ 4(\cos^2\frac{\pi}{mT} - \delta^2\sin^2\frac{\pi}{2(N+1)}), & mT \text{ is odd}, \end{cases}\end{aligned}$$

and

$$\lambda_{\min} = \min_{(k,l)\in[1,N]\times[1,mT]} \lambda_{kl} = -4\delta^2\sin^2\frac{N\pi}{2(N+1)}.$$

In this paper, we require $\sigma(L)\cap(0,\infty)\neq\emptyset$ and $\sigma(L)\cap(-\infty,0)\neq\emptyset$. Clearly, $\sigma(L)\cap(-\infty,0)\neq\emptyset$, thus, we only assume that the condition

$$\begin{cases} \delta\sin\frac{\pi}{2(N+1)} < 1, & mT \text{ is even}, \\ \delta\sin\frac{\pi}{2(N+1)} < \cos\frac{\pi}{mT}, & mT \text{ is odd}, \end{cases} \tag{18}$$

holds, where $\delta > 0$. On the other hand, we also need to suppose that the conditions

$$\sin^2\frac{(l-1)\pi}{mT} \neq \delta^2\sin^2\frac{k\pi}{2(N+1)} \tag{19}$$

hold for $(k,l)\in[1,N]\times[1,mT]$. In this case, we have $0\notin\sigma(L)$. Throughout this paper, we always assume that the conditions (18) and (19) hold.

The abstract critical point theorem plays an important role in proving our main results. Let E be a Hilbert space with norm $\|\cdot\|$ and inner product $\langle\cdot,\cdot\rangle$ and have an orthogonal decomposition $E = N\oplus N^{\perp}$, where $N\in E$ is a closed and separable subspace. Since N is separable, we can define a new norm $|v|_{\omega}$ satisfying $|v|_{\omega}\leq\|v\|$ for all $v\in N$ and such that the topology induced by this norm is equivalent to the weak topology of N on bounded subset of N. For $u = v + w\in E$ with $v\in N$ and $w\in N^{\perp}$, we define $|u|_{\omega}^2 = |v|_{\omega}^2 + \|w\|^2$, then $|u|_{\omega}\leq\|u\|$ for $u\in E$. Particularly, if $\{u_n = v_n + w_n\}_{n=1}^{\infty}\in E$ is $|\cdot|_{\omega}$-bounded and $u_n\to_{|\cdot|_{\omega}} u$, then $v_n\rightharpoonup v$ weakly in N, $w_n\to w$ strongly in $N^{\perp}$, $u_n\rightharpoonup v + w$ weakly in E (see [21]).

Let $E = E^-\oplus E^+$, $z_0\in E^+$ with $\|z_0\| = 1$. For any $u\in E$, we write $u = u^-\oplus sz_0\oplus w^+$ with $u^-\in E^-$, $s\in R$, $w^+\in(E^-\oplus Rz_0)^{\perp} := E_1^+$. For $R > 0$, let

$$Q = \{u = u^- + sz_0 \mid s\in R^+, u^-\in E^-, \|u\| < R\}$$

with $p_0 = s_0z_0\in Q$, $s_0 > 0$. We define

$$D = \{u = sz_0 + w^+ \mid s\geq 0, w^+\in E_1^+, \|sz_0 + w^+\| = s_0\}.$$

For $I\in C^1(E,R)$, define $h:[0,1]\times\overline{Q}\to E$ is $|\cdot|_{\omega}$ -continuous, $h(0,u) = u$, $I(h(s,u))\leq I(u)$ for $u\in\overline{Q}$, for any $(s_0,u_0)\in[0,1]\times\overline{Q}$, there is a $|\cdot|_{\omega}$-neighborhood $U_{(s_0,u_0)}$ such that

$$\{u - h(t,u) \mid (t,u)\in U_{(s_0,u_0)}\cap[0,1]\times\overline{Q}\}\subset E_{\text{fin}},$$

where E_{fin} denotes various finite-dimensional subspaces of E whose exact dimensions are irrelevant and depend on (s_0, u_0). Denote

$$\Gamma = \{h \mid h : [0,1] \times \overline{Q} \to E\},$$

then $\Gamma \neq \emptyset$ since $id \in \Gamma$.

The variant weak linking theorem is as follows.

Lemma 2 (see [21]) *The family of C^1-functional $\{H_\lambda\}$ has the form*

$$H_\lambda(u) = I(u) - \lambda K(u) \quad \text{for } \lambda \in [1,2]. \tag{20}$$

Assume that

(a) $K(u) \geq 0$, $u \in E$, $H_1 = H$;

(b) $I(u) \to \infty$ *or* $K(u) \to \infty$ *as* $\|u\| \to \infty$;

(c) H_λ *is* $|\cdot|_\omega$*-upper semicontinuous,* H'_λ *is weakly sequentially continuous on E. Moreover,* H_λ *maps bounded sets to bounded sets;*

(d) $\sup_{\partial Q} H_\lambda \leq \inf_D H_\lambda$ *for* $\lambda \in [1,2]$.

Then, for almost all $\lambda \in [1,2]$*, there exists a sequence* $\{u_n\}$ *such that*

$$\sup_n \|u_n\| < \infty, \qquad H'_\lambda(u_n) \to 0, \qquad H_\lambda(u_n) \to c_\lambda,$$

where

$$c_\lambda = \inf_{h \in \Gamma} \sup_{u \in Q} H_\lambda(h(1,u)) \in \left[\inf_D H_\lambda, \sup_{\overline{Q}} H\right].$$

3 Main result

First of all, we state the following conditions.

(i) $xf(n,x) \geq 0$ for $x \in R$ and $n \in Z$.

(ii) There exist $0 < a < 1$ and $r > 0$ such that

$$\int_0^x f(j,s)\, ds \leq \frac{a\sigma^+_{\min}}{2} x^2 \quad \text{for } |x| \leq r.$$

(iii) There exist $\rho > 0$ and $d > 1$ such that

$$\int_0^x f(j,s)\, ds \geq \frac{d\sigma^+_{\max}}{2} x^2 \quad \text{for } |x| > \rho,$$

where

$$\sigma^+_{\min} = \min\{\lambda_{kl} > 0, (k,l) \in [1,N] \times [1,mT]\}$$

and

$$\sigma^+_{\max} = \max\{\lambda_{kl} > 0, (k,l) \in [1,N] \times [1,mT]\}.$$

Theorem 1 *The function $f : Z \times R \to R$ is continuous about the second variable and there exists a positive integer T such that*

$$f(n+T,\cdot) = f(n,\cdot) \quad \text{for } n \in Z.$$

Suppose that the above conditions (i)-(iii), (18), *and* (19) *hold. Then for any positive integer m, for problem* (10)-(12) *there at least exists a non-zero time mT-periodic solution.*

Proof In view of Lemma 2, we need to prove that the conditions (a)-(d) hold. First of all, we give some symbols.

When the conditions (18) and (19) hold, we can denote

$$\sigma^+(L) = \{\lambda_{kl} > 0, (k,l) \in [1,N] \times [1,mT]\},$$
$$\sigma^-(L) = \{\lambda_{kl} < 0, (k,l) \in [1,N] \times [1,mT]\},$$
$$\sigma^+_{\min} = \min\{\lambda_{kl} > 0, (k,l) \in [1,N] \times [1,mT]\},$$
$$\sigma^+_{\max} = \max\{\lambda_{kl} > 0, (k,l) \in [1,N] \times [1,mT]\},$$
$$\sigma^-_{\min} = \min\{\lambda_{kl} < 0, (k,l) \in [1,N] \times [1,mT]\},$$

and

$$\sigma^-_{\max} = \max\{\lambda_{kl} < 0, (k,l) \in [1,N] \times [1,mT]\}.$$

Clearly, we have

$$\sigma^-_{\min} < \sigma^-_{\max} < 0 < \sigma^+_{\min} < \sigma^+_{\max},$$
$$\sigma(L) = \sigma^+(L) \cup \sigma^-(L),$$

and

$$\sigma^+(L) \cap \sigma^-(L) = \emptyset.$$

That is, the linear operator $L = -(\Delta^2 - \delta^2 A)$ has a sequence of eigenvalues

$$\sigma^-_{\min} = \lambda_{-p} \le \lambda_{-p+1} \le \cdots \le \lambda_{-1} = \sigma^-_{\max}$$
$$< 0 < \sigma^+_{\min} = \lambda_1 \le \lambda_2 \le \cdots \le \lambda_q = \sigma^+_{\max}$$

and the corresponding eigenvectors Υ_j for $j = -p, \ldots, -1, 1, \ldots, q$. (Clearly, we have $p + q = mTN$.)

Let

$$E^- = \left\{ \sum_{i=1}^{p} c_i \Upsilon_{-i} \,\middle|\, c_i \in R \right\} \quad \text{and} \quad E^+ = \left\{ \sum_{i=1}^{q} c_i \Upsilon_i \,\middle|\, c_i \in R \right\}.$$

Then $E_{mT} = E = E^- \oplus E^+$ and for any $U \in E_{mT}$ we have $U = U^- + U^+$, where $U^- \in E^-$ and $U^+ \in E^+$. Clearly, we have also $E^0 = \ker L = \{0\}$.

For any $U, V \in E_{mT}$, $U = U^+ + U^-$ and $V = V^+ + V^-$, we can define an equivalent new inner product $\langle\cdot\rangle$ and the corresponding norm $\|\cdot\|$ in E_{mT} by

$$\langle U, V\rangle = \langle LU^+, V^+\rangle_{mT} - \langle LU^-, V^-\rangle_{mT} \quad \text{and} \quad \|U\| = \langle U, U\rangle^{1/2},$$

see [25]. Therefore, H can be rewritten as

$$\begin{aligned} H(U) &= \frac{1}{2}\left(\|U^+\|^2 - \|U^-\|^2\right) - \Psi(U) \\ &= \frac{1}{2}\|U^+\|^2 - \left(\frac{1}{2}\|U^-\|^2 + \Psi(U)\right), \end{aligned}$$

where

$$\Psi(U) = \sum_{j=1}^{mT}\sum_{i=1}^{N}\int_0^{u_j^i} f(j,s)\,ds = \sum_{j=1}^{mT} H(j, U_j).$$

In order to apply Lemma 2, we consider the family of functional defined by

$$H_\lambda(U) = \frac{1}{2}\|U^+\|^2 - \lambda\left(\frac{1}{2}\|U^-\|^2 + \Psi(U)\right) \quad \text{for } \lambda \in [1,2]. \tag{21}$$

In the following, we give the proofs for the conditions (a)-(d) in Lemma 2.

(a) $K(U) \geq 0$, $U \in E_{mT}$, $H_1 = H$.

Let

$$K(U) = \frac{1}{2}\|U^-\|^2 + \Psi(U).$$

When $xf(n,x) \geq 0$, we find that (a) holds.

(b) $I(U) \to \infty$ or $K(U) \to \infty$ as $\|U\| \to \infty$. We will prove that $K(U) \to \infty$ as $\|U\| \to \infty$. In fact, this is clear when the condition (iii) holds.

(c) H_λ is $|\cdot|_\omega$-upper semicontinuous, H'_λ is weakly sequentially continuous on E. Moreover, H_λ maps bounded sets to bounded sets.

The condition (iii) implies that

$$\begin{aligned} H_\lambda(U) &= \frac{1}{2}\|U^+\|^2 - \lambda\left(\frac{1}{2}\|U^-\|^2 + \Psi(U)\right) \\ &\leq \frac{1}{2}\|U^+\|^2 - \frac{1}{2}\|U^-\|^2 - \Psi(U) \\ &\leq \frac{\sigma^+_{\max}}{2}\|U\|^2_{mT} - \frac{d\sigma^+_{\max}}{2}\|U\|^2_{mT} \\ &= -(d-1)\frac{\sigma^+_{\max}}{2}\|U\|^2_{mT} \to -\infty \quad \text{as } \|U\|_{mT} \to \infty. \end{aligned}$$

Thus, if $U_n \to_{|\cdot|_\omega} U$ and $H_\lambda(U_n) \geq a$, then $H_\lambda(U) \geq a$, which means that H_λ is $|\cdot|_\omega$-upper semicontinuous. The other cases are clear.

(d) $\sup_{\partial Q} H_\lambda \leq \inf_D H_\lambda$ for $\lambda \in [1,2]$.

Note by the condition (ii), letting $s_0 = r$, for $U \in S = \{U \mid U \in E^+, \|U\| = s_0\}$ we have

$$\begin{aligned} H_\lambda(U) &= \frac{1}{2}\left\|U^+\right\|^2 - \lambda\Psi(U) \\ &\geq \frac{\sigma^+_{\min}}{2}\|U\|^2_{mT} - 2\sum_{j=1}^{mT}\sum_{i=1}^{N}\int_0^{u^i_j} f(j,s)\,ds \\ &\geq \frac{\sigma^+_{\min}}{2}\|U\|^2_{mT} - \frac{a\sigma^+_{\min}}{2}\|U\|^2_{mT} \\ &= (1-a)\frac{\sigma^+_{\min}}{2}\|U\|^2_{mT}. \end{aligned}$$

Let $z_0 = \Upsilon_1/\|\Upsilon_1\|$ and

$$D = \left\{U \mid U = sz_0 + W^+, s \geq 0, W^+ \in E_1^+ \text{ and } \|U\| = s_0\right\},$$

we have $D \subset S$, which implies that

$$\inf_D H_\lambda(U) > 0.$$

Now, we choose $\rho > 0$ of the condition (ii) and let

$$Q = \left\{U \mid U = U^- + sz_0, s \geq 0, U^- \in E^- \text{ and } \|U\| < \rho\right\}.$$

For $U \in \partial Q$, we have

$$\begin{aligned} H_\lambda(U) &= \frac{1}{2}\left\|U^+\right\|^2 - \lambda\left(\frac{1}{2}\left\|U^-\right\|^2 + \Psi(U)\right) \\ &\leq -(d-1)\frac{\sigma^+_{\max}}{2}\|U\|^2_{mT} < 0. \end{aligned}$$

That is, the condition (d) holds.

In view of Lemma 2, we find that for almost all $\lambda \in [1,2]$ there exists a sequence $\{U^{(n)}\} \subset \overline{Q}$ such that

$$\sup_n \left\|U^{(n)}\right\| < \infty, \qquad H'_\lambda\left(U^{(n)}\right) \to 0$$

and

$$H_\lambda\left(U^{(n)}\right) \to c_\lambda \in \left[\frac{1}{2}(1-a)r\sigma^+_{\min}, \sup_{\overline{Q}} H_\lambda(U)\right]. \tag{22}$$

Note that the function $H_\lambda(U)$ is finite dimensional continuous, thus, there exist $\{U^{(n)}\} \subset \overline{Q}$ and $U^* \in \overline{Q}$ such that

$$\lim_{n\to\infty} H_1\left(U^{(n)}\right) = \lim_{n\to\infty} H\left(U^{(n)}\right) = \lim_{n\to\infty} H\left(U^*\right) = c_1 > 0.$$

The proof is complete. □

Remark 1 The symbols $\sigma_{\min}^{-}$ and $\sigma_{\max}^{-}$ have not been used. In fact, similarly, if we discuss the functional $-H(U)$, then the corresponding result can also be obtained. It is omitted.

Remark 2 From the proofs of (a)-(d), we can see that the conditions (ii)-(iii) need only to be hold locally. For convenience, we use the present state.

Remark 3 Our result is new; see Deng *et al.* [6] and Yan *et al.* [7]. Clearly, the sublinear case can also be established. It will be omitted.

Remark 4 For an mT-dimensional discrete system of the form

$$\Delta^2 U_{n-1} + BU_n + \nabla V(n, U_n) = 0, \quad n \in Z,$$

when B is a symmetric positive definite, negative definite, or infinite definite matrix, our method is also valid.

Remark 5 The superlinear condition (iii) cannot be used for the corresponding continuous wave equations or the continuous Hamiltonian systems because their eigenvalues are unbounded.

Remark 6 Our method is not suitable for the corresponding continuous wave equations. When ω^2 is irrational, the eigenvalues of the operator $\partial_{tt} - \partial_{xx}$ accumulate to 0. However, if ω^2 is rational, the number of 0-spectrum is infinite.

Remark 7 All conditions of Theorem 1 are easily satisfied. For the conditions (18) and (19), for example, let $\delta = 1$, $m = 1$ and $T = 2$, the condition (18) clearly holds for any N. At the same time, note that

$$\sin^2 \frac{(1-1)\pi}{2} = 0, \qquad \sin^2 \frac{(2-1)\pi}{2} = 1$$

and

$$0 < \sin^2 \frac{k\pi}{2(N+1)} < 1$$

for all $k \in [1, N]$. Thus, the condition (19) also holds. For the nonlinear term, the conditions (i)-(iii) are also easy satisfied. For example, for $\delta = 1$, $m = 1$, $T = 2$, and $N = 2$, we have

$$\sigma(L) = \left\{ -\frac{3}{4}, -\frac{1}{4}, 1, 3 \right\}, \qquad \sigma_{\min}^{+} = 1, \quad \text{and} \quad \sigma_{\max}^{+} = 3.$$

In this case, we let $f(-x) = -f(x)$ and

$$f(x) = \begin{cases} \frac{x}{4}, & 0 \le x < 1, \\ \frac{59}{8}x - \frac{57}{8}, & 1 \le x \le 3, \\ 5x, & x > 3. \end{cases}$$

Then all conditions of Theorem 1 are satisfied. However, such a function is not valid for the corresponding continuous wave equations or the continuous Hamiltonian systems.

Competing interests
The authors declare that they have no competing interests.

Authors' contributions
All authors contributed equally to the writing of this paper. All authors read and approved the final manuscript.

Author details
[1]School of Science, Tianjin University of Commerce, Tianjin, 300134, P.R. China. [2]Department of Computing & Information Systems, Department of Mathematics, Trent University, Peterborough, K9J 7B8, Canada. [3]School of Computer Science & Mathematics, Faculty of Science & Engineering, University of Chester, Thornton Science Park, Pool Lane, Ince, Chester, CH2 4NU, UK.

Acknowledgements
The authors would like to thank the anonymous referees for their careful review and helpful comments. This work was supported by the National Natural Science Foundation of China (No. 11371277). The project was completed while the first author was visiting Trent University and supported by a grant from the Natural Sciences and Engineering Research Council of Canada.

References
1. Vitt, VA: Distributed self-oscillating systems Zh. Tekh. Fiz. **4**, 144-157 (1934) (in Russian)
2. Rabinowitz, P: Periodic solutions of nonlinear hyperbolic partial differential equations. Commun. Pure Appl. Math. **20**, 145-205 (1967)
3. Brézis, H, Coron, JM: Periodic solutions of nonlinear wave equations and Hamiltonian systems. Am. J. Math. **103**(3), 559-570 (1981)
4. Berti, M, Bolle, P: Cantor families of periodic solutions for completely resonant nonlinear wave equations. Duke Math. J. **134**, 359-419 (2006)
5. Gentile, G, Mastropietro, V: Construction of periodic solutions of nonlinear wave equation with Dirichlet boundary conditions by the Lindstedt series method. J. Math. Pures Appl. **83**, 1019-1065 (2004)
6. Deng, XQ, Shi, HP, Xie, XL: Periodic solutions of second order discrete Hamiltonian systems with potential indefinite in sign. Appl. Math. Comput. **218**, 148-156 (2011)
7. Yan, SH, Wu, XP, Tang, CL: Multiple periodic solutions for second-order discrete Hamiltonian systems. Appl. Math. Comput. **234**, 142-149 (2014)
8. Jiang, LQ: Existence of periodic solutions for convex discrete Hamiltonian systems. Math. Sci. Res. J. **16**(3), 64-81 (2012)
9. Guo, ZM, Yu, JS: Multiplicity results for periodic solutions to second-order difference equations. J. Dyn. Differ. Equ. **18**(4), 943-960 (2006)
10. Poincaré, H: Les Méthodes Nouvelles de la Mécanique Céleste. Gauthier-Villars, Paris (1892)
11. Lyapunov, AM: Problème général de la stabilité du mouvement. Ann. Fac. Sci. Toulouse **2**, 203-474 (1907)
12. Birkhoff, GD, Lewis, DC: On the periodic motions near a given periodic motion of a dynamical system. Ann. Math. **12**, 117-133 (1933)
13. Moser, J: Periodic orbits near an equilibrium and a theorem by Alan Weinstein. Commun. Pure Appl. Math. **29**, 727-747 (1976)
14. Weinstein, A: Normal modes *f* or a nonlinear Hamiltonian system. Invent. Math. **20**, 47-57 (1973)
15. Rabinowitz, P: Periodic solutions of Hamiltonian systems. Commun. Pure Appl. Math. **31**, 157-184 (1978)
16. Ekeland, I, Lasry, JM: On the number of periodic trajectories for a Hamiltonian flow on a convex energy surface. Ann. Math. **112**(2), 283-319 (1980)
17. Pugh, CC, Robinson, RC: The C^1 closing lemma, including Hamiltonians. Ergod. Theory Dyn. Syst. **3**, 261-313 (1983)
18. Biasco, L, Gregorio, LD: Periodic solutions of Birkhoff-Lewis type for the nonlinear wave equation. Discrete Contin. Dyn. Syst. suppl., 102-109 (2007)
19. Long, YM: Multiple solutions of perturbed superquadratic second order Hamiltonian systems. Trans. Am. Math. Soc. **311**(2), 749-780 (1989)
20. Luan, S, Mao, A: Periodic solutions for a class of nonautonomous Hamiltonian systems. Nonlinear Anal. **61**(8), 1413-1426 (2005)
21. Schechter, M, Zou, W: Weak linking theorems and Schrödinger equations with critical Sobolev exponent. ESAIM Control Optim. Calc. Var. **9**, 601-619 (2003)
22. Chen, GW, Ma, SW: Periodic solutions for Hamiltonian systems without Ambrosetti-Rabinowitz condition and spectrum 0. J. Math. Anal. Appl. **379**, 842-851 (2011)
23. Bai, L, Zhang, G: Nontrivial solutions for a nonlinear discrete elliptic equation with periodic boundary conditions. Appl. Math. Comput. **210**, 321-333 (2009)
24. Cheng, SS: Partial Difference Equations. Taylor & Francis, London (2003)
25. Stuart, CA: Bifurcation into spectral gaps. Bull. Belg. Math. Soc. suppl., pp. 59 (1995)

Growth of solutions of some kinds of linear difference equations

Shun-Zhou Wu and Xiu-Min Zheng*

*Correspondence: zhengxiumin2008@sina.com
Institute of Mathematics and Information Science, Jiangxi Normal University, Nanchang, Jiangxi 330022, P.R. China

Abstract

In this paper, we investigate the growth and value distribution of solutions of some kinds of linear difference equations, where there may be more than one coefficient having the same maximal order and the same maximal type.

MSC: 30D35; 39A22

Keywords: linear difference equation; meromorphic solution; growth; exponent of convergence

1 Introduction

Throughout this paper, we use the standard notations and basic results of Nevanlinna's value distribution theory (see [1–4]). In addition, we use the notations $\sigma(f)$, $\sigma_2(f)$ and $\lambda(f-a)$ to denote, respectively, the order, the hyperorder, and the exponent of convergence of the sequence of a-points of a meromorphic function $f(z)$ in the complex plane, where $a \in \mathbb{C} \cup \{\infty\}$. Furthermore, we can get the definition of $\lambda(f-\varphi)$, when a is replaced by a meromorphic function $\varphi(z)$.

Recently, with the research and further development of difference analogs of Nevanlinna's theory, it has been applied more and more widely in the difference field. By this important tool, many scholars investigated the linear difference equation

$$A_k(z)f(z+c_k) + \cdots + A_1(z)f(z+c_1) + A_0(z)f(z) = 0, \tag{1.1}$$

where $k \in \mathbb{N}_+$ and c_j, $j = 1,\ldots,k$ are distinct nonzero complex constants, and obtained many results on the growth and the exponent of convergence of the sequence of zeros of meromorphic solutions of (1.1). For instance, in [5–7], the authors considered the case when there is exactly only one coefficient of (1.1) having the maximal order; in [8–10], the authors considered the case when there is exactly only one coefficient having the type strictly greater than the others among those having the maximal order; and in [5, 6, 11], the authors considered the case when the coefficients of (1.1) are polynomials.

Further, how about the case when there are more than one coefficient having the same maximal order and the same maximal type?

In 2013, Liu [12] considered the growth and the exponent of convergence of the sequence of small function value points of second order linear difference equations and obtained the following theorem.

Theorem A (see [12]) *Let c_1, c_2 $(\neq c_1)$, a be nonzero constants, $h_1(z)$ be a nonzero meromorphic functions with $\sigma(h_1) < 1$, $B(z)$ be a nonzero meromorphic function. If $B(z)$ satisfies any one of the following three conditions:*

(i) $\sigma(B) > 1$ *and* $\delta(\infty, B) > 0$;

(ii) $\sigma(B) < 1$;

(iii) $B(z) = h_0(z)e^{bz}$, *where b is a nonzero constant, $h_0(z)$ $(\not\equiv 0)$ is a meromorphic function with $\sigma(h_0) < 1$,*

then every meromorphic solution $f(z)$ $(\not\equiv 0)$ of the difference equation

$$f(z + c_2) + h_1(z)e^{az}f(z + c_1) + B(z)f(z) = 0$$

satisfies $\sigma(f) \geq \max\{\sigma(B), 1\} + 1$. Further, if $\varphi(z)$ $(\not\equiv 0)$ is a meromorphic function with $\sigma(\varphi) < \max\{\sigma(B), 1\} + 1$, then $\lambda(f - \varphi) = \sigma(f) \geq \max\{\sigma(B), 1\} + 1$.

Liu and Mao [13] considered the growth of meromorphic solutions of (1.1), where the coefficients may have the same order and the same type, and obtained the following theorem.

Theorem B (see [13]) *Let $A_j(z) = h_j(z)e^{P_j(z)} + D_j(z)$, $j = 0, 1, \ldots, k$, where $P_j(z) = a_{jn}z^n + \cdots + a_{j0}$ are polynomials with degree n (≥ 1), $h_j(z)$ $(\not\equiv 0)$, $D_j(z)$ are entire functions with order less than n. If a_{jn}, $j = 0, 1, \ldots, k$ are distinct complex numbers, then every meromorphic solution $f(z)$ $(\not\equiv 0)$ of (1.1) satisfies $\sigma(f) \geq \max_{0\leq j\leq k}\{\sigma(A_j)\} + 1$.*

In this paper, we are concerned with the more general problem than Theorems A and B, and obtain the following results, which extend and improve the previous results.

First, we consider the difference equation (1.1).

Theorem 1.1 *Let $k, n \in \mathbb{N}_+$, $A_j(z) = B_j(z)e^{P_j(z)} + D_j(z)e^{Q_j(z)} + R_j(z)$, $j = 0, 1, \ldots, k$, where $P_j(z) = a_{jn}z^n + \cdots + a_{j0}$, $Q_j(z) = b_{jn}z^n + \cdots + b_{j0}$, $j = 0, 1, \ldots, k$ are polynomials with degree n and satisfy $|a_{jn}| \geq |b_{jn}| > 0$, $j = 0, 1, \ldots, k$, $B_j(z)$, $D_j(z)$, $R_j(z)$, $j = 0, 1, \ldots, k$ are meromorphic functions and satisfy $\max_{0\leq j\leq k}\{\sigma(B_j), \sigma(D_j), \sigma(R_j)\} = \omega < n$, $A_j(z) - R_j(z) \not\equiv 0$, $j = 0, 1, \ldots, k$. Let c_j, $j = 1, \ldots, k$ be distinct nonzero complex constants. If there exists an $i \in \{0, 1, \ldots, k\}$ such that for all j $(\neq i)$, $|a_{in}| \geq |a_{jn}|$, and*

(i) $\arg a_{in} \neq \arg a_{jn}$, *or* $\arg a_{in} = \arg a_{jn}$, $|a_{in}| > |a_{jn}|$

and

(ii) $\arg a_{in} \neq \arg b_{jn}$, *or* $\arg a_{in} = \arg b_{jn}$, $|a_{in}| > |b_{jn}|$

hold simultaneously, then every meromorphic solution $f(z)$ $(\not\equiv 0)$ of (1.1) satisfies $\sigma(f) \geq n + 1$. Further, if $\varphi(z)$ $(\not\equiv 0)$ is a meromorphic function with $\sigma(\varphi) < n + 1$, then for every meromorphic solution $f(z)$ $(\not\equiv 0)$ of (1.1) with $\sigma_2(f) < 1$, we have $\lambda(f - \varphi) = \sigma(f) \geq n + 1$.

Remark 1.1 In Theorem 1.1, if there exist some j $(\neq i)$ such that $A_j(z) - R_j(z) \equiv 0$, or some of $B_j(z)$ $(j \neq i)$, $D_j(z)$, $R_j(z)$, $j = 0, 1, \ldots, k$, are equal to zero, then the corresponding result holds by using a similar proof to the one of Theorem 1.1.

Next, we consider difference operators instead of shift operators in (1.1).

For a nonzero complex constant c, the forward differences $\Delta^k f(z)$, $k \in \mathbb{N}_+$, are defined (see [14]) by

$$\Delta f(z) = \Delta^1 f(z) = f(z + c) - f(z), \qquad \Delta^{k+1} f(z) = \Delta^k f(z + c) - \Delta^k f(z), \quad k \in \mathbb{N}_+.$$

It is shown in [6] that

$$\Delta^k f(z) = \sum_{j=0}^{k} C_k^j (-1)^{k-j} f(z+jc), \qquad f(z+kc) = \sum_{j=0}^{k} C_k^j \Delta^j f(z), \quad k \in \mathbb{N}_+.$$

Then we can obtain the following theorem.

Theorem 1.2 *Suppose that $A_j(z), j = 0,1,\ldots,k$, satisfy the hypotheses of Theorem 1.1, and i is also defined as in Theorem 1.1. If $i = 0$, then every meromorphic solution $f(z)$ ($\not\equiv 0$) of the difference equation*

$$A_k(z)\Delta^k f(z) + \cdots + A_1(z)\Delta f(z) + A_0(z)f(z) = 0 \tag{1.2}$$

satisfies $\sigma(f) \geq n+1$. Further, if $\varphi(z)$ ($\not\equiv 0$) is a meromorphic function with $\sigma(\varphi) < n+1$, then for every meromorphic solution $f(z)$ ($\not\equiv 0$) of (1.2) with $\sigma_2(f) < 1$, we have $\lambda(f - \varphi) = \sigma(f) \geq n+1$.

In the end, we can easily get the following corollary.

Corollary 1.1 *Let $a_j \in \mathbb{C}, j = 0,1,\ldots,k$, such that $a_0 \neq a_j$, $|a_0| \geq |a_j| \geq 0$, $j = 1,\ldots,k$, and $h_j(z)$ ($\not\equiv 0$), $j = 0,1,\ldots,k$, be meromorphic functions with order less than n, then every meromorphic solution $f(z)$ ($\not\equiv 0$) of the difference equation*

$$\begin{aligned}&h_k(z)e^{a_k z^n}\Delta^k f(z) + h_{k-1}(z)e^{a_{k-1}z^n}\Delta^{k-1}f(z) + \cdots + h_1(z)e^{a_1 z^n}\Delta f(z)\\ &\quad + h_0(z)e^{a_0 z^n} f(z) = 0\end{aligned}$$

satisfies $\sigma(f) \geq n+1$. Further, if $\varphi(z)$ ($\not\equiv 0$) is a meromorphic function with $\sigma(\varphi) < n+1$, then for every meromorphic solution $f(z)$ ($\not\equiv 0$) with $\sigma_2(f) < 1$, we have $\lambda(f-\varphi) = \sigma(f) \geq n+1$.

2 Preliminary lemmas

Lemma 2.1 (see [6]) *Let η_1, η_2 be two arbitrary complex numbers, and $f(z)$ be a meromorphic function of finite order σ. Let $\varepsilon > 0$ be given, then there exists a set $E \subset (0,+\infty)$ with finite logarithmic measure such that for all $r \notin E \cup [0,1]$, we have*

$$\exp(-r^{\sigma-1+\varepsilon}) \leq \left|\frac{f(z+\eta_1)}{f(z+\eta_2)}\right| \leq \exp(r^{\sigma-1+\varepsilon}).$$

Remark 2.1 It follows from Lemma 2.1 that

$$\begin{aligned}\left|\frac{\Delta^i f(z)}{f(z)}\right| &= \left|\frac{\sum_{j=0}^{i} C_i^j(-1)^{i-j}f(z+jc)}{f(z)}\right| \leq \sum_{j=0}^{i} C_i^j \left|\frac{f(z+jc)}{f(z)}\right|\\ &\leq \sum_{j=0}^{i} C_i^j \exp\{r^{\sigma-1+\varepsilon}\} = 2^i \exp\{r^{\sigma-1+\varepsilon}\}, \quad i \in \mathbb{N}_+.\end{aligned}$$

Lemma 2.2 (see [15]) *Let $f(z)$ be a meromorphic function with $\sigma(f) = \beta < +\infty$, then for any given $\varepsilon > 0$, there exists a set $E \subset [0,+\infty)$ with $mE < +\infty$ such that for all z with $|z| =$*

$r \notin [0,1] \cup E$, $r \to \infty$, *we have*

$$\exp\{-r^{\beta+\varepsilon}\} \le |f(z)| \le \exp\{r^{\beta+\varepsilon}\}.$$

Lemma 2.3 (see [16]) *Suppose that* $P(z) = (\alpha + \beta i)z^n + \cdots$ (α, β *are real numbers such that* $|\alpha| + |\beta| \neq 0$) *is a polynomial with degree* n (≥ 1), *and* $\omega(z)$ ($\not\equiv 0$) *is a meromorphic function with* $\sigma(\omega) < n$. *Set* $g(z) = \omega(z)e^{P(z)}$, $z = re^{i\theta}$, $\delta(P,\theta) = \alpha \cos n\theta - \beta \sin n\theta$. *Then for any given* $\varepsilon > 0$, *there exists a set* $H_0 \subset [0,2\pi)$ *with linear measure zero, such that for any* $\theta \in [0,2\pi)\backslash(H_0 \cup H_1)$, *there exists* $r_0 = r_0(\theta,\varepsilon)$ (> 0) *such that for* $|z| = r > r_0$, *we have*

(i) *if* $\delta(P,\theta) > 0$, *then*

$$\exp\{(1-\varepsilon)\delta(P,\theta)r^n\} < |g(re^{i\theta})| < \exp\{(1+\varepsilon)\delta(P,\theta)r^n\};$$

(ii) *if* $\delta(P,\theta) < 0$, *then*

$$\exp\{(1+\varepsilon)\delta(P,\theta)r^n\} < |g(re^{i\theta})| < \exp\{(1-\varepsilon)\delta(P,\theta)r^n\},$$

where $H_1 = \{\theta \in [0,2\pi) : \delta(P,\theta) = 0\}$ *is a finite set.*

Remark 2.2 Let $P(z) = az^n + \cdots$ be a polynomial with degree n and $z = re^{i\theta}$, we denote $\delta(P,\theta) = |a| \cos(\arg a + n\theta)$.

Lemma 2.4 (see [17]) *Let* $f(z)$ *be a nonconstant meromorphic function,* $\varepsilon > 0$, *and* $c \in \mathbb{C}\backslash\{0\}$. *If* $\zeta = \sigma_2(f) < 1$, *then*

$$m\left(r, \frac{f(z+c)}{f(z)}\right) = o\left(\frac{T(r,f)}{r^{1-\zeta-\varepsilon}}\right)$$

for all r *outside of a set of finite logarithmic measure.*

Lemma 2.5 *Let* c_j, $j = 1,\ldots,k$ *be distinct nonzero complex constants,* $A_j(z)$, $j = 0,1,\ldots,k$, $F(z)$ *be meromorphic functions such that* $A_k(z)A_0(z)F(z) \not\equiv 0$. *If* $f(z)$ *is a meromorphic solution of the difference equation*

$$A_k(z)f(z+c_k) + \cdots + A_1(z)f(z+c_1) + A_0(z)f(z) = F(z) \tag{2.1}$$

and satisfies $\max\{\sigma(F),\sigma(A_j), j = 0,1,\ldots,k\} = \omega < \sigma(f) = \sigma$ ($0 < \sigma \leq \infty$), $\sigma_2(f) < 1$, *then we have*

$$\lambda(f) = \sigma(f).$$

Proof We use a similar proof to the one in [12] here. We rewrite (2.1) as

$$\frac{1}{f(z)} = \frac{1}{F(z)}\left(A_k(z)\frac{f(z+c_k)}{f(z)} + \cdots + A_1(z)\frac{f(z+c_1)}{f(z)} + A_0(z)\right). \tag{2.2}$$

By Lemma 2.4, there exists a set $E \subset (1,+\infty)$ of finite logarithmic measure such that for all z satisfying $|z| = r \notin E$, we have

$$m\left(r, \frac{f(z+c_j)}{f(z)}\right) = o(T(r,f)), \quad j = 1,\ldots,k.$$

Thus, (2.2) implies that

$$\begin{aligned} T(r,f) &= N\left(r,\frac{1}{f}\right) + m\left(r,\frac{1}{f}\right) + O(1) \\ &\le N\left(r,\frac{1}{f}\right) + m\left(r,\frac{1}{F}\right) + \sum_{j=0}^{k} m(r,A_j) + \sum_{j=1}^{k} m\left(r,\frac{f(z+c_j)}{f(z)}\right) + O(1) \\ &\le N\left(r,\frac{1}{f}\right) + T(r,F) + \sum_{j=0}^{k} T(r,A_j) + o(T(r,f)), \quad r \notin E. \end{aligned} \tag{2.3}$$

Set $m_l E = \log \delta < \infty$. Since $\sigma(f) = \sigma$, there exists a sequence $\{r'_n\}_{n=1}^{\infty}$ tending to ∞ such that $(\delta + 2)r'_n < r'_{n+1}$ and

$$\lim_{r'_n \to \infty} \frac{\log T(r'_n, f)}{\log r'_n} = \sigma.$$

We may choose $r_n \in [r'_n, (\delta+2)r'_n]\backslash E$, $n = 1, 2, \dots$. Since

$$\frac{\log T(r_n, f)}{\log r_n} \ge \frac{\log T(r'_n, f)}{\log(\delta+2)r'_n} = \frac{\log T(r'_n, f)}{\log r'_n (1 + \frac{\log(\delta+2)}{\log r'_n})},$$

we have

$$\varliminf_{r_n \to \infty} \frac{\log T(r_n, f)}{\log r_n} \ge \lim_{r'_n \to \infty} \frac{\log T(r'_n, f)}{\log r'_n (1 + \frac{\log(\delta+2)}{\log r'_n})} = \sigma,$$

that is,

$$\lim_{r_n \to \infty} \frac{\log T(r_n, f)}{\log r_n} = \sigma.$$

Then for sufficiently small ε $(0 < \varepsilon < \frac{\sigma-\omega}{2})$ and sufficiently large r_n, we have

$$T(r_n, f) \ge r_n^{\sigma-\varepsilon}$$

and

$$T(r_n, F) \le r_n^{\omega+\varepsilon}, \qquad T(r_n, A_j) \le r_n^{\omega+\varepsilon}, \quad j = 0, 1, \dots, k.$$

Hence,

$$\frac{T(r_n, F)}{T(r_n, f)} \le r_n^{\omega-\sigma+2\varepsilon} \to 0, \quad r_n \to \infty$$

and

$$\frac{T(r_n, A_j)}{T(r_n, f)} \le r_n^{\omega-\sigma+2\varepsilon} \to 0, \quad r_n \to \infty, j = 0, 1, \dots, k,$$

hold. Then, for sufficiently large r_n, we have

$$T(r_n,F) \le \frac{1}{k+4}T(r_n,f), \qquad o(T(r_n,f)) \le \frac{1}{k+4}T(r_n,f),$$
$$T(r_n,A_j) \le \frac{1}{k+4}T(r_n,f), \quad j=0,1,\ldots,k.$$

It follows from (2.3) that for sufficiently large r_n,

$$T(r_n,f) \le N\left(r_n,\frac{1}{f}\right) + \frac{k+3}{k+4}T(r_n,f),$$

that is,

$$T(r_n,f) \le (k+4)N\left(r_n,\frac{1}{f}\right).$$

Therefore,

$$\sigma = \lim_{r_n\to\infty}\frac{\log T(r_n,f)}{\log r_n} \le \overline{\lim_{r_n\to\infty}}\frac{\log N(r_n,\frac{1}{f})}{\log r_n} \le \lambda(f),$$

that is, $\lambda(f) = \sigma(f)$. □

Remark 2.3 Noting that

$$m\left(r,\frac{\Delta^i f}{f}\right) \le O\left(\sum_{j=0}^{i} m\left(r,\frac{f(z+jc)}{f(z)}\right)\right), \quad i\in\mathbb{N}_+,$$

we see that if $f(z+c_i)$, $i=1,\ldots,k$, in Lemma 2.5 are replaced, respectively, by $\Delta^i f$, $i = 1,\ldots,k$, the corresponding result holds.

3 Proofs of Theorems 1.1 and 1.2

Proof of Theorem 1.1 Suppose that $f(z)$ ($\not\equiv 0$) is a meromorphic solution of (1.1). If $\sigma(f) = \infty$, then $\sigma(f) \ge n+1$ holds obviously. Now, we suppose $\sigma(f) = \sigma < \infty$. Set

$$I_1 = \{j \ne i \mid \arg a_{in} \ne \arg a_{jn} \text{ and } \arg a_{in} \ne \arg b_{jn}\},$$
$$I_2 = \{j \ne i \mid \arg a_{in} = \arg b_{jn} \ne \arg a_{jn} \text{ and } |a_{in}| > |b_{jn}|\},$$
$$I_3 = \{j \ne i \mid \arg a_{in} = \arg a_{jn} \ne \arg b_{jn} \text{ and } |a_{in}| > |a_{jn}|\},$$
$$I_4 = \{j \ne i \mid \arg a_{in} = \arg b_{jn} = \arg a_{jn} \text{ and } |a_{in}| > |a_{jn}| \ge |b_{jn}|\}.$$

It is clear that I_1, I_2, I_3, I_4 do not intersect with each other, and $I = I_1 \cup I_2 \cup I_3 \cup I_4 = \{0,1,\ldots,k\}\setminus\{i\}$.

Now, we may choose $\theta_0 \in (0,2\pi)$ such that $\cos(\arg a_{in} + n\theta_0) = 1$. (If $n=1$ and $\arg a_{in} = 0$, then we replace $[0,2\pi)$ in Lemma 2.3 by $[-\frac{\pi}{2},\frac{3\pi}{2})$; if $n \ge 2$, this kind of θ_0 can always be chosen.) Set $z = re^{i\theta}$, $\theta\in[0,2\pi)$.

For $j \in I_1$, there exists sufficiently small ε_1 (> 0) such that for all $\theta \in (\theta_0 - \varepsilon_1, \theta_0 + \varepsilon_1) \subset (0, 2\pi)$, we have

$$\cos(\arg a_{in} + n\theta) > \max_{j\in I_1}\left\{\cos(\arg a_{jn} + n\theta), \cos(\arg b_{jn} + n\theta), 0\right\}.$$

Since $|a_{in}| \geq |a_{jn}| \geq |b_{jn}| > 0$, $j \neq i$, we see that

$$\delta(P_i, \theta) > \max_{j\in I_1}\left\{\delta(P_j, \theta), \delta(Q_j, \theta), 0\right\}.$$

For $j \in I_2$, there exists sufficiently small ε_2 (> 0) such that for all $\theta \in (\theta_0 - \varepsilon_2, \theta_0 + \varepsilon_2) \subset (0, 2\pi)$, we have

$$\cos(\arg a_{in} + n\theta) = \cos(\arg b_{jn} + n\theta) > \max_{j\in I_2}\left\{\cos(\arg a_{jn} + n\theta), 0\right\}.$$

Since $|a_{in}| > |b_{jn}|$, $j \neq i$, and $|a_{in}| \geq |a_{jn}|$, we see that

$$\delta(P_i, \theta) > \max_{j\in I_2}\left\{\delta(P_j, \theta), \delta(Q_j, \theta)\right\} > 0.$$

For $j \in I_3$, there exists sufficiently small ε_3 (> 0) such that for all $\theta \in (\theta_0 - \varepsilon_3, \theta_0 + \varepsilon_3) \subset (0, 2\pi)$, we have

$$\cos(\arg a_{in} + n\theta) = \cos(\arg a_{jn} + n\theta) > \max_{j\in I_3}\left\{\cos(\arg b_{jn} + n\theta), 0\right\}.$$

Since $|a_{in}| > |a_{jn}| \geq |b_{jn}|$, $j \neq i$, we see that

$$\delta(P_i, \theta) > \max_{j\in I_3}\left\{\delta(P_j, \theta), \delta(Q_j, \theta)\right\} > 0.$$

For $j \in I_4$, there exists sufficiently small ε_4 (> 0) such that for all $\theta \in (\theta_0 - \varepsilon_4, \theta_0 + \varepsilon_4) \subset (0, 2\pi)$, we have

$$\cos(\arg a_{in} + n\theta) = \cos(\arg b_{jn} + n\theta) = \cos(\arg a_{jn} + n\theta) > 0.$$

Since $|a_{in}| > |a_{jn}| \geq |b_{jn}|$, we see that

$$\delta(P_i, \theta) > \max_{j\in I_4}\left\{\delta(P_j, \theta), \delta(Q_j, \theta)\right\} > 0.$$

Set $\varepsilon_0' = \min\{\varepsilon_1, \varepsilon_2, \varepsilon_3, \varepsilon_4\}$, then for any $\theta \in (\theta_0 - \varepsilon_0', \theta_0 + \varepsilon_0') \subset (0, 2\pi)$, we have

$$\delta(P_i, \theta) = \delta_1 > \max_{j\in I}\left\{\delta(P_j, \theta), \delta(Q_j, \theta), 0\right\} = \delta_2. \tag{3.1}$$

If $|b_{in}| < |a_{in}|$ and $\arg b_{in} = \arg a_{in}$, then for all $\theta \in (\theta_0 - \varepsilon_0', \theta_0 + \varepsilon_0') \subset (0, 2\pi)$, we have

$$\delta(P_i, \theta) > \delta(Q_i, \theta) > 0.$$

If $\arg a_{in} \neq \arg b_{in}$ and $|b_{in}| \leq |a_{in}|$, then there exists ε_0 $(0 < \varepsilon_0 \leq \varepsilon_0')$ such that for all $\theta \in (\theta_0 - \varepsilon_0, \theta_0 + \varepsilon_0) \subset (0, 2\pi)$, we have

$$\cos(\arg a_{in} + n\theta) > \max\{\cos(\arg b_{in} + n\theta), 0\}$$

and, correspondingly,

$$\delta(P_i, \theta) > \max\{\delta(Q_i, \theta), 0\}.$$

Therefore, if $b_{in} \neq a_{in}$, and $|b_{in}| \leq |a_{in}|$, then there exists ε_0 (> 0) such that for all $\theta \in (\theta_0 - \varepsilon_0, \theta_0 + \varepsilon_0) \subset (0, 2\pi)$, we have

$$\delta(P_i, \theta) = \delta_1 > \max\{\delta(Q_i, \theta), 0\} = \delta_3. \tag{3.2}$$

If $b_{in} = a_{in}$, then $A_i(z) = \tilde{B}_i(z)e^{a_{in}z^n} + R_i(z)$, where $\tilde{B}_i(z)$ is a meromorphic function and satisfies $\sigma(\tilde{B}_i) \leq \max\{\omega, n-1\}$. Since $A_i(z) - R_i(z) \not\equiv 0$, we see that $\tilde{B}_i(z) \not\equiv 0$. Here, $D_i(z) \equiv 0$, and

$$\delta(a_{in}z^n, \theta) = \delta(P_i, \theta) = \delta_1 > 0. \tag{3.3}$$

By Lemma 2.2, for any given ε $(0 < \varepsilon < \frac{1}{2}\min\{\frac{\delta_1 - \delta_2}{\delta_1 + \delta_2}, \frac{\delta_1 - \delta_3}{\delta_1 + \delta_3}, n - \omega\})$, there exists a set $E_1 \subset (0, +\infty)$ with finite linear measure, such that for all z satisfying $|z| = r \notin E_1$, $j = 0, 1, \ldots, k$ and $r \to \infty$, we have

$$|B_j(z)| \leq \exp\{r^{\omega+\varepsilon}\}, \qquad |D_j(z)| \leq \exp\{r^{\omega+\varepsilon}\}, \qquad |R_j(z)| \leq \exp\{r^{\omega+\varepsilon}\}. \tag{3.4}$$

By Lemma 2.3 and (3.1)-(3.4), for the above $\varepsilon > 0$, there exists $H_0 \subset [0, 2\pi)$ with linear measure zero and a finite set $H_1 = \bigcup_{i=0}^{k}\{\theta \in [0, 2\pi) | \delta(P_i, \theta) = 0 \text{ or } \delta(Q_i, \theta) = 0\}$, such that for all $\theta \in (\theta_0 - \varepsilon_0, \theta_0 + \varepsilon_0) \backslash (H_0 \cup H_1)$, there exists $r_0 = r_0(\theta, \varepsilon) > 0$, such that for $r > r_0$, we have

$$\begin{aligned} |A_j(z)| &\leq |B_j(z)e^{P_j(z)}| + |D_j(z)e^{Q_j(z)}| + |R_j(z)| \\ &\leq 2\exp\{(1+\varepsilon)\delta_2 r^n\} + \exp\{r^{\omega+\varepsilon}\} \\ &\leq 3\exp\{(1+\varepsilon)\delta_2 r^n\}\exp\{r^{\omega+\varepsilon}\}, \quad j \neq i \end{aligned} \tag{3.5}$$

and

$$\begin{aligned} |A_i(z)| &\geq |B_i(z)e^{P_i(z)}| - |D_i(z)e^{Q_i(z)}| - |R_i(z)| \\ &\geq \exp\{(1-\varepsilon)\delta_1 r^n\} - \exp\{(1+\varepsilon)\delta_3 r^n\} - \exp\{r^{\omega+\varepsilon}\} \\ &\geq \exp\{(1-\varepsilon)\delta_1 r^n\} - 2\exp\{(1+\varepsilon)\delta_3 r^n\}\exp\{r^{\omega+\varepsilon}\}. \end{aligned} \tag{3.6}$$

By Lemma 2.1, for the above $\varepsilon > 0$, there exists a set $E_2 \subset (1, +\infty)$ with finite logarithmic measure, such that for all z satisfying $|z| = r \notin E_2 \cup [0, 1]$, we have

$$\left|\frac{f(z + c_j)}{f(z + c_i)}\right| \leq \exp\{r^{\sigma-1+\varepsilon}\}, \quad j \neq i. \tag{3.7}$$

Equation (1.1) gives

$$-A_i(z) = A_k(z)\frac{f(z+c_k)}{f(z+c_i)} + \cdots + A_{i+1}(z)\frac{f(z+c_{i+1})}{f(z+c_i)} + A_{i-1}(z)\frac{f(z+c_{i-1})}{f(z+c_i)} + \cdots + A_0(z)\frac{f(z)}{f(z+c_i)}$$

and, correspondingly,

$$\left|A_i(z)\right| \le \left|A_k(z)\right|\left|\frac{f(z+c_k)}{f(z+c_i)}\right| + \cdots + \left|A_{i+1}(z)\right|\left|\frac{f(z+c_{i+1})}{f(z+c_i)}\right| + \left|A_{i-1}(z)\right|\left|\frac{f(z+c_{i-1})}{f(z+c_i)}\right| + \cdots + \left|A_0(z)\right|\left|\frac{f(z)}{f(z+c_i)}\right|. \tag{3.8}$$

It follows from (3.5)-(3.8) that for all $z = re^{i\theta}$, where $\theta \in (\theta_0 - \varepsilon_0, \theta_0 + \varepsilon_0)\backslash(H_0 \cup H_1)$, $r \notin [0,1] \cup E_1 \cup E_2$ and $r \to \infty$, we have

$$\exp\{(1-\varepsilon)\delta_1 r^n\} - 2\exp\{(1+\varepsilon)\delta_3 r^n\}\exp\{r^{\omega+\varepsilon}\} \le 3k\exp\{r^{\sigma-1+\varepsilon}\}\exp\{(1+\varepsilon)\delta_2 r^n\}\exp\{r^{\omega+\varepsilon}\}. \tag{3.9}$$

Since $0 < \varepsilon < \frac{1}{2}\min\{\frac{\delta_1-\delta_2}{\delta_1+\delta_2}, \frac{\delta_1-\delta_3}{\delta_1+\delta_3}, n-\omega\}$, we have

$$(1+\varepsilon)\delta_2 < (1-2\varepsilon)\delta_1, \qquad (1+\varepsilon)\delta_3 < (1-2\varepsilon)\delta_1, \qquad \omega+\varepsilon < n-\varepsilon$$

and

$$\frac{2\exp\{(1+\varepsilon)\delta_3 r^n\}\exp\{r^{\omega+\varepsilon}\}}{\exp\{(1-\varepsilon)\delta_1 r^n\}} < 2\exp\{r^{\omega+\varepsilon} - \varepsilon\delta_1 r^n\} \to 0, \quad r \notin [0,1] \cup E_1 \cup E_2, r \to \infty.$$

Then from (3.9), for sufficiently large r, we have

$$\frac{1}{2}\exp\{(1-\varepsilon)\delta_1 r^n\} \le 3k\exp\{r^{\sigma-1+\varepsilon}\}\exp\{(1+\varepsilon)\delta_2 r^n\}\exp\{r^{\omega+\varepsilon}\},$$

i.e.,

$$\exp\{(1-\varepsilon)\delta_1 r^n - (1+\varepsilon)\delta_2 r^n - r^{\omega+\varepsilon}\} \le 6k\exp\{r^{\sigma-1+\varepsilon}\},$$

i.e.,

$$\exp\left\{\frac{1}{2}\varepsilon\delta_1 r^n\right\} \le 6k\exp\{r^{\sigma-1+\varepsilon}\}. \tag{3.10}$$

Then (3.10) implies $n \le \sigma - 1 + \varepsilon$. Since ε is arbitrary, we have $\sigma(f) = \sigma \ge n+1$.

Therefore, every meromorphic solution $f(z)$ ($\not\equiv 0$) satisfies $\sigma(f) \ge n+1$.

Set $g(z) = f(z) - \varphi(z)$, then $g(z)$ solves the equation

$$A_k(z)g(z+c_k) + \cdots + A_1(z)g(z+c_1) + A_0(z)g(z) = -A_k(z)\varphi(z+c_k) - \cdots - A_1(z)\varphi(z+c_1) - A_0(z)\varphi(z).$$

Since $\sigma(\varphi) < n+1$, $\varphi(z)$ $(\not\equiv 0)$ does not solve (1.1), that is,

$$A_k(z)\varphi(z+c_k) + \cdots + A_1(z)\varphi(z+c_1) + A_0(z)\varphi(z) \not\equiv 0$$

and

$$\begin{aligned}&\sigma\left(A_k(z)\varphi(z+c_k) + \cdots + A_1(z)\varphi(z+c_1) + A_0(z)\varphi(z)\right)\\ &\quad \le \max\left\{\sigma(A_j), j = 0,1,\ldots,k, \sigma(\varphi)\right\}\\ &\quad < n+1 \le \sigma(f).\end{aligned}$$

Therefore, by Lemma 2.5, we have $\lambda(g) = \sigma(g)$, *i.e.*, $\lambda(f-\varphi) = \sigma(f)$.

The proof of Theorem 1.1 is completed. □

Proof of Theorem 1.2 Equation (1.2) gives

$$-A_0(z) = A_k(z)\frac{\Delta^k f(z)}{f(z)} + \cdots + A_1(z)\frac{\Delta f(z)}{f(z)}. \tag{3.11}$$

By Remark 2.1, we have

$$\left|\frac{\Delta^j f(z)}{f(z)}\right| \le O\left(\exp\left\{r^{\sigma-1+\varepsilon}\right\}\right), \quad j = 1,\ldots,k. \tag{3.12}$$

By combining (3.5), (3.6), (3.12) with (3.11), for all $z = re^{i\theta}$, where $\arg z = \theta \in (\theta_0 - \varepsilon_0, \theta_0 + \varepsilon_0)\backslash(H_0 \cup H_1)$, $|z| = r \notin [0,1] \cup E_1 \cup E_2$ and for sufficiently large r, we have

$$\begin{aligned}&\exp\left\{(1-\varepsilon)\delta_1 r^n\right\} - 2\exp\left\{(1+\varepsilon)\delta_3 r^n\right\}\exp\left\{r^{\omega+\varepsilon}\right\}\\ &\quad \le O\left(\exp\left\{r^{\sigma-1+\varepsilon}\right\}\right)\exp\left\{(1+\varepsilon)\delta_2 r^n\right\}\exp\left\{r^{\omega+\varepsilon}\right\}.\end{aligned}$$

By using a similar method to the one in the proof of Theorem 1.1, we have $\sigma(f) = \sigma \ge n+1$.

Further, set $g(z) = f(z) - \varphi(z)$, then $g(z)$ solves the equation

$$\begin{aligned}&A_k(z)\Delta^k g(z) + \cdots + A_1(z)\Delta g(z) + A_0(z)g(z)\\ &\quad = -A_k(z)\Delta^k\varphi(z) - \cdots - A_1(z)\Delta\varphi(z) - A_0(z)\varphi(z).\end{aligned}$$

Since $\sigma(\varphi) < n+1$, $\varphi(z)$ $(\not\equiv 0)$ does not solve (1.2), that is,

$$A_k(z)\Delta^k\varphi(z) + \cdots + A_1(z)\Delta\varphi(z) + A_0(z)\varphi(z) \not\equiv 0$$

and

$$\sigma\left(A_k\Delta^k\varphi + \cdots + A_1\Delta\varphi + A_0\varphi\right) \le \max\left\{\sigma(A_j), j = 0,1,\ldots,k, \sigma(\varphi)\right\} < n+1 \le \sigma(f).$$

Therefore, by Lemma 2.5 and Remark 2.3, we have $\lambda(g) = \sigma(g)$, *i.e.*, $\lambda(f-\varphi) = \sigma(f)$.

The proof of Theorem 1.2 is completed. □

Competing interests

The authors declare that they have no competing interests.

Authors' contributions

All authors drafted the manuscript, and they read and approved the final manuscript.

Acknowledgements

This work was supported by the National Natural Science Foundation of China (11301233, 11171119), the Natural Science Foundation of Jiangxi Province in China (20151BAB201004), and the Youth Science Foundation of Education Bureau of Jiangxi Province in China (GJJ14271).

References

1. Hayman, WK: Meromorphic Functions. Clarendon, Oxford (1964)
2. Laine, I: Nevanlinna Theory and Complex Differential Equations. de Gruyter, Berlin (1993)
3. Yang, L: Value Distribution Theory and Its New Research. Science Press, Beijing (1982) (in Chinese)
4. Yi, HX, Yang, CC: Uniqueness Theory of Meromorphic Functions. Mathematics and Its Applications, vol. 557. Kluwer Academic, Dordrecht (2003)
5. Chen, ZX, Shon, KH: On growth of meromorphic solutions for linear difference equations. Abstr. Appl. Anal. **2013**, Article ID 619296 (2013)
6. Chiang, YK, Feng, SJ: On the Nevanlinna characteristic of $f(z+\eta)$ and difference equations in the complex plane. Ramanujan J. **16**(1), 105-129 (2008)
7. Li, S, Gao, ZS: Finite order meromorphic solutions of linear difference equations. Proc. Jpn. Acad., Ser. A, Math. Sci. **87**(5), 73-76 (2011)
8. Laine, I, Yang, CC: Clunie theorems for difference and q-difference polynomials. J. Lond. Math. Soc. **76**(2), 556-566 (2007)
9. Latreuch, Z, Belaïdi, B: Growth and oscillation of meromorphic solutions of linear difference equations. Mat. Vesn. **66**(2), 213-222 (2014)
10. Zheng, XM, Tu, J: Growth of meromorphic solutions of linear difference equations. J. Math. Anal. Appl. **384**(2), 349-356 (2011)
11. Chen, ZX: Zeros of entire solutions to complex linear difference equations. Acta Math. Sci., Ser. B **32**(3), 1141-1148 (2012)
12. Liu, YX: On growth of meromorphic solutions for linear difference equations with meromorphic coefficients. Adv. Differ. Equ. **2013**, 60 (2013)
13. Liu, HF, Mao, ZQ: On the meromorphic solutions of some linear difference equations. Adv. Differ. Equ. **2013**, 133 (2013)
14. Bergweiler, W, Langley, JK: Zeros of differences of meromorphic functions. Math. Proc. Camb. Philos. Soc. **142**(1), 133-147 (2007)
15. Chen, ZX: The growth of solutions of a type of second order differential equations with entire coefficients. Chin. Ann. Math., Ser. A **20**(1), 7-14 (1999) (in Chinese)
16. Cao, CL, Chen, ZX: On the orders and zeros of the solutions of certain linear differential equations with entire coefficients. Acta Math. Appl. Sin. **25**(1), 123-131 (2002) (in Chinese)
17. Halburd, RG, Korhonen, RJ, Toghe, K: Holomorphic curves with shift-invariant hyperplane preimages. Trans. Am. Math. Soc. **336**(8), 4267-4298 (2014)

Oscillation criteria for a class of nonlinear neutral differential equations

Shuhui Wu[1*], Pargat Singh Calay[2] and Zhanyuan Hou[2]

*Correspondence: s.wu@zust.edu.cn
[1]School of Sciences, Zhejiang University of Science and Technology, 318 Liuhe Road, Hangzhou, 310023, China
Full list of author information is available at the end of the article

Abstract

In this paper, we deal with the oscillation of the second order nonlinear neutral differential equations of the form

$$(a(t)(x(t)+\delta p(t)x(t-\tau))')' + f(t,x(t-\sigma)) - g(t,x(t-\rho)) = 0.$$

The oscillation criteria for these equations have been obtained. Furthermore, examples are given to illustrate the criteria, respectively.

Keywords: neutral differential equations; bounded oscillation; almost oscillation; bounded almost oscillation

1 Introduction

The differential equations that we study describe many phenomena and dynamical processes in various fields, and they have attracted a great deal of attention of researchers in physical sciences, mathematics, biology, and economy. In addition, these equations play an important role in numerical simulations of nonlinear partial differential equations, queuing problems, and discretization in solid state and quantum physics. For the application, please see [1].

In this paper, we consider the oscillation of second order nonlinear neutral differential equations with mixed type term of the form

$$\big(a(t)\big(x(t)+\delta p(t)x(t-\tau)\big)'\big)' + f\big(t,x(t-\sigma)\big) - g\big(t,x(t-\rho)\big) = 0, \tag{1.1}$$

where $\delta = +1$ or -1, $t \geq t_0$, $a(t)$ is a continuously differentiable function, $p(t)$ is a continuous bounded function with $a(t) > 0$, $p(t) \geq 0$, $f(t,u)$ and $g(t,v)$ are continuous functions, the constants $\tau,\sigma,\rho \in [0,\infty)$. Denote $\lambda = \max\{\tau,\sigma,\rho\}$, $t_1 = t_0 + \lambda$, $L^1[t_0,\infty) = \{x(t) \mid \int_{t_0}^{\infty} |x(s)|\,ds < \infty\}$.

The following conditions will be assumed throughout this paper:

(H_1) $\int_t^{\infty} \frac{1}{a(s)}\,ds = \infty$ for all $t \geq t_0$,
(H_2) $\frac{f(t,u)}{u} \geq q(t-\sigma) > 0$ for $u \neq 0$ and $0 < \frac{g(t,v)}{v} \leq r(t-\rho)$ for $v \neq 0$,
(H_3) $0 < \frac{f(t,u)}{u} \leq q(t-\sigma)$ for $u \neq 0$ and $\frac{g(t,v)}{v} \geq r(t-\rho) > 0$ for $v \neq 0$,
(H_4) $\frac{1}{r(t)-q(t)}$ is bounded, where $q, r \in C([t_0,\infty), R^+)$.

We also assume that $x(t)$ is a nontrivial solution of (1.1). The investigation of oscillatory behavior of solutions of various types of differential equations done by many researchers is motivated by many application problems in physics, biology, ecology, and so on. In particular, an increasing interest in obtaining oscillation criteria for different classes of differential and functional differential equations has been manifested recently. Please see [1–16].

The paper is organized as follows. We will first present criteria for (1.1) when $\delta = +1$ in Section 2 and then for (1.1) when $\delta = -1$ in Section 3. Some examples will be given to illustrate the obtained criteria, respectively. The proofs of the main results are left to Section 4.

2 Statement of the main results when $\delta = +1$

We first rewrite (1.1) as

$$\left(a(t)\left(x(t)+p(t)x(t-\tau)\right)'\right)' + f\left(t, x(t-\sigma)\right) - g\left(t, x(t-\rho)\right) = 0. \tag{2.1}$$

In this section, four oscillatory criteria will be presented and some illustrated examples will be given.

Theorem 2.1 *Suppose that conditions* (H_1), (H_2) *and* (H_4) *hold,* $q(t) > r(t)$, $r(t)$ *is bounded and* $\sigma \geq \rho$. *Then* (2.1) *is bounded oscillatory.*

Remark 2.2 In [16], the authors considered

$$\left(x(t)+p(t)x\left(\tau(t)\right)\right)^{(n)} + q(t)f\left(x\left(\sigma(t)\right)\right) = 0$$

and established the criteria for the solution to be oscillatory when $0 \leq p(t) < 1$. Even though this result is about the higher order equations, the generality of our results is not robbed since our equations include a larger class of equations.

Example 1 Consider the differential equation

$$\left(\left(1+\frac{1}{t}\right)\left(x(t)+2x(t-2\pi)\right)'\right)' + 3\left(1+\frac{1}{t}\right)x(t-2\pi)$$
$$+ 3\left(1+\frac{1}{t}\right)x^5(t-2\pi) - \frac{3}{t^2}x\left(t-\frac{\pi}{2}\right) = 0. \tag{2.2}$$

Viewing (2.2) as (2.1), we have $a(t) = 1 + (1/t)$, $p(t) = 2 > 0$, and

$$q(t) = 3\left(1+\frac{1}{t+2\pi}\right) > r(t) = \frac{3}{(t+\frac{\pi}{2})^2}.$$

Moreover, $\tau = 2\pi$, $\sigma = 2\pi > \rho = \pi/2$, and $r(t)$ is bounded for $t \geq 2\pi$. Note that conditions (H_1), (H_2) and (H_4) are satisfied and, by Theorem 2.1, (2.2) is bounded oscillatory.

Theorem 2.3 *Suppose that conditions* (H_1), (H_2) *and* (H_4) *hold,* $q(t) > r(t)$, $q(t)$, $1/a(t)$ *are bounded and* $\sigma < \rho$. *Then* (2.1) *is almost oscillatory.*

Example 2 Consider the differential equation

$$\left(t\left(x(t)+x(t-\pi)\right)'\right)' + \frac{t\pi}{t-2\pi}x(t-2\pi) - \frac{\pi}{t-\frac{7}{2}\pi}x\left(t-\frac{7}{2}\pi\right) = 0. \tag{2.3}$$

Viewing (2.3) as (2.1), we have $a(t) = t$, $p(t) = 1 > 0$, and

$$q(t) = \frac{(t+2\pi)\pi}{t} > r(t) = \frac{\pi}{t}.$$

Moreover, $\tau = \pi, \sigma = 2\pi < \rho = 7\pi/2$, $q(t)$ is bounded for $t \geq 4\pi$. Note that conditions (H_1), (H_2) and (H_4) are satisfied. By Theorem 2.3, (2.3) is almost oscillatory. Indeed, $x(t) = t\sin t$ is an unbounded oscillatory solution of (2.3).

Theorem 2.4 *Suppose that conditions* (H_1), (H_3) *and* (H_4) *hold,* $q(t) < r(t)$, $r(t)$, $1/a(t)$ *are bounded and* $\sigma \geq \rho$. *Then* (2.1) *is bounded almost oscillatory.*

Theorem 2.5 *Suppose that conditions* (H_1), (H_3) *and* (H_4) *hold,* $q(t) < r(t)$, $q(t)$ *is bounded and* $\sigma < \rho$. *Then* (2.1) *is bounded almost oscillatory.*

Example 3 Consider the differential equation

$$\begin{aligned}&\left(\frac{t+1-3\pi}{t-3\pi}\left(x(t)+\frac{t-\pi}{t(t+1-\pi)}x(t-\pi)\right)'\right)'\\ &\quad+\frac{2t+2-3\pi}{(t-3\pi)(2t-3\pi)}x\left(t-\frac{3\pi}{2}\right)-x(t-3\pi)=0.\end{aligned} \tag{2.4}$$

Viewing (2.4) as (2.1), we have

$$\begin{aligned}a(t) &= \frac{t+1-3\pi}{t-3\pi},\\ p(t) &= \frac{t-\pi}{t(t+1-\pi)} > 0,\\ q(t) &= \frac{2t+2}{t(2t-3\pi)} < r(t) = 1.\end{aligned}$$

Also,

$$\tau = \pi, \sigma = \frac{3\pi}{2} < \rho = 3\pi \text{ and } q(t) \text{ is bounded for } t \geq 3\pi.$$

We note that conditions (H_1), (H_3) and (H_4) are satisfied and, by Theorem 2.5, (2.4) is bounded almost oscillatory. In fact, $x(t) = (1+(1/t))\sin t$ is a bounded oscillatory solution of (2.4).

3 Statement of the main results when $\delta = -1$

In this section, we consider (1.1) when $\delta = -1$. So (1.1) becomes

$$\left(a(t)\left(x(t)-p(t)x(t-\tau)\right)'\right)' + f\left(t,x(t-\sigma)\right) - g\left(t,x(t-\rho)\right) = 0. \tag{3.1}$$

Four oscillation criteria have been obtained. In addition, some example will be given to demonstrate the obtained results.

Theorem 3.1 *Suppose that conditions* (H_1), (H_3) *and* (H_4) *hold,* $p(t) \geq 1$, $q(t) < r(t)$, $\sigma \leq \rho$ *and* $r(t)$ *is bounded. Then* (3.1) *is bounded oscillatory.*

Example 4 Consider the differential equation

$$\left(\left(1-\frac{1}{t}\right)(x(t)-2x(t-\pi))'\right)' + \frac{3}{t^2}x\left(t-\frac{\pi}{2}\right) - 3x(t-\pi) = 0. \tag{3.2}$$

Viewing (3.2) as (3.1), we have $\tau = \pi$, $\sigma = \frac{\pi}{2} < \rho = \pi$,

$$a(t) = 1 - \frac{1}{t},$$

$$p(t) = 2 > 0,$$

$$q(t) = \frac{3}{(t+\frac{\pi}{2})^2} < r(t) = 3$$

for $t \geq \pi$. We note that conditions (H_1), (H_3) and (H_4) are satisfied and, by Theorem 3.1, (3.2) is bounded oscillatory.

Theorem 3.2 *Suppose that conditions* (H_1), (H_3) *and* (H_4) *hold,* $q(t) < r(t)$, $\sigma \geq \rho$, $0 \leq p(t) \leq p_1 < 1$ *or* $1 < p_2 \leq p(t)$, $r(t)$ *and* $1/a(t)$ *are bounded. Then* (3.1) *is bounded almost oscillatory.*

Theorem 3.3 *Suppose that conditions* (H_1), (H_2) *and* (H_4) *hold,* $q(t) > r(t)$, $\sigma < \rho$, $0 \leq p(t) \leq p_1 < 1$ *or* $1 < p_2 \leq p(t)$, $q(t)$ *and* $1/a(t)$ *are bounded. Then* (3.1) *is bounded almost oscillatory.*

Example 5 Consider the differential equation

$$\left(\frac{t}{t+1}\left(x(t) - \frac{2(t-2\pi)}{t}x(t-2\pi)\right)'\right)' + \frac{(t^3+t^2-2t-1)(t-\pi)}{2t^2(t+1)^2}x(t-\pi)$$
$$-\frac{(2t+1)(t-\frac{3\pi}{2})}{2t(t+1)^2}x\left(t-\frac{3\pi}{2}\right) = 0. \tag{3.3}$$

Viewing (3.3) as (3.1), we have $\tau = 2\pi$, $\sigma = \pi < \rho = 3\pi/2$,

$$a(t) = \frac{t}{t+1},$$

$$p(t) = \frac{2(t-2\pi)}{t} \geq 1.2 \quad \text{for } t \geq 5\pi,$$

$$q(t) = \frac{t((t+\pi)^3 + (t+\pi)^2 - 2t - 2\pi - 1)}{2(t+\pi)^2(t+\pi+1)^2},$$

$$r(t) = \frac{t(2t+3\pi+1)}{2(t+3\pi/2)(t+3\pi/2+1)^2}.$$

Clearly, $q(t) > r(t)$ for large t and $1/a(t)$ and $q(t)$ are bounded. We note that conditions (H_1), (H_2) and (H_4) are satisfied and, by Theorem 3.3, (3.3) is bounded almost oscillatory.

Theorem 3.4 *Suppose that conditions* (H_1), (H_2) *and* (H_4) *hold,* $q(t) > r(t)$, $\sigma \geq \rho$, $q(t)$ *is bounded,* $0 \leq p(t) \leq p_1 < 1$, *or* $1/a(t)$ *is bounded and* $1 < p_2 \leq p(t)$. *Then* (3.1) *is bounded almost oscillatory.*

Example 6 Consider the differential equation

$$\left(\frac{t}{t+1}\left(x(t) - \frac{t-2\pi}{2t}x(t-2\pi)\right)'\right)' + \frac{(t^3+t^2-2t-1)(t-2\pi)}{2t^2(t+1)^2}x(t-2\pi)$$
$$- \frac{(2t+1)(t-\pi/2)}{2t(t+1)^2}x(t-\pi/2) = 0. \tag{3.4}$$

Regarding (3.4) as (3.1), we have $\tau = 2\pi$, $\sigma = 2\pi > \rho = \frac{\pi}{2}$,

$$a(t) = \frac{t}{t+1},$$
$$0 < p(t) = \frac{t-2\pi}{2t} \leq \frac{1}{2} < 1, \quad t \geq 3\pi,$$
$$q(t) = \frac{t((t+2\pi)^3 + (t+2\pi)^2 - 2t - 4\pi - 1)}{2(t+2\pi)^2(t+2\pi+1)^2},$$
$$r(t) = \frac{t(2t+\pi+1)}{2(t+\pi/2)(t+\pi/2+1)^2}.$$

Clearly, $q(t) > r(t)$ for large enough t and $q(t)$ is bounded. Notice that (H_1), (H_2) and (H_4) are satisfied therefore, by Theorem 3.4, (3.4) is bounded almost oscillatory.

4 Proofs of the main results

Here we will give the proofs of the main results.

Proof of Theorem 2.1 Let $x(t)$ be a bounded non-oscillatory solution. Suppose $x(t)$ is an eventually positive solution. Then there exists $t_2 \geq t_1$ such that $x(t) > 0$ and $x(t-\lambda) > 0$ for $t \geq t_2$. Let

$$z(t) = a(t)\big(x(t) + p(t)x(t-\tau)\big)' - \int_{t-\sigma}^{t-\rho} r(s)x(s)\,ds. \tag{4.1}$$

From (2.1) and (H_2) it follows that

$$z'(t) \leq \big(r(t-\sigma) - q(t-\sigma)\big)x(t-\sigma) < 0, \quad t \geq t_2. \tag{4.2}$$

So $z(t)$ is decreasing, and

$$-\infty \leq \lim_{t\to\infty} z(t) = c < \infty.$$

If $c = -\infty$, from (4.1) and the boundedness of $x(t)$ and $r(t)$, we have

$$\lim_{t\to\infty} a(t)\big(x(t) + p(t)x(t-\tau)\big)' = -\infty.$$

Then there exist $l_1 > 0$ and $t_3 \geq t_2$ such that

$$\big(x(t) + p(t)x(t-\tau)\big)' \leq -\frac{l_1}{a(t)}, \quad t \geq t_3.$$

Integrating both sides of the above inequality, according to (H_1), we obtain

$$\lim_{t\to\infty} \big(x(t) + p(t)x(t-\tau)\big) = -\infty,$$

which contradicts the boundedness of $x(t)$ and $p(t)$. This contradiction shows that $|c| < \infty$, *i.e.*, $z(t)$ is bounded.

From (4.2) it follows that

$$x(t-\sigma) \leq \frac{1}{r(t-\sigma) - q(t-\sigma)} z'(t). \tag{4.3}$$

So $x \in L^1[t_0, \infty)$ by (H_4).

(i) If $c > 0$, from (4.1) we have

$$z(t) \leq a(t)\big(x(t) + p(t)x(t-\tau)\big)', \quad t \geq t_2.$$

Therefore, since $z(t) \to c$ as $t \to \infty$,

$$\big(x(t) + p(t)x(t-\tau)\big)' \geq \frac{c}{a(t)}, \quad t \geq t_2.$$

From (H_1) we have $\lim_{t\to\infty}(x(t) + p(t)x(t-\tau)) = \infty$, which contradicts the boundedness of $x(t)$.

(ii) If $c < 0$, in view of $x \in L^1[t_0, \infty)$, we have

$$\lim_{t\to\infty} \int_{t-\sigma}^{t-\rho} r(s)x(s)\, ds = 0.$$

Then, since $z(t) \to c$ as $t \to \infty$, there exist $\epsilon \in (0, -c)$ and $t_4 \geq t_2$ such that

$$a(t)\big(x(t) + p(t)x(t-\tau)\big)' \leq c + \epsilon < 0, \quad t \geq t_4.$$

Hence, by (H_1) again, we obtain

$$\lim_{t\to\infty} \big(x(t) + p(t)x(t-\tau)\big) = -\infty,$$

a contradiction to the boundedness of $x(t)$ and $p(t)$.

(iii) If $c = 0$, in view of $z'(t) < 0$, we have $z(t) > 0$. So

$$a(t)\big(x(t) + p(t)x(t-\tau)\big)' > \int_{t-\sigma}^{t-\rho} r(s)x(s)\, ds > 0, \quad t \geq t_2.$$

Since $x(t) + p(t)x(t-\tau)$ is positive and increasing, the integral

$$\int_{t_0}^{\infty} \big(x(t) + p(t)x(t-\tau)\big)\, dt$$

is divergent, a contradiction to $x \in L^1[t_0, \infty)$. The contradictions obtained in the above three cases show that (2.1) has no bounded eventually positive solution. Now suppose that $x(t)$ is a bounded eventually negative solution. Then $x(t-\lambda) < 0$ for some $t_2 > t_1$ and all $t \geq t_2$. From (2.1), (4.1) and (H_2), we have

$$z'(t) \geq \big(r(t-\sigma) - q(t-\sigma)\big)x(t-\sigma) > 0, \quad t \geq t_2. \tag{4.4}$$

So $z(t)$ is increasing and $-\infty < \lim_{t\to\infty} z(t) = c \leq \infty$. Then an argument parallel to the above also leads to contradictions. Therefore, every bounded solution of (2.1) is oscillatory. □

Proof of Theorem 2.3 Without loss of generality, suppose that $x(t)$ is an eventually positive solution. Take $t_2 \geq t_1$ such that $x(t-\lambda) > 0$ for all $t \geq t_2$. Let

$$z(t) = a(t)\big(x(t) + p(t)x(t-\tau)\big)' + \int_{t-\rho}^{t-\sigma} q(s)x(s)\,ds. \tag{4.5}$$

From (2.1) it follows that

$$z'(t) \leq \big(r(t-\rho) - q(t-\rho)\big)x(t-\rho) < 0, \quad t \geq t_2. \tag{4.6}$$

So $z(t)$ is decreasing and

$$-\infty \leq \lim_{t\to\infty} z(t) = c < \infty.$$

If $c = -\infty$, then

$$\lim_{t\to\infty} a(t)\big(x(t) + p(t)x(t-\tau)\big)' = -\infty.$$

By (H_1), we obtain $\lim_{t\to\infty}(x(t) + p(t)x(t-\tau)) = -\infty$, which contradicts $x(t) + p(t)x(t-\tau) > 0$. Therefore $|c| < \infty$ so $z(t)$ is bounded.

From (4.6) we have

$$x(t-\rho) \leq \frac{1}{r(t-\rho) - q(t-\rho)} z'(t), \quad t \geq t_2 \tag{4.7}$$

so, by (H_4), $x \in L^1[t_0, \infty)$ and $\lim_{t\to\infty} \int_{t-\rho}^{t-\sigma} q(s)x(s)\,ds = 0$. Since $1/a(t)$ is bounded, by (4.5), $(x(t) + p(t)x(t-\tau))'$ is bounded. This implies that $x(t) + p(t)x(t-\tau)$ is uniformly continuous on $[t_1, \infty)$. Note that the property $x \in L^1[t_0, \infty)$ and the boundedness of $p(t)$ imply that $x(t) + p(t)x(t-\tau) \in L^1[t_0, \infty)$. Hence $\lim_{t\to\infty}(x(t) + p(t)x(t-\tau)) = 0$, so $\lim_{t\to\infty} x(t) = 0$. Therefore, every solution x of (2.1) which is not in the class of $o(1)$ as $t \to \infty$ is oscillatory. □

Proof of Theorem 2.4 Without loss of generality, assume that $x(t)$ is a bounded eventually positive solution and $z(t)$ is defined by (4.1). Take $t_2 \geq t_1$ such that $x(t-\lambda) > 0$ for $t \geq t_2$. From (2.1) and (H_3), we have

$$z'(t) \geq \big(r(t-\sigma) - q(t-\sigma)\big)x(t-\sigma) > 0, \quad t \geq t_2. \tag{4.8}$$

So $z(t)$ is increasing. Then

$$-\infty < \lim_{t\to\infty} z(t) = d \le \infty.$$

If $\lim_{t\to\infty} z(t) = \infty$, then from (4.1) and the boundedness of $x(t)$ and $r(t)$, we obtain

$$\lim_{t\to\infty} a(t)\big(x(t) + p(t)x(t-\tau)\big)' = \infty.$$

Then there exist $l_2 > 0$ and $t_3 \ge t_2$ such that

$$a(t)\big(x(t) + p(t)x(t-\tau)\big)' \ge l_2, \quad t \ge t_3.$$

From (H_1) it follows that

$$\lim_{t\to\infty} \big(x(t) + p(t)x(t-\tau)\big) = \infty,$$

a contradiction to the boundedness of $x(t)$ and $p(t)$. So $|d| < \infty$ and $z(t)$ is bounded. From (4.8) we have

$$x(t-\sigma) \le \frac{1}{r(t-\sigma) - q(t-\sigma)} z'(t), \quad t \ge t_2.$$

Therefore, by (H_4), $x \in L^1[t_0, \infty)$. By the same reasoning as that used in the proof of Theorem 2.3, we have $\lim_{t\to\infty} x(t) = 0$. Therefore, every bounded solution x of (2.1) which is not in the class of $o(1)$ as $t \to \infty$ must be oscillatory. □

Proof of Theorem 2.5 Without loss of generality, suppose that $x(t)$ is a bounded eventually positive solution. Let $z(t)$ be defined by (4.5). Take $t_2 \ge t_1$ such that $x(t-\lambda) > 0$ for $t \ge t_2$. From (2.1) and (H_3), we have

$$z'(t) \ge \big(r(t-\rho) - q(t-\rho)\big)x(t-\rho) > 0, \quad t \ge t_2. \tag{4.9}$$

Hence $z(t)$ is increasing and

$$-\infty < \lim_{t\to\infty} z(t) = d \le \infty.$$

By using the method similar to that used in the proof of Theorem 2.4, we have $-\infty < d < \infty$. Therefore $z(t)$ is bounded. From (4.9) it follows that

$$x(t-\rho) \le \frac{1}{r(t-\rho) - q(t-\rho)} z'(t), \quad t \ge t_2.$$

Thus, by (H_4), $x \in L^1[t_0, \infty)$ and $\lim_{t\to\infty} \int_{t-\rho}^{t-\sigma} q(s)x(s)\,ds = 0$. Then it follows from (4.5) that

$$\lim_{t\to\infty} a(t)\big(x(t) + p(t)x(t-\tau)\big)' = d.$$

(i) If $d > 0$, then there exists $t_5 \ge t_2$ such that

$$a(t)\big(x(t) + p(t)x(t-\tau)\big)' \ge \frac{d}{2}, \quad t \ge t_5.$$

From (H_1) we have

$$\lim_{t\to\infty}\big(x(t)+p(t)x(t-\tau)\big)=\infty,$$

which contradicts the boundedness of $x(t)$ and $p(t)$.

(ii) If $d<0$, similar to the case (i), we have

$$\lim_{t\to\infty}\big(x(t)+p(t)x(t-\tau)\big)=-\infty,$$

a contradiction to the boundedness of $x(t)$ and $p(t)$ again. Hence $d=0$, *i.e.*, $\lim_{t\to\infty} z(t)=0$. On the other hand, from (4.9) and $\lim_{t\to\infty} z(t)=0$, we have $z(t)<0$. In view of (4.5), $(x(t)+p(t)x(t-\tau))'<0$, which implies that $x(t)+p(t)x(t-\tau)$ is decreasing. From $x(t)+p(t)x(t-\tau)\in L^1[t_0,\infty)$, we have $\lim_{t\to\infty}(x(t)+p(t)x(t-\tau))=0$. Thus $\lim_{t\to\infty} x(t)=0$. Therefore, every bounded solution x of (2.1) which is not in the class of $o(1)$ as $t\to\infty$ must be oscillatory. □

Proof of Theorem 3.1 Suppose that $x(t)$ is a bounded non-oscillatory solution. Without loss of generality, we assume that $x(t)$ is an eventually positive solution. Let

$$z(t)=a(t)\big(x(t)-p(t)x(t-\tau)\big)'+\int_{t-\rho}^{t-\sigma} q(s)x(s)\,ds. \tag{4.10}$$

From a proof similar to that of Theorem 2.4, we obtain

$$z'(t)>0,\ \lim_{t\to\infty} z(t)=c,\ |c|<\infty,\ \text{and } x\in L^1[t_0,\infty).$$

(i) If $c>0$, from (4.10) it follows that

$$\lim_{t\to\infty} a(t)\big(x(t)-p(t)x(t-\tau)\big)'=c>\frac{c}{2}.$$

So, for large enough t,

$$\big(x(t)-p(t)x(t-\tau)\big)'\geq\frac{c}{2a(t)}.$$

Hence $\lim_{t\to\infty}(x(t)-p(t)x(t-\tau))=\infty$ by (H_1), which contradicts the boundedness of $x(t)$ and $p(t)$.

(ii) If $c<0$, in view of $\lim_{t\to\infty} z(t)=c$ and $x\in L^1[t_0,\infty)$, there exists $t_6\geq t_1$ such that

$$a(t)\big(x(t)-p(t)x(t-\tau)\big)'\leq\frac{c}{2}<0,\quad t\geq t_6.$$

Hence $\lim_{t\to\infty}(x(t)-p(t)x(t-\tau))=-\infty$ by (H_1), a contradiction to the boundedness of $x(t)$ and $p(t)$ again.

(iii) If $c=0$, in view of $z'(t)>0$, we have $z(t)<0$. Further,

$$\big(x(t)-p(t)x(t-\tau)\big)'<0.$$

We show that $x(t)-p(t)x(t-\tau)>0$. In fact, if there exists $t_7 \geq t_1$ such that $x(t_7)-p(t_7)x(t_7-\tau)<0$, then, for all $t \geq t_7$,

$$x(t)-p(t)x(t-\tau) \leq x(t_7)-p(t_7)x(t_7-\tau)<0.$$

This contradicts $x(t)-p(t)x(t-\tau) \in L^1[t_0,\infty)$. Hence $x(t)-p(t)x(t-\tau)>0$ for all large $t \geq t_1$. From this and the assumption on p, we have $x(t) \geq p(t)x(t-\tau) \geq x(t-\tau)$, which contradicts $x \in L^1[t_0,\infty)$. Thus (3.1) is bounded oscillatory. □

Proof of Theorem 3.2 Without loss of generality, assume that $x(t)$ is a bounded eventually positive solution. By a proof similar to that of Theorem 2.4, we obtain $\lim_{t\to\infty}(x(t)-p(t)x(t-\tau))=0$. Suppose

$$\limsup_{t\to\infty} x(t)=l>0.$$

So there exists a sequence $\{t_k\}$ such that $t_k \to \infty$ as $k \to \infty$ and

$$\lim_{k\to\infty} x(t_k)=l>0.$$

(i) If $0 \leq p(t) \leq p_1 < 1$, then we have $(1-p_1)l \leq 0$, which contradicts $l>0$ and $1-p_1>0$.
(ii) If $1 < p_2 \leq p(t)$, then we have $0 \leq (1-p_2)l$, which contradicts $l>0$ and $p_2-1>0$. Therefore, we must have

$$\limsup_{t\to\infty} x(t)=0.$$

Then $\lim_{t\to\infty} x(t)=0$ as $x(t)$ is eventually positive. This shows that (3.1) is bounded almost oscillatory. □

Proof of Theorem 3.3 Without loss of generality, suppose that $x(t)$ is a bounded eventually positive solution. As in the proof of Theorem 2.3, we obtain

$$\lim_{t\to\infty}\big(x(t)-p(t)x(t-\tau)\big)=0.$$

Then the rest follows from the proof of Theorem 3.2. □

Proof of Theorem 3.4 Without loss of generality, suppose that $x(t)$ is a bounded eventually positive solution. Let

$$z(t)=a(t)\big(x(t)-p(t)x(t-\tau)\big)'-\int_{t-\sigma}^{t-\rho} r(s)x(s)\,ds. \tag{4.11}$$

By the reasoning similar to that used in the proof of Theorem 2.1, we have $x \in L^1[t_0,\infty)$, $\lim_{t\to\infty} z(t)=c=0$ and $z'(t)<0$. So $z(t)>0$, $(x(t)-p(t)x(t-\tau))'>0$ and $x(t)-p(t)x(t-\tau)$ is increasing. We claim that $x(t)-p(t)x(t-\tau)<0$ for $t \geq t_1$. In fact, if there exists $t_8 \geq t_1$ such that $x(t_8)-p(t_8)x(t_8-\tau) \geq 0$, then $x(t)-p(t)x(t-\tau) \geq x(t_8+1)-p(t_8+1)x(t_8+1-\tau)>0$ for $t \geq t_8+1$, which contradicts $x(t)-p(t)x(t-\tau) \in L^1[t_1,\infty)$. Hence $x(t)-p(t)x(t-\tau)<0$

for all $t \geq t_1$. If $0 \leq p(t) \leq p < 1$ is satisfied, then $x(t) < px(t-\tau)$ for all $t \geq t_1$. This implies that $\lim_{t\to\infty} x(t) = 0$.

If $1/a(t)$ is bounded and $1 < p_2 \leq p(t)$, from the proof of Theorem 3.2, we have $\lim_{t\to\infty}(x(t) - p(t)x(t-\tau)) = 0$ and thus $\lim_{t\to\infty} x(t) = 0$. Therefore, (3.1) is bounded almost oscillatory. □

Competing interests

The authors declare that they have no competing interests.

Authors' contributions

All authors carried out the proof. All authors conceived of the study and participated in its design and coordination. All authors read and approved the final manuscript.

Author details

[1]School of Sciences, Zhejiang University of Science and Technology, 318 Liuhe Road, Hangzhou, 310023, China. [2]Faculty of Computing, London Metropolitan University, 166-220 Holloway Road, London, N7 8DB, UK.

Acknowledgements

This research is supported by the NNSF of China via Grant 11171306 and the interdisciplinary research funding from Zhejiang University of Science and Technology via Grant 2012JC09Y.

References

1. Gepreel, K, Shehata, AR: Rational Jacobi elliptic solutions for nonlinear difference differential lattice equation. Appl. Math. Lett. **25**, 1173-1178 (2012)
2. Al-Hamouri, R, Zein, A: Oscillation criteria for certain even order neutral delay differential equations. Int. J. Differ. Equ. **2014**, Article ID 437278 (2014)
3. Baculíková, B, Džurina, J: Oscillation theorems for second order neutral differential equations. Comput. Math. Appl. **61**, 94-99 (2011)
4. Elabbay, EM, Hegazi, AS, Saker, SH: Oscillation of solutions to delay differential equations with positive and negative coefficients. Electron. J. Differ. Equ. **2000**, 13 (2000)
5. Fu, XL, Li, TX, Zhang, CH: Oscillation of second-order damped differential equations. Adv. Differ. Equ. **2013**, Article ID 326 (2013)
6. Gai, MJ, Shi, B, Zhang, DC: Oscillations criteria for second order nonlinear differential equations of neutral type. Appl. Math. J. Chin. Univ. Ser. B **16**(2), 122-126 (2001)
7. Grammatikopoulos, MK, Ladas, G, Meimaridou, A: Oscillation and asymptotic behavior of second order neutral differential equations. Ann. Math. Pures Appl. **148**, 29-40 (1987)
8. Hasanbulli, M, Rogovchenko, YV: Oscillation criteria for second order nonlinear neutral differential equations. Appl. Math. Comput. **215**, 4392-4399 (2010)
9. Karpuz, B, Öcalan, Ö, Özturk, S: Comparison theorem on the oscillation and asymptotics behaviour of higher-order neutral differential equations. Glasg. Math. J. **52**, 107-114 (2010)
10. Li, WT, Agarwal, RP: Interval oscillations criteria for second-order nonlinear differential equations with damping. Comput. Math. Appl. **40**, 217-230 (2000)
11. Li, TX, Rogovchenko, YV, Zhang, CH: Oscillation results for second-order nonlinear neutral differential equations. Adv. Differ. Equ. **2013**, Article ID 336 (2013)
12. Li, HJ, Yeh, CC: Oscillations criteria for second-order neutral delay difference equations. Comput. Appl. Math. **36**(10-12), 123-132 (1998)
13. Qin, HZ, Shang, N, Lu, YM: A note on oscillation criteria of second order nonlinear neutral delay differential equations. Comput. Math. Appl. **56**, 2987-2992 (2008)
14. Ruan, S: Oscillations of second order neutral differential equations. Can. Math. Bull. **36**(4), 485-496 (1993)
15. Xing, GJ, Li, TX, Zhang, CH: Oscillation of higher-order quasi-linear neutral differential equations. Adv. Differ. Equ. **2011**, Article ID 45 (2011)
16. Zhang, QX, Yan, JR, Gao, L: Oscillation behavior of even-order nonlinear neutral differential equations with variable coefficients. Comput. Math. Appl. **59**, 426-430 (2010)

7

Improved results on reachable set bounding for linear systems with discrete and distributed delays

Hao Chen[1,2*], Jun Cheng[3], Shouming Zhong[2], Jinxiang Yang[4] and Wei Kang[2]

*Correspondence: chh0308@126.com
[1]School of Mathematical Sciences, Huaibei Normal University, Huaibei, Anhui 235000, P.R. China
[2]School of Mathematical Sciences, University of Electronic Science and Technology of China, Chengdu, Sichuan 611731, P.R. China
Full list of author information is available at the end of the article

Abstract

This paper investigates the bound of a reachable set for linear systems with discrete and distributed delays. By utilizing the Lyapunov-Krasovskii functional, delay decomposition technique, reciprocally convex method and free-weighting matrix approach, some new results in the form of linear matrix inequalities are derived. Finally, a tighter bound of the reachable set is obtained. Three numerical examples are given to illustrate the effectiveness and advantage of the proposed results compared with the existing criteria.

Keywords: reachable set; distributed delay; Lyapunov-Krasovskii functional; linear matrix inequality

1 Introduction

The bounding of a reachable set is of practical importance in the design of the controller for ellipsoid dynamic systems with disturbance. It plays an important role for state estimation and parameter estimation in control theory [1–4]. Therefore, the reachable set estimation and its related fields have been investigated by many researchers. For a dynamic system, the reachable set is a set which bounds all the states starting from the origin by inputs with peak values. The problem of reachable set bounding for time-delay systems has received considerable attention in recent years, for instance, [1–11] and the references therein. However, time delays cannot be avoided during practice, and they cause undesirable dynamic network behaviors such as oscillation and instability [12–30]. Then, it is natural to ask what about the reachable set of systems with time delays.

Boyd researched linear systems without time-delay, and an LMI condition for an ellipsoid that bounds the reachable set was given in [31]. In [9], Fridman and Shaked studied the uncertain linear systems with time-varying delays and bounded peak input, and LMIs criteria of an ellipsoid that bounds the reachable set based on the Razumikhin theory were firstly obtained. In [3], Kim improved the condition by using the modified Lyapunov-Razumikhin functional. More recently, Nam and Pathirana obtained a smaller reachable set bound by the delay decomposition technique [8]. The maximal Lyapunov functional, combined with the Razumikhin methodology, was utilized to give a non-ellipsoidal description of the reachable set in [10]. Moreover, the authors focused on reachable set bounding for linear systems with discrete and distributed delays in [1, 2]. However, up to now, there have been only few literature works available about linear systems with mixed delays. Hence, it is necessary to make further study on linear systems with mixed delays.

The paper [8] considered non-differentiable time-varying delays, and differentiable time-varying delays were considered in [3–7, 9–11]. In [3], the derivative of time delays was assumed to be less than 1. As is well known, a large value of the derivative of time delays may yield bigger reachable set bounding. In fact, the constraint for time-varying delays may be relaxed, that is, a value of the derivative of delays may not necessarily be less than 1.

Motivated by the above discussions, in this paper we aim to study the reachable set bounding for linear systems with discrete and distributed delays. The constraint of delay is relaxed. The value of derivatives of time delay is not necessary to be less than 1. We construct Lyapunov functionals, combined with the delay decomposition technique, reciprocally convex approach and free-weighting matrix method to derive a more accurate description of the reachable set bound. To the best of our knowledge, it is the first time to introduce triple integrations for reachable set bounds of linear systems with discrete and distributed delays. Numerical examples are given to illustrate the effectiveness of the obtained results, and reachable set bound is tighter than the ones in [1–4, 6–8].

Notations: The following notations are used in our paper except where otherwise specified. R^n is the n-dimension Euclidean space, $R^{n\times m}$ denotes the set of $n \times m$-dimension real matrices; real matrix $P > 0$ (≥ 0) means that P is a symmetric positive definite (positive semi-definite) matrix. Superscript 'T' denotes the transposition of a vector or a matrix; $\star$ represents the elements below the main diagonal of a symmetric block matrix; I denotes an identity matrix; '-' in tables represents no feasible solution for matrix inequality.

2 Preliminaries

Consider the following delayed linear systems with disturbances:

$$\dot{z}(t) = Az(t) + Bz\big(t - \tau(t)\big) + D\int_{t-\sigma}^{t} z(s)\,ds + Ew(t), \qquad z(t) = 0, \quad t \in [-d, 0], \tag{1}$$

where $z(t) \in R^n$ is the state vector, $w(t) \in R^m$ is the disturbance, $\tau(t)$ is discrete time delay and σ is distributed time delay, $A \in R^{n\times n}$, $B \in R^{n\times n}$, $D \in R^{n\times n}$ and $E \in R^{n\times m}$, A, D, B and E are known constant matrices.

$\tau(t)$ is time-varying discrete delay satisfying

$$0 \leq \tau(t) \leq \tau, \qquad \dot{\tau}(t) \leq \mu,$$

where τ and μ are constants. Moreover, $d = \max\{\tau, \sigma\}$.

The disturbance $w(t) \in R^m$ is the input with bounded peak value, that is,

$$w^T(t)w(t) \leq w_m^2, \tag{2}$$

where w_m is a constant.

The following lemmas are useful in deriving the criteria.

Lemma 2.1 *The following relation is known as the Leibniz rule*:

$$\frac{d}{dt}\int_{a(t)}^{b(t)} f(t,s)\,ds = \dot{b}(t)f\big[t, b(t)\big] - \dot{a}(t)f\big[t, a(t)\big] + \int_{a(t)}^{b(t)} \frac{\partial}{\partial t} f(t,s)\,ds.$$

Lemma 2.2 [29] *Given constant symmetric matrices R_1, R_2, R_3, where $R_1 = R_1^T, R_2 = R_2^T > 0$, then $R_1 + R_3^T R_2^{-1} R_3 < 0$ if and only if*

$$\begin{pmatrix} R_1 & R_3^T \\ R_3 & -R_2 \end{pmatrix} < 0,$$

or

$$\begin{pmatrix} -R_2 & R_3 \\ R_3^T & R_1 \end{pmatrix} < 0.$$

Lemma 2.3 [13] *For any constant matrix $P = P^T > 0$ and $h_2 > h_1 \geq 0$ such that the following integrations are well defined, then*

$$-(h_2 - h_1) \int_{t-h_2}^{t-h_1} z^T(s) P z(s)\, ds \leq -\left(\int_{t-h_2}^{t-h_1} z(s)\, ds \right)^T P \left(\int_{t-h_2}^{t-h_1} z(s)\, ds \right).$$

Lemma 2.4 [30] *For any constant matrix $R > 0$, scalars $h_2 > h_1 \geq 0$ such that the following integrations are well defined, then*

$$-\frac{1}{2}(h_2 - h_1)^2 \int_{-h_2}^{-h_1} \int_{t+\theta}^{t-h_1} z^T(s) R z(s)\, ds\, d\theta \leq -\left(\int_{-h_2}^{-h_1} \int_{t+\theta}^{t-h_1} z^T(s)\, ds\, d\theta \right) \cdot R \left(\int_{-h_2}^{-h_1} \int_{t+\theta}^{t-h_1} z(s)\, ds\, d\theta \right).$$

Lemma 2.5 [18] *Let $f_1, f_2, \ldots, f_N : R^m \mapsto R$ have positive values in an open subset D of R^m. Then the reciprocally convex combination of f_i over D satisfies*

$$\min_{\{\alpha_i | \alpha_i > 0, \sum_{i=1}^{N} \alpha_i = 1\}} \sum_{i=1}^{N} \frac{1}{\alpha_i} f_i(t) + \max_{g_{i,j}(t)} \sum_{i \neq j} g_{i,j}(t),$$

subject to

$$\left\{ g_{i,j}(t) : R^m \mapsto R, g_{j,i}(t) = g_{i,j}(t), \begin{bmatrix} f_i(t) & g_{i,j}(t) \\ g_{i,j}(t) & f_j(t) \end{bmatrix} \geq 0 \right\}.$$

Lemma 2.6 [26] *For any vectors x_1, x_2, constant matrices T_i $(i = 1, 2, 3, 4)$, S and scalars $\alpha > 0$, $\beta > 0$ satisfying $\alpha + \beta = 1$, then the following inequality holds:*

$$-\frac{1}{\alpha} x_1^T T_1 x_1 - \frac{1}{\beta} x_2^T T_2 x_2 - \frac{\beta}{\alpha} x_1^T T_3 x_1 - \frac{\alpha}{\beta} x_2^T T_4 x_2 \leq -\begin{bmatrix} x_1 \\ x_2 \end{bmatrix}^T \begin{bmatrix} T_1 & S \\ S^T & T_2 \end{bmatrix} \begin{bmatrix} x_1 \\ x_2 \end{bmatrix},$$

subject to

$$\begin{bmatrix} T_1 + T_3 & S \\ S^T & T_2 + T_4 \end{bmatrix} \geq 0.$$

Lemma 2.7 [5] *Let V be a Lyapunov function for system* (1)-(2). *If*

$$\dot{V} + \alpha V - \frac{\alpha}{w_m^2} w^T(t)w(t) \le 0,$$

then $V \le 1$.

3 Main results

Theorem 3.1 *If there exist matrices $P > 0$, $Q > 0$, $R > 0$, $K_1 > 0$, $K_2 > 0$, $M_1 > 0$, $M_2 > 0$, S, N with appropriate dimensions, and a scalar $\alpha > 0$ such that the following inequalities holds*:

$$\begin{bmatrix} R + K_1 & S \\ S^T & R + K_2 \end{bmatrix} \ge 0, \tag{3}$$

$$\Phi = \begin{bmatrix} \Phi_{11} & \Phi_{12} & \Phi_{13} & \Phi_{14} & 0 & \Phi_{16} & \Phi_{17} & \Phi_{18} \\ \star & \Phi_{22} & \Phi_{23} & \Phi_{24} & \Phi_{25} & \Phi_{26} & 0 & 0 \\ \star & \star & \Phi_{33} & 0 & \Phi_{35} & 0 & 0 & 0 \\ \star & \star & \star & \Phi_{44} & 0 & 0 & 0 & 0 \\ \star & \star & \star & \star & \Phi_{55} & 0 & 0 & 0 \\ \star & \star & \star & \star & \star & \Phi_{66} & \Phi_{67} & \Phi_{68} \\ \star & \star & \star & \star & \star & \star & \Phi_{77} & 0 \\ \star & \star & \star & \star & \star & \star & \star & -\frac{\alpha}{w_m^2} I \end{bmatrix} \le 0, \tag{4}$$

where

$$\Phi_{11} = \alpha P + PA + A^T P + Q + \sigma M_1 + M_2 - 2e^{-\alpha\tau} K_1 - e^{-\alpha\tau} R,$$

$$\Phi_{12} = PB + e^{-\alpha\tau} R - e^{-\alpha\tau} S, \qquad \Phi_{13} = e^{-\alpha\tau} S, \qquad \Phi_{14} = 2e^{-\alpha\tau} K_1,$$

$$\Phi_{16} = A^T N^T, \qquad \Phi_{17} = \sigma PD, \qquad \Phi_{18} = PE,$$

$$\Phi_{22} = -(1-\mu)e^{-\alpha\tau} M_2 - 2e^{-\alpha\tau} K_1 - 2e^{-\alpha\tau} K_2 - 2e^{-\alpha\tau} R + e^{-\alpha\tau}\left(S^T + S\right),$$

$$\Phi_{23} = e^{-\alpha\tau} R - e^{-\alpha\tau} S, \qquad \Phi_{24} = 2e^{-\alpha\tau} K_2,$$

$$\Phi_{25} = 2e^{-\alpha\tau} K_1, \qquad \Phi_{26} = B^T N^T,$$

$$\Phi_{33} = -e^{-\alpha\tau} Q - 2e^{-\alpha\tau} K_2 - e^{-\alpha\tau} R, \qquad \Phi_{35} = 2e^{-\alpha\tau} K_2,$$

$$\Phi_{44} = -2e^{-\alpha\tau} K_1 - 2e^{-\alpha\tau} K_2, \qquad \Phi_{55} = -2e^{-\alpha\tau} K_1 - 2e^{-\alpha\tau} K_2,$$

$$\Phi_{66} = \tau^2 R + \frac{1}{2}\tau^2 K_1 + \frac{1}{2}\tau^2 K_2 - N - N^T, \qquad \Phi_{67} = \sigma ND, \qquad \Phi_{68} = NE,$$

$$\Phi_{77} = -\sigma e^{-\alpha\sigma} M_1.$$

Then the reachable sets of system (1) *are bounded by a ball* $B(0, r) = \{z \in R^n \mid \|z\| \le r\}$ *with*

$$r = \frac{1}{\sqrt{\lambda_{\min}(P)}}. \tag{5}$$

Proof Construct the following Lyapunov-Krasovskii functional:

$$V(z_t) = \sum_{i=1}^{6} V_i(z_t),$$

where

$$V_1(z_t) = z^T(t)Pz(t),$$

$$V_2(z_t) = \int_{t-\tau}^{t} e^{\alpha(s-t)} z^T(s)Qz(s)\,ds,$$

$$V_3(z_t) = \tau \int_{-\tau}^{0}\int_{t+\theta}^{t} e^{\alpha(s-t)} \dot{z}^T(s)R\dot{z}(s)\,ds,$$

$$V_4(z_t) = \int_{-\tau}^{0}\int_{\eta}^{0}\int_{t+\theta}^{t} e^{\alpha(s-t)} \dot{z}^T(s)K_1\dot{z}(s)\,ds\,d\theta\,d\eta,$$

$$V_5(z_t) = \int_{-\tau}^{0}\int_{-\tau}^{\eta}\int_{t+\theta}^{t} e^{\alpha(s-t)} \dot{z}^T(s)K_2\dot{z}(s)\,ds\,d\theta\,d\eta,$$

$$V_6(z_t) = \int_{t-\sigma}^{t} e^{\alpha(s-t)}(\sigma - t + s) z^T(s)M_1z(s)\,ds + \int_{t-\tau(t)}^{t} e^{\alpha(s-t)} z^T(s)M_2z(s)\,ds.$$

Taking the time derivative of $V(z_t)$ along the trajectory of system (1), we obtain

$$\begin{aligned}\dot{V}_1(z_t) &= 2z^T(t)P\dot{z}(t) = -\alpha V_1(z_t) + 2z^T(t)P\dot{z}(t) + \alpha z^T(t)Pz(t)\\ &= -\alpha V_1(z_t) + \alpha z^T(t)Pz(t) + 2z^T(t)P\Big(Az(t) + Bz(t-\tau(t))\\ &\quad + D\int_{t-\sigma}^{t} z(s)\,ds + Ew(t)\Big),\end{aligned} \tag{6}$$

$$\dot{V}_2(z_t) = -\alpha V_2(z_t) + z^T(t)Qz(t) - e^{-\alpha\tau} z^T(t-\tau)Qz(t-\tau), \tag{7}$$

$$\begin{aligned}\dot{V}_3(z_t) &= -\alpha V_3(z_t) + \tau^2\dot{z}^T(t)R\dot{z}(t) - \tau\int_{t-\tau}^{t} e^{\alpha(s-t)}\dot{z}^T(s)R\dot{z}(s)\,ds\\ &\le -\alpha V_3(z_t) + \tau^2\dot{z}^T(t)R\dot{z}(t) - e^{-\alpha\tau}\tau\int_{t-\tau(t)}^{t} \dot{z}^T(s)R\dot{z}(s)\,ds\\ &\quad - e^{-\alpha\tau}\tau\int_{t-\tau}^{t-\tau(t)} \dot{z}^T(s)R\dot{z}(s)\,ds,\end{aligned} \tag{8}$$

$$\begin{aligned}\dot{V}_4(z_t) &= -\alpha V_4(z_t) + \frac{1}{2}\tau^2\dot{z}(t)K_1\dot{z}(t) - \int_{-\tau}^{0}\int_{t+\theta}^{t} e^{\alpha(s-t)}\dot{z}^T(s)K_1\dot{z}(s)\,ds\,d\theta\\ &\le -\alpha V_4(z_t) + \frac{1}{2}\tau^2\dot{z}(t)K_1\dot{z}(t) - e^{-\alpha\tau}\int_{-\tau}^{0}\int_{t+\theta}^{t} \dot{z}^T(s)K_1\dot{z}(s)\,ds\,d\theta\\ &= -\alpha V_4(z_t) + \frac{1}{2}\tau^2\dot{z}(t)K_1\dot{z}(t) - e^{-\alpha\tau}\int_{-\tau(t)}^{0}\int_{t+\theta}^{t} \dot{z}^T(s)K_1\dot{z}(s)\,ds\,d\theta\\ &\quad - e^{-\alpha\tau}\int_{-\tau}^{-\tau(t)}\int_{t+\theta}^{t-\tau(t)} \dot{z}^T(s)K_1\dot{z}(s)\,ds\,d\theta\\ &\quad - e^{-\alpha\tau}\int_{-\tau}^{-\tau(t)}\int_{t-\tau(t)}^{t} \dot{z}^T(s)K_1\dot{z}(s)\,ds\,d\theta,\end{aligned} \tag{9}$$

$$
\begin{aligned}
\dot{V}_5(z_t) &= -\alpha V_5(z_t) + \frac{1}{2}\tau^2\dot{z}(t)K_2\dot{z}(t) - \int_{-\tau}^{0}\int_{t-\tau}^{t+\theta} e^{\alpha(s-t)}\dot{z}^T(s)K_2\dot{z}(s)\,ds\,d\theta \\
&\leq -\alpha V_5(z_t) + \frac{1}{2}\tau^2\dot{z}(t)K_2\dot{z}(t) - e^{-\alpha\tau}\int_{-\tau}^{0}\int_{t-\tau}^{t+\theta} \dot{z}^T(s)K_2\dot{z}(s)\,ds\,d\theta \\
&= -\alpha V_5(z_t) + \frac{1}{2}\tau^2\dot{z}(t)K_2\dot{z}(t) - e^{-\alpha\tau}\int_{-\tau(t)}^{0}\int_{t-\tau(t)}^{t+\theta} \dot{z}^T(s)K_2\dot{z}(s)\,ds\,d\theta \\
&\quad - e^{-\alpha\tau}\int_{-\tau(t)}^{0}\int_{t-\tau}^{t-\tau(t)} \dot{z}^T(s)K_2\dot{z}(s)\,ds\,d\theta \\
&\quad - e^{-\alpha\tau}\int_{-\tau}^{-\tau(t)}\int_{t-\tau}^{t+\theta} \dot{z}^T(s)K_2\dot{z}(s)\,ds\,d\theta, \qquad (10)
\end{aligned}
$$

$$
\begin{aligned}
\dot{V}_6(z_t) &= -\alpha V_6(z_t) - \int_{t-\sigma}^{t} e^{\alpha(s-t)}z^T(s)M_1z(s)\,ds + \sigma z^T(t)M_1z(t) \\
&\quad + z^T(t)M_2z(t) - \big(1-\dot{\tau}(t)\big)z^T\big(t-\tau(t)\big)e^{-\alpha\tau(t)}M_2z\big(t-\tau(t)\big) \\
&\leq -\alpha V_6(z_t) - e^{-\alpha\sigma}\int_{t-\sigma}^{t} z^T(s)M_1z(s)\,ds + \sigma z^T(t)M_1z(t) \\
&\quad + z^T(t)M_2z(t) - (1-\mu)e^{-\alpha\tau}z^T\big(t-\tau(t)\big)M_2z\big(t-\tau(t)\big). \qquad (11)
\end{aligned}
$$

Using Lemma 2.2, we have

$$
\begin{aligned}
\dot{V}_3(z_t) &\leq -\alpha V_3(z_t) + \tau^2\dot{z}^T(t)R\dot{z}(t) \\
&\quad - e^{-\alpha\tau}\frac{\tau}{\tau(t)}\big(z^T(t) - z^T\big(t-\tau(t)\big)\big)R\big(z(t) - z\big(t-\tau(t)\big)\big) \\
&\quad - e^{-\alpha\tau}\frac{\tau}{\tau-\tau(t)}\big(z^T\big(t-\tau(t)\big) - z^T(t-\tau)\big)R\big(z\big(t-\tau(t)\big) - z(t-\tau)\big) \\
&= -\alpha V_3(z_t) + \tau^2\dot{z}^T(t)R\dot{z}(t) \\
&\quad - e^{-\alpha\tau}\frac{1}{\beta_1}\big(z^T(t) - z^T\big(t-\tau(t)\big)\big)R\big(z(t) - z\big(t-\tau(t)\big)\big) \\
&\quad - e^{-\alpha\tau}\frac{1}{\beta_2}\big(z^T\big(t-\tau(t)\big) - z^T(t-\tau)\big)R\big(z\big(t-\tau(t)\big) - z(t-\tau)\big), \qquad (12)
\end{aligned}
$$

where $\beta_1 = \frac{\tau(t)}{\tau}$, $\beta_2 = \frac{\tau-\tau(t)}{\tau}$ and $\beta_1 + \beta_2 = 1$.

Using Lemma 2.3, we get

$$
\begin{aligned}
\dot{V}_4(z_t) &\leq -\alpha V_4(z_t) + \frac{1}{2}\tau^2\dot{z}(t)K_1\dot{z}(t) \\
&\quad - 2e^{-\alpha\tau}\frac{1}{(\tau(t))^2}\bigg(\int_{-\tau(t)}^{0}\int_{t+\theta}^{t}\dot{z}^T(s)\,ds\,d\theta\bigg)K_1\bigg(\int_{-\tau(t)}^{0}\int_{t+\theta}^{t}\dot{z}(s)\,ds\,d\theta\bigg) \\
&\quad - 2e^{-\alpha\tau}\frac{1}{(\tau-\tau(t))^2}\bigg(\int_{-\tau}^{-\tau(t)}\int_{t+\theta}^{t-\tau(t)}\dot{z}^T(s)\,ds\,d\theta\bigg) \\
&\quad \cdot K_1\bigg(\int_{-\tau}^{-\tau(t)}\int_{t+\theta}^{t-\tau(t)}\dot{z}(s)\,ds\,d\theta\bigg) \\
&\quad - e^{-\alpha\tau}\frac{\tau-\tau(t)}{\tau(t)}\bigg(\int_{t-\tau(t)}^{t}\dot{z}^T(s)\,ds\,d\theta\bigg)K_1\bigg(\int_{t-\tau(t)}^{t}\dot{z}^T(s)\,ds\,d\theta\bigg)
\end{aligned}
$$

$$
\begin{aligned}
&= -\alpha V_4(z_t) + \frac{1}{2}\tau^2 \dot{z}(t) K_1 \dot{z}(t) \\
&\quad - 2e^{-\alpha\tau}\left(z^T(t) - \frac{1}{\tau(t)}\int_{t-\tau(t)}^{t} z^T(s)\,ds\right) K_1 \left(z(t) - \frac{1}{\tau(t)}\int_{t-\tau(t)}^{t} z(s)\,ds\right) \\
&\quad - 2e^{-\alpha\tau}\left(z(t-\tau(t)) - \frac{1}{\tau - \tau(t)}\int_{t-\tau}^{t-\tau(t)} z(s)\,ds\right)^T K_1 \left(z(t-\tau(t))\right. \\
&\quad \left. - \frac{1}{\tau - \tau(t)}\int_{t-\tau}^{t-\tau(t)} z(s)\,ds\right) \\
&\quad - e^{-\alpha\tau}\frac{\beta_2}{\beta_1}\left(z^T(t) - z^T(t-\tau(t))\right) K_1 \left(z(t) - z(t-\tau(t))\right), \qquad (13)
\end{aligned}
$$

$$
\begin{aligned}
\dot{V}_5(z_t) &\leq -\alpha V_5(z_t) + \frac{1}{2}\tau^2 \dot{z}(t) K_2 \dot{z}(t) \\
&\quad - 2e^{-\alpha\tau}\frac{1}{(\tau(t))^2}\left(\int_{-\tau(t)}^{0}\int_{t-\tau(t)}^{t+\theta} \dot{z}^T(s)\,ds\,d\theta\right) K_2 \left(\int_{-\tau(t)}^{0}\int_{t-\tau(t)}^{t+\theta} \dot{z}^T(s)\,ds\,d\theta\right) \\
&\quad - e^{-\alpha\tau}\frac{\tau(t)}{\tau - \tau(t)}\left(\int_{t-\tau}^{t-\tau(t)} \dot{z}^T(s)\,ds\,d\theta\right) K_2 \left(\int_{t-\tau}^{t-\tau(t)} \dot{z}(s)\,ds\,d\theta\right) \\
&\quad - 2e^{-\alpha\tau}\frac{1}{(\tau - \tau(t))^2}\left(\int_{-\tau}^{-\tau(t)}\int_{t-\tau}^{t+\theta} \dot{z}^T(s)\,ds\,d\theta\right) K_2 \left(\int_{-\tau}^{-\tau(t)}\int_{t-\tau}^{t+\theta} \dot{z}(s)\,ds\,d\theta\right) \\
&= -\alpha V_5(z_t) + \frac{1}{2}\tau^2 \dot{z}(t) K_2 \dot{z}(t) \\
&\quad - 2e^{-\alpha\tau}\left(\frac{1}{\tau(t)}\int_{t-\tau(t)}^{t} z^T(s)\,ds - z(t-\tau(t))\right) \\
&\quad \cdot K_2 \left(\frac{1}{\tau(t)}\int_{t-\tau(t)}^{t} z(s)\,ds - z(t-\tau(t))\right) \\
&\quad - e^{-\alpha\tau}\frac{\beta_1}{\beta_2}\left(z^T(t-\tau(t)) - z^T(t-\tau)\right) K_2 \left(z(t-\tau(t)) - z(t-\tau)\right) \\
&\quad - 2e^{-\alpha\tau}\left(\frac{1}{\tau - \tau(t)}\int_{t-\tau}^{t-\tau(t)} z^T(s)\,ds - z^T(t-\tau)\right) \\
&\quad \cdot K_2 \left(\frac{1}{\tau - \tau(t)}\int_{t-\tau}^{t-\tau(t)} z(s)\,ds - z(t-\tau)\right). \qquad (14)
\end{aligned}
$$

From Lemma 2.5 and inequality (6), one can obtain

$$
\begin{aligned}
&-e^{-\alpha\tau}\frac{1}{\beta_1}\left(z^T(t) - z^T(t-\tau(t))\right) R \left(z(t) - z(t-\tau(t))\right) \\
&\quad - e^{-\alpha\tau}\frac{1}{\beta_2}\left(z^T(t-\tau(t)) - z^T(t-\tau)\right) R \left(z(t-\tau(t)) - z(t-\tau)\right) \\
&\quad - e^{-\alpha\tau}\frac{\beta_2}{\beta_1}\left(z^T(t) - z^T(t-\tau(t))\right) K_1 \left(z(t) - z(t-\tau(t))\right) \\
&\quad - e^{-\alpha\tau}\frac{\beta_1}{\beta_2}\left(z^T(t-\tau(t)) - z^T(t-\tau)\right) K_2 \left(z(t-\tau(t)) - z(t-\tau)\right) \\
&\leq -e^{-\alpha\tau}\begin{bmatrix} z(t) - z(t-\tau(t)) \\ z(t-\tau(t)) - z(t-\tau) \end{bmatrix}^T \begin{bmatrix} R & S \\ S^T & R \end{bmatrix} \begin{bmatrix} z(t) - z(t-\tau(t)) \\ z(t-\tau(t)) - z(t-\tau) \end{bmatrix}. \qquad (15)
\end{aligned}
$$

It is clear that the following equation holds:

$$2\dot{z}^T(t)N\left(-\dot{z}(t) + Az(t) + Bz\big(t-\tau(t)\big) + D\int_{t-\sigma}^{t} z(s)\,ds + Ew(t)\right) = 0. \tag{16}$$

Combining (9)-(19), one gets

$$\dot{V}(z_t) + \alpha V(z_t) - \frac{\alpha}{w_m^2}w^T(t)w(t) \le \frac{1}{\sigma}\int_{t-\sigma}^{t} \xi^T(t)\Phi\xi(t)\,ds, \tag{17}$$

where

$$\xi^T(t) = \Bigg[z^T(t),\quad z^T\big(t-\tau(t)\big),\quad z^T(t-\tau),\quad \frac{1}{\tau(t)}\int_{t-\tau(t)}^{t} z^T(s)\,ds,$$
$$\frac{1}{\tau-\tau(t)}\int_{t-\tau}^{t-\tau(t)} z^T(s)\,ds,\quad \dot{z}^T(t),\quad z^T(s),\quad w^T(t)\Bigg].$$

Since (6) and (7) hold, we can conclude that $\dot{V} + \alpha V - \frac{\alpha}{w_m^2}w^T(t)w(t) \le 0$.

Therefore, one can obtain $V(z_t) \le 1$ by Lemma 2.7.

Using the spectral properties of a symmetric positive definite matrix P, the following inequality holds:

$$\lambda_{\min}(P)\left\|z(t)\right\|^2 \le V(z_t). \tag{18}$$

This further implies that $\|z(t)\| \le r = \frac{1}{\sqrt{\lambda_{\min}(P)}}$ due to (19). This completes the proof. □

If $D = 0$ in system (1), the following corollary is true.

Corollary 3.1 *If there exist matrices $P > 0$, $Q > 0$, $R > 0$, $K_1 > 0$, $K_2 > 0$, $M > 0$, S, N with appropriate dimensions, and a scalar $\alpha > 0$ such that the following inequalities holds:*

$$\begin{bmatrix} R+K_1 & S \\ S^T & R+K_2 \end{bmatrix} \ge 0, \tag{19}$$

$$\Phi = \begin{bmatrix} \Phi_{11} & \Phi_{12} & \Phi_{13} & \Phi_{14} & 0 & \Phi_{16} & \Phi_{17} \\ \star & \Phi_{22} & \Phi_{23} & \Phi_{24} & \Phi_{25} & \Phi_{26} & 0 \\ \star & \star & \Phi_{33} & 0 & \Phi_{35} & 0 & 0 \\ \star & \star & \star & \Phi_{44} & 0 & 0 & 0 \\ \star & \star & \star & \star & \Phi_{55} & 0 & 0 \\ \star & \star & \star & \star & \star & \Phi_{66} & 0 \\ \star & \star & \star & \star & \star & \star & -\frac{\alpha}{w_m^2}I \end{bmatrix} \le 0, \tag{20}$$

where

$$\Phi_{11} = \alpha P + PA + A^TP + Q + M - 2e^{-\alpha\tau}K_1 - e^{-\alpha\tau}R,$$

$$\Phi_{12} = PB + e^{-\alpha\tau}R - e^{-\alpha\tau}S, \qquad \Phi_{13} = e^{-\alpha\tau}S,$$

$$\Phi_{14} = 2e^{-\alpha\tau}K_1, \qquad \Phi_{16} = A^T N^T, \qquad \Phi_{17} = PE,$$

$$\Phi_{22} = -(1-\mu)e^{-\alpha\tau}M - 2e^{-\alpha\tau}K_1 - 2e^{-\alpha\tau}K_2 - 2e^{-\alpha\tau}R + e^{-\alpha\tau}\left(S^T + S\right),$$

$$\Phi_{23} = e^{-\alpha\tau}R - e^{-\alpha\tau}S, \qquad \Phi_{24} = 2e^{-\alpha\tau}K_2,$$

$$\Phi_{25} = 2e^{-\alpha\tau}K_1, \qquad \Phi_{26} = B^T N^T,$$

$$\Phi_{33} = -e^{-\alpha\tau}Q - 2e^{-\alpha\tau}K_2 - e^{-\alpha\tau}R, \qquad \Phi_{35} = 2e^{-\alpha\tau}K_2,$$

$$\Phi_{44} = -2e^{-\alpha\tau}K_1 - 2e^{-\alpha\tau}K_2, \qquad \Phi_{55} = -2e^{-\alpha\tau}K_1 - 2e^{-\alpha\tau}K_2,$$

$$\Phi_{66} = \tau^2 R + \frac{1}{2}\tau^2 K_1 + \frac{1}{2}\tau^2 K_2 - N - N^T.$$

Then the reachable sets of system (1) *are bounded by a ball* $B(0, r) = \{z \in R^n \mid \|z\| \leq r\}$ *with*

$$r = \frac{1}{\sqrt{\lambda_{\min}(P)}}. \tag{21}$$

Proof By setting $D = 0$, $M_1 = 0$ in the proof of Theorem 3.1, one can easily get the conclusion in the corollary. □

Furthermore, consider the following uncertain polytopic time-delayed linear systems with disturbances:

$$\begin{aligned} \dot{z}(t) &= (A + \Delta A)z(t) + (B + \Delta B)z\big(t - \tau(t)\big) + (E + \Delta E)w(t), \\ z(t) &= 0, \quad t \in [-\tau_M, 0]. \end{aligned} \tag{22}$$

The uncertainties are expressed as a linear convex-hull of known matrices A_i, B_i and E_i,

$$\Delta A = \sum_{i=1}^{N} \theta_i(t)A_i, \qquad \Delta B = \sum_{i=1}^{N} \theta_i(t)B_i, \qquad \Delta E = \sum_{i=1}^{N} \theta_i(t)E_i$$

with $\theta_i(t) \in [0, 1]$ and $\sum_{i=1}^{N} \theta_i(t) = 1$, $\forall t > 0$.

As for system (22), the following corollary is true.

Corollary 3.2 *If there exist matrices* $P > 0$, $Q > 0$, $R > 0$, $K_1 > 0$, $K_2 > 0$, $M > 0$, S, N *with appropriate dimensions, and a scalar* $\alpha > 0$ *satisfying the following inequalities for all* $i = 1, 2, \ldots, N$:

$$\begin{bmatrix} R + K_1 & S \\ S^T & R + K_2 \end{bmatrix} \geq 0, \tag{23}$$

$$\Phi_i = \begin{bmatrix} \Phi_{11i} & \Phi_{12i} & \Phi_{13i} & \Phi_{14i} & 0 & \Phi_{16i} & \Phi_{17i} \\ \star & \Phi_{22i} & \Phi_{23i} & \Phi_{24i} & \Phi_{25i} & \Phi_{26i} & 0 \\ \star & \star & \Phi_{33i} & 0 & \Phi_{35i} & 0 & 0 \\ \star & \star & \star & \Phi_{44i} & 0 & 0 & 0 \\ \star & \star & \star & \star & \Phi_{55i} & 0 & 0 \\ \star & \star & \star & \star & \star & \Phi_{66i} & 0 \\ \star & \star & \star & \star & \star & \star & -\frac{\alpha}{w_m^2}I \end{bmatrix} \leq 0, \tag{24}$$

where

$$\Phi_{11i} = \alpha P + P(A + A_i) + (A + A_i)^T P + Q + M - 2e^{-\alpha\tau} K_1 - e^{-\alpha\tau} R,$$

$$\Phi_{12i} = P(B + B_i) + e^{-\alpha\tau} R - e^{-\alpha\tau} S, \qquad \Phi_{13i} = e^{-\alpha\tau} S,$$

$$\Phi_{14i} = 2e^{-\alpha\tau} K_1, \Phi_{16i} = (A + A_i)^T N^T, \qquad \Phi_{17i} = P(E + E_i),$$

$$\Phi_{22i} = -(1 - \mu)e^{-\alpha\tau} M - 2e^{-\alpha\tau} K_1 - 2e^{-\alpha\tau} K_2 - 2e^{-\alpha\tau} R + e^{-\alpha\tau}\left(S^T + S\right),$$

$$\Phi_{23i} = e^{-\alpha\tau} R - e^{-\alpha\tau} S, \qquad \Phi_{24i} = 2e^{-\alpha\tau} K_2,$$

$$\Phi_{25i} = 2e^{-\alpha\tau} K_1, \qquad \Phi_{26i} = (B + B_i)^T N^T,$$

$$\Phi_{33i} = -e^{-\alpha\tau} Q - 2e^{-\alpha\tau} K_2 - e^{-\alpha\tau} R, \qquad \Phi_{35i} = 2e^{-\alpha\tau} K_2,$$

$$\Phi_{44i} = -2e^{-\alpha\tau} K_1 - 2e^{-\alpha\tau} K_2, \qquad \Phi_{55i} = -2e^{-\alpha\tau} K_1 - 2e^{-\alpha\tau} K_2,$$

$$\Phi_{66i} = \tau^2 R + \frac{1}{2}\tau^2 K_1 + \frac{1}{2}\tau^2 K_2 - N - N^T.$$

Then the reachable sets of system (25) *are bounded by a ball* $B(0, r) = \{z \in R^n \mid \|z\| \leq r\}$ *with*

$$r = \frac{1}{\sqrt{\lambda_{\min}(P)}}. \tag{25}$$

Proof Replacing A, B, E with $\sum_{i=1}^{N} \theta_i(t)(A + A_i)$, $\sum_{i=1}^{N} \theta_i(t)(B + B_i)$, $\sum_{i=1}^{N} \theta_i(t)(E + E_i)$ in the proof of Corollary 3.1, respectively, one can easily obtain the conclusion. □

Remark 1 In this paper, the delay decomposition technique, free-weighting matrix approach and reciprocally convex method are used to construct a Lyapunov functional. Triple integrals are introduced in the Lyapunov functional for the first time to investigate bounding of a reachable set for linear systems with discrete and distributed delays, which may lead to tighter bounding. It will be verified by the following numerical examples.

Remark 2 In order to guarantee negative definite, μ is required to be less than 1 in [3]. It should be noted that the values of derivatives of time delays are not necessary to be less than 1 in Theorem 3.1 because the term $\Phi_{22} = -(1 - \mu)e^{-\alpha\tau} M_2 - 2e^{-\alpha\tau} K_1 - 2e^{-\alpha\tau} K_2 - 2e^{-\alpha\tau} R + e^{-\alpha\tau}(S^T + S)$ can be negative definite by choosing appropriate K_1, K_2, R, S, M_2 when $\mu > 1$. Obviously, the results in this paper have a wider scope of application than the ones in [3].

Remark 3 The reachable set of system (1) can be minimized by solving the following optimization problem for a scalar $\delta > 0$:

$$\begin{aligned} &\min \bar{\delta} \quad \left(\bar{\delta} = \frac{1}{\delta}\right) \\ &\text{s.t.} \begin{cases} \text{(a)} \quad P \geq \delta I, \\ \text{(b)} \quad \text{(6)-(7) or (22)-(23) or (26)-(27).} \end{cases} \end{aligned} \tag{26}$$

Remark 4 It should be noted that the matrix inequalities in Theorem 3.1 and Corollaries 3.1-3.2 cannot be simplified to LMIs. However, when α is fixed, the matrix inequalities reduced to LMIs, and Matlab's Toolbox is employed to solve the matrix inequalities in Theorem 3.1 and Corollaries 3.1-3.2.

4 Examples

In this section, two numerical examples are presented to show the validity of the main results derived in this paper.

Example 1 Consider the following time-delayed system with parameters:

$$\dot{z}(t) = \begin{bmatrix} 0 & 1 \\ -2 & -3 \end{bmatrix} z(t) + \begin{bmatrix} 0 & -0.1 \\ 0.2 & 0.3 \end{bmatrix} z\big(t - \tau(t)\big) + \begin{bmatrix} 0.2 & 0.2 \\ 0 & -0.2 \end{bmatrix} \int_{t-\sigma}^{t} z(s)\,ds + \begin{bmatrix} 1 \\ 1 \end{bmatrix} w(t), \tag{27}$$

and $w^T(t)w(t) \le 1$.

In order to compare with previous results, r's for different values of σ with $\tau = 0.2$, $\mu = 0.5$ are listed in Table 1. r's for different values of τ with $\sigma = 0.1$, $\mu = 0.5$ are listed in Table 2. r's for different values of μ with $\tau = 0.2$, $\sigma = 0.1$ are listed in Table 3. It is shown that tighter bounds are obtained than the ones in [1, 2] by the proposed method in this paper.

Solving system (1) with parameters $\sigma = 0.1$, $\mu = 0.5$, $\tau = 0.2$ by Theorem 3.1, we have

$$P = \begin{bmatrix} 0.8599 & 0.7400 \\ 0.7400 & 2.1083 \end{bmatrix}, \qquad r = \sqrt{1.9380}. \tag{28}$$

To give a direct comparison, we plotted the ellipsoid of the reachable set in Figure 1. The solid line is the result computed by the proposed method in this paper and the dotted line is the one obtained by [2].

Table 1 Computed r's of Example 1 for different values of σ with $\tau = 0.2$, $\mu = 0.5$

σ	0.2	0.4	0.6	0.7	0.8
[1]	$\sqrt{10.0502}$	$\sqrt{11.5762}$	$\sqrt{13.7431}$	$\sqrt{14.9492}$	$\sqrt{16.3729}$
[2]	$\sqrt{5.8645}$	$\sqrt{6.4724}$	$\sqrt{7.3010}$	$\sqrt{7.7887}$	$\sqrt{8.3381}$
Theorem 3.1	$\sqrt{2.2624}$	$2\sqrt{.9329}$	$\sqrt{3.0799}$	$\sqrt{3.2658}$	$\sqrt{3.3009}$

Table 2 Computed r's of Example 1 for different values of τ with $\sigma = 0.1$, $\mu = 0.5$

τ	0.1	0.3	0.5	0.7	0.9
[1]	$\sqrt{9.3269}$	$\sqrt{9.5769}$	$\sqrt{9.8531}$	$\sqrt{10.1593}$	$\sqrt{10.5000}$
[2]	$\sqrt{5.5456}$	$\sqrt{5.7058}$	$\sqrt{5.9109}$	$\sqrt{6.1423}$	$\sqrt{6.4505}$
Theorem 3.1	$\sqrt{3.0762}$	$\sqrt{3.9978}$	$\sqrt{4.1181}$	$\sqrt{4.4765}$	$\sqrt{5.0025}$

Table 3 Computed r's of Example 1 for different values of μ with $\sigma = 0.1$, $\tau = 0.2$

μ	0.0	0.1	0.5	0.8	0.9	2
[1]	$\sqrt{8.0664}$	$\sqrt{8.2338}$	$\sqrt{9.4488}$	$\sqrt{13.4135}$	$\sqrt{21.2778}$	-
[2]	$\sqrt{5.5885}$	$\sqrt{5.5992}$	$\sqrt{5.6198}$	$\sqrt{5.6198}$	$\sqrt{5.6198}$	$\sqrt{5.6198}$
Theorem 3.1	$\sqrt{4.0411}$	$\sqrt{4.0928}$	$\sqrt{4.2011}$	$\sqrt{4.4049}$	$\sqrt{4.5205}$	$\sqrt{5.1675}$

Figure 1 Bounding ellipsoids computed by different methods.

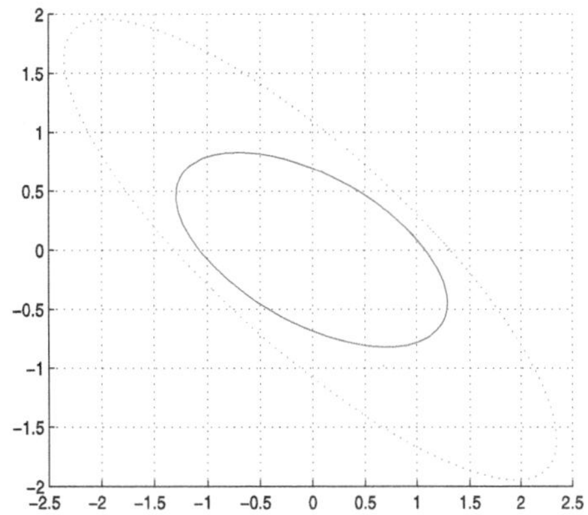

Table 4 Computed r's of Example 2 for different values of τ with $\mu = 0$

τ	0.1	0.3	0.5	0.7	0.9
[3]	$\sqrt{0.83}$	$\sqrt{1.28}$	$\sqrt{1.94}$	$\sqrt{2.90}$	$\sqrt{4.46}$
[8]	$\sqrt{0.74}$	$\sqrt{0.92}$	$\sqrt{1.36}$	$\sqrt{2.30}$	$\sqrt{3.51}$
[4]	$\sqrt{0.68}$	$\sqrt{0.80}$	$\sqrt{0.97}$	$\sqrt{1.64}$	$\sqrt{3.22}$
[6]	$\sqrt{0.66}$	$\sqrt{0.75}$	$\sqrt{0.94}$	$\sqrt{1.61}$	$\sqrt{3.14}$
[7]	$\sqrt{0.66}$	$\sqrt{0.75}$	$\sqrt{0.94}$	$\sqrt{1.61}$	$\sqrt{3.14}$
[1]	$\sqrt{0.66}$	$\sqrt{0.75}$	$\sqrt{0.94}$	$\sqrt{1.61}$	$\sqrt{3.14}$
Corollary 3.1	$\sqrt{0.41}$	$\sqrt{0.71}$	$\sqrt{0.92}$	$\sqrt{1.25}$	$\sqrt{2.74}$

Example 2 Consider the following uncertain time-delayed system with parameters:

$$\dot{z}(t) = \begin{bmatrix} -2 & 0 \\ 0 & -0.7 \end{bmatrix} z(t) + \begin{bmatrix} -1 & 0 \\ -1 & -0.9 \end{bmatrix} z\big(t - \tau(t)\big) + \begin{bmatrix} -0.5 \\ 1 \end{bmatrix} w(t), \tag{29}$$

and $w^T(t)w(t) \leq 1$.

By employing the method of Corollary 3.1 in this paper, r's for different values of τ with $\mu = 0$ are listed in Table 4. It is clear that the bounds obtained in this paper are better than the ones of [1, 3, 4, 6–8].

Example 3 Consider the following uncertain time-delayed system with parameters:

$$\begin{aligned} A + A_1 &= \begin{bmatrix} -2 & 0 \\ 0 & -0.7 \end{bmatrix}, \quad & A + A2 &= \begin{bmatrix} -2 & 0 \\ 0 & -1.1 \end{bmatrix}, \\ B + B_1 &= \begin{bmatrix} -1 & 0 \\ -1 & -0.9 \end{bmatrix}, \quad & B + B_2 &= \begin{bmatrix} -1 & 0 \\ -1 & -1.1 \end{bmatrix}, \\ E + E_1 &= \begin{bmatrix} -0.5 \\ 1 \end{bmatrix} = E + E_2, \quad & w^T(t)w(t) &\leq 1. \end{aligned} \tag{30}$$

By solving optimization problems (26), computed $\bar{\delta}$'s for the case $0 \leq \tau(t) \leq 0.75$ with different values of μ are listed in Table 5. It is clear that the proposed method in this paper yields tighter bounds than the one in [3, 6].

Table 5 Computed $\bar{\delta}$'s of Example 3 for the case $0 \le \tau(t) \le 0.75$, $\dot{\tau}(t) \le \mu$

μ	0	0.1	0.2	0.3	0.4	0.5	0.6	0.9	2
[6]	3.34	3.79	4.53	5.88	8.85	18.36	127.70	-	-
[3]	2.28	2.35	2.45	2.57	2.68	2.85	4.62	5.57	13.39
Corollary 3.2	2.03	2.33	2.34	2.45	2.60	2.72	4.37	4.92	8.73

5 Conclusions

In this paper, the problem of reachable set bounding for linear systems with both discrete and distributed delays has been investigated. By using Lyapunov-Krasovskii functional theory, delay decomposition technique, reciprocally convex method and free-weighting matrix approach, new reachable set bounds are obtained. Triple integrations are introduced for the first time in Lyapunov functionals to study reachable set bounding for linear systems with mixed delays. Meanwhile, delay decomposition technique is employed, which leads to tighter reachable set bounds. The results have been given to illustrate the advantages over the ones in [1–4, 6–8]. The foregoing results have great potential to be useful for further study in this area, such as linear neutral systems and nonlinear systems. Meanwhile, it is expected that the approach will be used for practical applications in the future, for instance, finite-time boundedness of state estimation for neural networks.

Competing interests

The authors declare that they have no competing interests.

Authors' contributions

All authors contributed equally to the writing of this paper. All authors read and approved the final manuscript.

Author details

[1]School of Mathematical Sciences, Huaibei Normal University, Huaibei, Anhui 235000, P.R. China. [2]School of Mathematical Sciences, University of Electronic Science and Technology of China, Chengdu, Sichuan 611731, P.R. China. [3]School of Electrical and Information Technology, Yunnan Minzu University, Kunming, Yunnan 650500, P.R. China. [4]School of Computer Science and Technology, Southwest University for Nationalities, Chengdu, Sichuan 610041, P.R. China.

Acknowledgements

This work was supported by the Anhui Provincial Natural Science Foundation under Grant No. 1408085MA14, the Anhui Provincial Natural Science Foundation under Grant No. 1508085MA14, the Key Natural Science Foundation of Universities of Anhui Province under Grant No. KJ2014A223, the Fundamental Research Funds for the Central Universities under Grant Nos. 12NZYQN17, 12NZYQN21 and 13NZYBS07, the State Ethnic Affairs Commission Project under Grant No. 12XNZ2003.

References

1. Zuo, Z, Ho, D, Wang, Y: Reachable set estimation for linear systems in the presence of both discrete and distributed delays. IET Control Theory Appl. **5**(15), 1808-1812 (2011)
2. Zuo, Z, Wang, Y: Results on reachable set estimation for linear systems with both discrete and distributed delays. IET Control Theory Appl. **6**(14), 2346-2350 (2012)
3. Kim, J: Improved ellipsoidal bound of reachable sets for time-delayed linear systems with disturbances. Automatica **44**(11), 2940-2943 (2008)
4. Kwon, O, Lee, S, Park, J: On reachable set bounding of uncertain dynamic systems with time-varying delays and disturbances. Inf. Sci. **181**(17), 3735-3748 (2011)
5. Zuo, Z, Chen, Y, Wang, Y, Ho, D, Chen, M, Li, H: A note on reachable set bounding for delayed systems with polytopic uncertainties. J. Franklin Inst. **350**(7), 1827-1835 (2013)
6. Zuo, Z, Ho, D, Wang, Y: Reachable set bounding for delayed systems with polytopic uncertainties: the maximal Lyapunov-Krasovskii functional approach. Automatica **46**(5), 949-952 (2010)
7. Shen, C, Zhong, S: The ellipsoidal bound of reachable sets for linear neutral systems with disturbances. J. Franklin Inst. **348**(9), 2570-2585 (2011)
8. Nam, P, Pathirana, P: Further result on reachable set bounding for linear uncertain polytopic systems with interval time-varying delays. Automatica **47**(8), 1838-1841 (2011)
9. Fridman, E, Shaked, U: On reachable sets for linear systems with delay and bounded peak inputs. Automatica **39**(11), 2005-2010 (2003)
10. Zuo, Z, Wang, Z, Chen, Y, Wang, Y: A non-ellipsoidal reachable set estimation for uncertain neural networks with time-varying delay. Commun. Nonlinear Sci. Numer. Simul. **19**(4), 1097-1106 (2014)

11. Zuo, Z, Chen, Y, Wang, Y: New criteria of reachable set estimation for time delay systems subject to polytopic uncertainties. In: The 7th IFAC Symposium on Robust Control Design, Aalborg, Denmark, June 2012, pp. 231-235 (2012)
12. Xiao, N, Jia, Y: New approaches on stability criteria for neural networks with two additive time-varying delay components. Neurocomputing **118**(10), 150-156 (2013)
13. Shi, L, Zhu, H, Zhong, S, Hou, L: Globally exponential stability for neural networks with time-varying delays. Appl. Math. Comput. **219**(21), 10487-10498 (2013)
14. Li, H, Yue, D: Synchronization stability of general complex dynamical networks with time-varying delays: a piecewise analysis method. J. Comput. Appl. Math. **232**(2), 149-158 (2009)
15. Kwon, O, Park, M, Lee, S, Park, J, Cha, E: Stability for neural networks with time-varying delays via some new approaches. IEEE Trans. Neural Netw. Learn. Syst. **24**(2), 181-193 (2013)
16. Xiao, J, Zeng, Z, Wu, A: New criteria for exponential stability of delayed recurrent neural networks. Neurocomputing **134**(6), 182-188 (2014)
17. Lakshmanan, S, Senthilkmar, T, Balasubramaniam, P: Improved results on robust stability of neutral systems with mixed time-varying delays and nonlinear perturbations. Appl. Math. Model. **35**(11), 5355-5368 (2011)
18. Tian, J, Zhong, S, Wang, Y: Improved exponential stability criteria for neural networks with time-varying delays. Neurocomputing **97**(11), 164-173 (2012)
19. Shao, J, Huang, T: A new result on global exponential robust stability of neural networks with time-varying delays. J. Control Theory Appl. **7**(3), 315-320 (2009)
20. Cheng, J, Xiong, L, Wang, B, Yang, J: Robust finite-time boundedness of H_∞ filtering for switched systems with time-varying delay. Optim. Control Appl. Methods (2015). doi:10.1002/oca.2165
21. Cheng, J, Xiong, L: Improved integral inequality approach on stabilization for continuous-time systems with time-varying input delay. Neurocomputing **160**, 274-280 (2015)
22. Wang, Y, Wang, C, Zuo, Z: On exponential stability analysis for neural networks with time-varying delays and general activation functions. Commun. Nonlinear Sci. Numer. Simul. **17**(3), 1447-1459 (2012)
23. Aeunkumar, A, Sakthivel, R, Mathiyalagan, K: Robust state estimation for discrete-time BAM neural networks with time-varying delay. Neurocomputing **131**(5), 171-178 (2014)
24. Zhang, F, Zhang, Y: State estimation of neural networks with both time-varying delays and norm-bounded parameter uncertainties via a delay decomposition approach. Commun. Nonlinear Sci. Numer. Simul. **18**, 3517-3529 (2013)
25. Huang, H, Huang, T, Chen, X: A mode-dependent approach to state estimation of recurrent neural networks with Markovian jumping parameters and mixed delays. Neural Netw. **46**, 50-61 (2013)
26. Lee, W, Lee, S, Par, P: Improved criteria on robust stability and h_∞ performance for linear systems with interval time-varying delays via new triple integral functionals. Appl. Math. Comput. **243**, 570-577 (2014)
27. Cheng, J, Zhu, H, Zhong, S, Zeng, Y: Finite-time h_∞ control for a class of Markovian jump systems with mode-dependent time-varying delays via new Lyapunov functionals. ISA Trans. **52**(6), 768-774 (2013)
28. Cheng, J, Zhong, S, Zhong, Q, Zhu, H, Du, Y: Finite-time boundedness of state estimation for neural networks with time-varying delays. Neurocomputing **129**, 257-264 (2014)
29. Wu, H, Liao, X, Feng, W, Guo, S, Zhang, W: Robust stability analysis of uncertain systems with two additive time-varying delay components. Appl. Math. Model. **33**, 4345-4353 (2009)
30. Sun, J, Liu, J, Chen, J: Improved delay-range-dependent stability criteria for linear systems with time-varying delays. Automatica **46**, 466-470 (2010)
31. Boyd, S, Ghaoui, L, Feron, E, Balakrishnan, V: Linear Matrix Inequalities in Systems and Control Theory. SIAM, Philadelphia (1994)

8

Dissipativity of the backward Euler method for nonlinear Volterra functional differential equations in Banach space

Siqing Gan*

*Correspondence:
sqgan@csu.edu.cn
School of Mathematics and Statistics, Central South University, Changsha, Hunan 410083, China

Abstract

This paper concerns the dissipativity of nonlinear Volterra functional differential equations (VFDEs) in Banach space and their numerical discretization. We derive sufficient conditions for the dissipativity of nonlinear VFDEs. The general results provide a unified theoretical treatment for dissipativity analysis to ordinary differential equations (ODEs), delay differential equations (DDEs), integro-differential equations (IDEs) and VFDEs of other type appearing in practice. Then the dissipativity property of the backward Euler method for VFDEs is investigated. It is shown that the method can inherit the dissipativity of the underlying system. The close relationship between the absorbing set of the numerically discrete system generated by the backward Euler method and that of the underlying system is revealed.

Keywords: dissipativity; Volterra functional differential equation; Banach space; backward Euler method

1 Introduction

Many dynamical systems are characterized by the property of possessing a bounded absorbing set where all trajectories enter in a finite time and thereafter remain inside. Such systems are called dissipative. Dissipativity means that the eventual time evolution of solutions is confined to a bounded absorbing set. In the study of numerical methods, it is natural to ask whether those discrete systems preserve the dissipativity of the continuous system.

Since the 1990s considerable process has been made in dissipativity analysis of numerical methods. The papers [1–5] focus on the numerical methods for ordinary differential equations. For the delay differential equations (DDEs) with constant delay, sufficient conditions for the dissipativity of analytical and numerical solutions are presented in [6–8]. Since that, the analysis is extended to DDEs with variable lags [9, 10] and Volterra functional differential equations [11–16].

The dissipativity analysis of numerical methods for VFDEs in the literature was limited in Euclidean spaces or Hilbert spaces. The aim of this paper is to investigate the dissipativity of nonlinear VFDEs in Banach space and their numerical discretization.

The main contributions of this paper could be summarized as follows.

(a) Sufficient conditions for the dissipativity of nonlinear VFDEs in Banach space are derived. The general results provide a unified theoretical treatment for dissipativity analysis to ordinary differential equations (ODEs), delay differential equations (DDEs), integro-differential equations (IDEs) and VFDEs of other type appearing in practice. In particular, the theory covers the existing dissipativity results of DDEs with a wide variety of delay arguments such as constant delays, bounded and unbounded vary delays, discrete and distributed delays and so on.

(b) It is proved that the backward Euler method can inherit the dissipativity of the underlying system. Theorem 4.1 and Theorem 4.2 show the close relationship between the absorbing set of VFDEs and that of the numerically discrete system generated by the backward Euler method. It implies that the radius of the absorbing set of the discrete system approaches to that of the underlying system as the stepsize approaches to zero. On the contrary, most of the existing dissipativity results of numerical methods for VFDEs are independent of the size of the absorbing set of the underlying system.

This paper is organized as follows. In Section 2, some basic concepts for nonlinear VFDEs in Banach space are presented. In Section 3, some sufficient conditions for the dissipativity of nonlinear VFDEs are given. In Section 4, it is shown that the backward Euler method can inherit the dissipativity of the underlying system.

2 Some concepts

Let X be a real or complex Banach space with the norm $\|\cdot\|$. For any given closed interval $I \subset \mathbb{R}$, let the symbol $C_X(I)$ denote a Banach space consisting of all continuous mappings $x : I \to X$, on which the norm is defined by $\|x\|_\infty = \max_{t\in I} \|x(t)\|$.

Consider the following initial value problem (IVP) [17]:

$$\begin{cases} y'(t) = f(t, y(t), y), & t \geq a, \\ y(t) = \varphi(t), & a - \tau \leq t \leq a, \end{cases} \tag{2.1}$$

where a, τ are constants, $0 \leq \tau \leq +\infty$, $\varphi \in C_X[a - \tau, a]$ is a given initial function, $f : [a, +\infty) \times X \times C_X[a - \tau, +\infty) \to X$ is a given continuous mapping satisfying the conditions

$$\begin{aligned} &\big(1 - \lambda\alpha(t)\big)D_f(0, t, u, \psi) \\ &\quad \leq D_f(\lambda, t, u, \psi) + \lambda\Big(\gamma(t) + \beta(t) \max_{t-\mu_2(t)\leq\xi\leq t-\mu_1(t)} \big\|\psi(\xi)\big\|^2\Big), \\ &\quad \forall\lambda \geq 0, t \in [a, +\infty), u \in X, \psi \in C_X[a - \tau, +\infty), \end{aligned} \tag{2.2}$$

where

$$D_f(\lambda, t, u, \psi) = \big\|u - \lambda f(t, u, \psi)\big\|^2. \tag{2.3}$$

Here $\alpha(t)$, $\beta(t)$, $\gamma(t)$ are continuous functions, $\mu_1(t)$ and $\mu_2(t)$ satisfy

$$0 \leq \mu_1(t) \leq \mu_2(t) \leq t - a + \tau, \quad \forall t \in [a, +\infty). \tag{2.4}$$

Define

$$\mu_1^{(0)} = \inf_{a \le t < \infty} \mu_1(t) \ge 0, \qquad \mu_2^{(0)}(\xi_1, \xi_2) = \inf_{\xi_1 \le t \le \xi_2} \big(t - \mu_2(t)\big) \ge a - \tau,$$
$$\forall \xi_1, \xi_2 : a \le \xi_1 \le \xi_2 < +\infty. \tag{2.5}$$

We always assume that problem (2.1) has a unique solution on the interval $[a - \tau, +\infty)$. Condition (2.2) implies that the mapping $f(t, \psi(t), \psi)$ is independent of the values of the function $\psi(\xi)$ with $t < \xi \le b$, *i.e.*, $f(t, \psi(t), \psi)$ is a Volterra functional.

For simplicity, we use the symbol $\mathcal{A}(\alpha, \beta, \gamma, \mu_1, \mu_2)$ to denote the problem class consisting of all problems (2.1) satisfying condition (2.2).

For the special case where X is a Hilbert space with the inner product $\langle \cdot, \cdot \rangle$ and the corresponding norm $\| \cdot \|$, condition (2.2) is equivalent to

$$2\,\mathrm{Re}\big\langle u, f(t, u, \psi)\big\rangle \le \gamma(t) + \alpha(t)\|u\|^2 + \beta(t) \max_{t-\mu_2(t) \le \xi \le t-\mu_1(t)} \|\psi(\xi)\|^2,$$
$$\forall t \in [a, +\infty), u \in X, \psi \in C_X[a - \tau, +\infty). \tag{2.6}$$

The dissipativity analysis of (2.6) can be found in [14, 16].

3 Dissipativity of nonlinear Volterra functional differential equations

Definition 3.1 [6] The evolutionary equation (2.1) is said to be dissipative in X if there is a bounded set $\mathcal{B} \subset X$ such that for all bounded sets $\Psi \subset X$ there is a time $t_0 = t_0(\Psi)$ such that for all initial functions $\varphi(t)$ contained in Ψ, the corresponding solution $y(t)$ is contained in $\mathcal{B}$ for all $t \ge t_0$. $\mathcal{B}$ is called an absorbing set in X.

Lemma 3.2 *Equation* (2.2) *implies that* $\gamma(t) \ge 0$ *and* $\beta(t) \ge 0$.

Proof Setting $u \equiv 0$ in (2.2) we obtain

$$0 \le \lambda \|f(t, 0, \psi)\|^2 + \gamma(t) + \beta(t) \max_{t-\mu_2(t) \le \xi \le t-\mu_1(t)} \|\psi(\xi)\|^2,$$
$$\forall \lambda > 0, t \in [a, +\infty), \psi \in C_X[a - \tau, +\infty),$$

which yields

$$0 \le \gamma(t) + \beta(t) \max_{t-\mu_2(t) \le \xi \le t-\mu_1(t)} \|\psi(\xi)\|^2, \quad \forall t \in [a, +\infty), \psi \in C_X[a - \tau, +\infty) \tag{3.1}$$

as $\lambda \to 0$. Let $\psi(t) \equiv 0$. It follows from (3.1) that $\gamma(t) \ge 0$. If there is $\tilde{t} \ge a$ such that $\beta(\tilde{t}) < 0$, it is easy to find a function $\psi \in C_X[a - \tau, +\infty)$ which satisfies

$$\gamma(\tilde{t}) + \beta(\tilde{t}) \max_{\tilde{t}-\mu_2(\tilde{t}) \le \xi \le \tilde{t}-\mu_1(\tilde{t})} \|\psi(\xi)\|^2 < 0,$$

which contradicts (3.1). Therefore, $\beta(t) \ge 0$.

For a continuous real-valued function $y(t)$ of a real variable, the Dini derivatives $D^+y(t)$ and $D_-y(t)$ are defined as

$$D^+y(t) = \limsup_{\delta \searrow 0} \frac{y(t+\delta) - y(t)}{\delta} \quad \text{and} \quad D_-y(t) = \liminf_{\delta \nearrow 0} \frac{y(t+\delta) - y(t)}{\delta}. \qquad \square$$

Lemma 3.3 *If* $u(t) \geq 0$, $t \in (-\infty, +\infty)$, *and*

$$u'(t) \leq \gamma(t) + \alpha(t)u(t) + \beta(t) \sup_{t-\tau(t)\leq\xi\leq t} u(\xi), \quad t \geq a, \tag{3.2}$$

$$u(t) = |\psi(t)|, \quad t \leq a,$$

where $\psi(t)$ *is bounded and continuous for* $t \leq a$, *continuous functions* $\gamma(t) \geq 0$, $\beta(t) \geq 0$ *and* $\alpha(t) < 0$ *for* $t \in [a, +\infty)$, $\tau(t) \geq 0$ *and*

$$\lim_{t\to+\infty} \left(t - \tau(t)\right) \to +\infty, \qquad \sup_{a\leq t<+\infty} \alpha(t) < 0, \qquad \sup_{a\leq t<+\infty} \frac{\beta(t)}{|\alpha(t)|} < 1. \tag{3.3}$$

Then, for any given $\epsilon > 0$, *there exists* $\hat{t} = \hat{t}(G, \epsilon) > a$ *such that*

$$u(t) \leq \frac{\gamma^*}{\sigma} + \epsilon, \quad t \geq \hat{t}, \tag{3.4}$$

where

$$\gamma^* = \sup_{a\leq t<+\infty} \gamma(t), \qquad G = \sup_{-\infty<\xi\leq a} |\psi(\xi)|.$$

Proof The last two inequalities of (3.3) imply that (2.10) and (2.12) of [16] hold. The conclusion follows from Theorem 2.3 of [16] directly. □

Theorem 3.4 *Suppose problem* (2.1) $\in \mathcal{A}(\alpha, \beta, \gamma, \mu_1, \mu_2)$ *and that*

$$\lim_{t\to+\infty} \left(t - \mu_2(t)\right) \to +\infty, \qquad \sup_{a\leq t<+\infty} \alpha(t) < 0, \qquad \sup_{a\leq t<+\infty} \frac{\beta(t)}{|\alpha(t)|} < 1. \tag{3.5}$$

Then, for any given $\epsilon > 0$, *there exists* $\check{t} = \check{t}(\bar{\varphi}, \epsilon)$ *such that*

$$\|y(t)\|^2 \leq \frac{\gamma^*}{\sigma} + \epsilon, \quad t > \check{t}, \tag{3.6}$$

where $\gamma^* = \sup_{t\geq a} \gamma(t)$, $\bar{\varphi} = \sup_{t\leq a} \|\varphi(t)\|^2$. *Hence the system is dissipative with an absorbing set* $B = B(0, \sqrt{\gamma^*/\sigma + \epsilon})$.

Proof By the definition of Dini derivative, we have

$$\begin{aligned} D_-\left(\|y(t)\|^2\right) &= \liminf_{\delta\to-0} \frac{\|y(t+\delta)\|^2 - \|y(t)\|^2}{\delta} \\ &= \liminf_{\delta\to-0} \frac{\|y(t+\delta)\| - \|y(t)\|}{\delta} \left(\|y(t+\delta)\| + \|y(t)\|\right), \\ D^+\left(\|y(t)\|^2\right) &= \liminf_{\delta\to+0} \frac{\|y(t+\delta)\|^2 - \|y(t)\|^2}{\delta} \\ &= \liminf_{\delta\to+0} \frac{\|y(t+\delta)\| - \|y(t)\|}{\delta} \left(\|y(t+\delta)\| + \|y(t)\|\right). \end{aligned} \tag{3.7}$$

Applying Lemma 4.6.2 of [18], we see that the limits

$$\lim_{\delta \to -0} \frac{\|y(t) + \delta y'(t)\| - \|y(t)\|}{\delta} \quad \text{and} \quad \lim_{\delta \to +0} \frac{\|y(t) + \delta y'(t)\| - \|y(t)\|}{\delta}$$

exist. Lemma 4.6.3 of [18] tells us that

$$\begin{aligned} \liminf_{\delta \to -0} \frac{\|y(t+\delta)\| - \|y(t)\|}{\delta} &= \lim_{\delta \to -0} \frac{\|y(t) + \delta y'(t)\| - \|y(t)\|}{\delta}, \\ \liminf_{\delta \to +0} \frac{\|y(t+\delta)\| - \|y(t)\|}{\delta} &= \lim_{\delta \to +0} \frac{\|y(t) + \delta y'(t)\| - \|y(t)\|}{\delta}. \end{aligned} \tag{3.8}$$

Then (3.7) and (3.8) together imply that $D_-(\|y(t)\|^2) = D^+(\|y(t)\|^2)$. Let $u(t) = \|y(t)\|^2$. Therefore, $u'(t)$ exists, and

$$\begin{aligned} u'(t) &= \lim_{\delta \to -0} \frac{\|y(t+\delta)\|^2 - \|y(t)\|^2}{\delta} \\ &= \lim_{\delta \to -0} \frac{\|y(t) + \delta y'(t) + o(\delta)\|^2 - \|y(t) + \delta y'(t)\|^2 + \|y(t) + \delta y'(t)\|^2 - \|y(t)\|^2}{\delta} \\ &= \lim_{\delta \to -0} \frac{\|y(t) + \delta y'(t)\|^2 - \|y(t)\|^2}{\delta} \\ &= \lim_{\delta \to -0} \frac{\|y(t) + \delta f(t, y(t), y)\|^2 - \|y(t)\|^2}{\delta}. \end{aligned}$$

Using condition (2.2), we have

$$\begin{aligned} u'(t) &\leq \lim_{\delta \to -0} \frac{(1 + \delta\alpha(t))\|y(t)\|^2 + \delta(\gamma(t) + \beta(t)\max_{t-\mu_2(t) \leq \xi \leq t-\mu_1(t)} \|y(\xi)\|^2) - \|y(t)\|^2}{\delta} \\ &= \gamma(t) + \alpha(t)\|y(t)\|^2 + \beta(t) \max_{t-\mu_2(t) \leq \xi \leq t-\mu_1(t)} \|y(\xi)\|^2, \end{aligned}$$

that is,

$$u'(t) \leq \gamma(t) + \alpha(t)u(t) + \beta(t) \max_{t-\mu_2(t) \leq \xi \leq t-\mu_1(t)} u(\xi). \tag{3.9}$$

The desired result follows from Theorem 3.3. The proof is complete. □

Remark 3.5 Specializing Theorem 3.4 to Hilbert spaces, we can obtain the corresponding result which is in accordance with that obtained in [16].

Remark 3.6 Specializing Theorem 3.4 to DDEs in Hilbert spaces with constant delays and $\alpha(t) \equiv \alpha$, $\beta(t) \equiv \beta$, $\gamma(t) \equiv \gamma$, we can obtain the corresponding result which is in accordance with that obtained in [6].

Remark 3.7 Specializing Theorem 3.4 to ODEs in Euclidean spaces with $\alpha(t) \equiv \alpha$, $\gamma(t) \equiv \gamma$, we can obtain the corresponding result which is in accordance with that obtained in [4].

Remark 3.8 Theorem 3.4 covers most of the existing dissipativity results of DDEs with a wide variety of delay arguments such as constant delays [6], bounded varying delays

[10] and unbounded varying delays [9], discrete and distributed delays [11, 12] and so on. In brief, Theorem 3.4 provides a unified theoretical treatment for dissipativity analysis to ordinary differential equations (ODEs), delay differential equations (DDEs), integro-differential equations (IDEs) and VFDEs of other type appearing in practice.

4 Dissipativity of the backward Euler method

For simplicity, from now on we assume that

$$\gamma(t) \equiv \gamma, \qquad \alpha(t) \equiv \alpha, \qquad \beta(t) \equiv \beta, \quad t \in [a, +\infty).$$

Theorem 3.4 can be rewritten as follows.

Theorem 4.1 *Suppose problem* (2.1) $\in \mathcal{A}(\alpha, \beta, \gamma, \mu_1, \mu_2)$ *and that*

$$\lim_{t\to+\infty} \big(t - \mu_2(t)\big) \to +\infty, \quad \alpha + \beta < 0.$$

Then, for any given $\epsilon > 0$, *there exist* $\check{t} = \check{t}(\bar{\varphi}, \epsilon)$, $\bar{\varphi} = \sup_{t\le a} \|\varphi(t)\|^2$ *such that*

$$\|y(t)\| \le \sqrt{\frac{\gamma}{-(\alpha+\beta)} + \epsilon}, \quad t > \check{t}.$$

The backward Euler method applied to (2.1) gives

$$\begin{cases} y_{n+1} = y_n + hf(t_{n+1}, y_{n+1}, y^h(\cdot)), & n = 0, 1, 2, \ldots, \\ y^h(t) = \pi^h(t, \varphi, y_1, y_2, \ldots, y_{n+1}), & a - \tau \le t \le t_{n+1}, \end{cases} \tag{4.1}$$

where π^h is an appropriate interpolation operator which approximates to the exact solution $y(t)$ on the interval $[a - \tau, b]$, $h > 0$ is the stepsize, y_n is an approximation to the exact solution $y(t_n)$ with $t_n = a + nh$.

Noting that the backward Euler method for ODEs is of order one, we can use the following piecewise linear interpolation:

$$y^h(t) = \begin{cases} \frac{1}{h}[(t_{i+1} - t)y_i + (t - t_i)y_{i+1}], & t_i \le t \le t_{i+1}, \\ \varphi(t), & a - \tau \le t \le a. \end{cases} \tag{4.2}$$

Theorem 4.2 *Assume that problem* (2.1) $\in \mathcal{A}(\alpha, \beta, \gamma, \mu_1, \mu_2)$ *and that*

$$\lim_{t\to+\infty} \big(t - \mu_2(t)\big) \to +\infty, \quad \alpha + \beta < 0.$$

Let $\{y_n\}$ *be the sequence of numerical solutions obtained by* (4.1)-(4.2). *Then, for any given* $\epsilon > 0$, *there exists* $n_0 = n_0(\bar{\varphi}, \epsilon)$ *such that*

$$\|y_n\| \le \sqrt{\frac{\gamma}{-(\alpha+\beta)} \cdot \frac{1 - h\alpha}{1 - h(\alpha+\beta)} + \epsilon}, \quad n \ge n_0.$$

Proof It follows from (4.1) that

$$\left\|y_{n+1} - hf\big(t_{n+1}, y_{n+1}, y^h(\cdot)\big)\right\|^2 = \|y_n\|^2. \tag{4.3}$$

Using (2.2), we have

$$\|y_{n+1} - hf(t_{n+1}, y_{n+1}, y^h(\cdot))\|^2$$
$$\geq (1 - h\alpha)\|y_{n+1}\|^2 - h\Big(\gamma + \beta \max_{t_{n+1}-\mu_2(t_{n+1})\leq\xi\leq t_{n+1}-\mu_1(t_{n+1})} \|y^h(\xi)\|^2\Big). \tag{4.4}$$

It follows from (4.3) and (4.4) that

$$(1 - h\alpha)\|y_{n+1}\|^2 \leq \|y_n\|^2 + h\Big(\gamma + \beta \max_{t_{n+1}-\mu_2(t_{n+1})\leq\xi\leq t_{n+1}-\mu_1(t_{n+1})} \|y^h(\xi)\|^2\Big). \tag{4.5}$$

In view of (4.2), we have

$$\max_{t_{n+1}-\mu_2(t_{n+1})\leq\xi\leq t_{n+1}-\mu_1(t_{n+1})} \|y^h(\xi)\|^2 \leq \max\Big\{\max_{1\leq i\leq n+1} \|y_i\|^2, \max_{a-\tau\leq t\leq a} \|\varphi(t)\|^2\Big\}, \tag{4.6}$$

where we used the following inequality:

$$\|(1-\delta)y_i + \delta y_{i+1}\|^2 \leq (1-\delta)\|y_i\|^2 + \delta\|y_{i+1}\|^2 \leq \max\{\|y_i\|^2, \|y_{i+1}\|^2\}.$$

A combination of (4.5) and (4.6) leads to

$$(1 - h\alpha)\|y_{n+1}\|^2 \leq \|y_n\|^2 + h\Big(\gamma + \beta \max\Big\{\max_{1\leq i\leq n+1} \|y_i\|^2, \max_{a-\tau\leq t\leq a} \|\varphi(t)\|^2\Big\}\Big). \tag{4.7}$$

For simplicity, for any given nonnegative integer n, we write

$$Q_n = \max\Big\{\max_{1\leq i\leq n} \|y_i\|^2, \max_{a-\tau\leq t\leq a} \|\varphi(t)\|^2\Big\}, \quad n \geq 1,$$
$$Q_0 = \max_{a-\tau\leq t\leq a} \|\varphi(t)\|^2.$$

We now consider two cases:

(a) $\max\Big\{\max_{1\leq i\leq n+1} \|y_i\|^2, \max_{a-\tau\leq t\leq a} \|\varphi(t)\|^2\Big\} = \|y_{n+1}\|^2,$

(b) $\max\Big\{\max_{1\leq i\leq n+1} \|y_i\|^2, \max_{a-\tau\leq t\leq a} \|\varphi(t)\|^2\Big\} \neq \|y_{n+1}\|^2.$

In the case of (a), it follows from (4.7) that

$$\|y_{n+1}\|^2 \leq \frac{h\gamma}{1 - h(\alpha+\beta)} + \frac{1}{1 - h(\alpha+\beta)}\|y_n\|^2. \tag{4.8}$$

In the case of (b), it follows from (4.7) that

$$\|y_{n+1}\|^2 \leq \frac{h\gamma}{1 - h\alpha} + \frac{1}{1 - h\alpha}(\|y_n\|^2 + h\beta Q_n). \tag{4.9}$$

To summarize both of the two cases, we have shown that

$$\|y_{n+1}\|^2 \leq \max\Big\{\frac{h\gamma}{1 - h(\alpha+\beta)}, \frac{h\gamma}{1 - h\alpha}\Big\} + \max\Big\{\frac{1}{1 - h(\alpha+\beta)}, \frac{1 + h\beta}{1 - h\alpha}\Big\}Q_n,$$

which yields

$$\|y_{n+1}\|^2 \le \frac{h\gamma}{1-h(\alpha+\beta)} + \frac{1+h\beta}{1-h\alpha} Q_n =: d_0 + d_1 Q_n, \tag{4.10}$$

where

$$d_0 = \frac{h\gamma}{1-h(\alpha+\beta)}, \qquad d_1 = \frac{1+h\beta}{1-h\alpha} < 1.$$

Considering $Q_n = \|y_n\|^2$ or $Q_n \neq \|y_n\|^2$ and inserting (4.10) repeatedly, we obtain

$$\begin{aligned} \|y_n\|^2 &\le d_0\left(1 + d_1 + \cdots + d_1^{n-1}\right) + d_1^n Q_0 \\ &\le \frac{d_0}{1-d_1} + d_1^n Q_0 = \frac{\gamma}{-(\alpha+\beta)} \frac{1-h\alpha}{1-h(\alpha+\beta)} + d_1^n Q_0. \end{aligned} \tag{4.11}$$

Therefore, for any given $\epsilon > 0$, there exists $n_0 = n_0(\bar{\varphi}, \epsilon)$ such that

$$\|y_n\| \le \sqrt{\frac{\gamma}{-(\alpha+\beta)} \cdot \frac{1-h\alpha}{1-h(\alpha+\beta)} + \epsilon}, \quad n \ge n_0.$$

This completes the proof of the theorem. □

Remark 4.3 Theorem 4.1 and Theorem 4.2 show the close relationship between the absorbing set of the underlying system and that of the numerically discrete system generated by the backward Euler method. On the contrary, most of the existing dissipativity results of numerical methods for VFDEs are independent of the size of the absorbing set of the underlying system. It is obvious that the radius of the absorbing set of the discrete system is longer than that of the underlying system because of $\frac{1-h\alpha}{1-h(\alpha+\beta)} \ge 1$. Furthermore, Theorem 4.2 implies the following facts.

- In the case of ODEs, that is $\beta = 0$, for any given $\epsilon > 0$, the ball $B(0, \sqrt{\frac{\gamma}{-\alpha} + \epsilon})$ is an absorbing set of the discrete system as well as the underlying system. The absorbing set is independent of the stepsize of the backward Euler method.
- For fixed $\beta > 0$, $h > 0$, if $|\alpha|$ is sufficiently large, then the difference between the radius of the absorbing set of the numerically discrete system and that of the underlying system is sufficiently small.
- Notice that $\frac{1-h\alpha}{1-h(\alpha+\beta)} \to 1$ as $h \to 0$, hence given any $\epsilon > 0$, there exists $h_0 = h_0(\epsilon)$ such that for $h < h_0$ the ball $B(0, \sqrt{\frac{\gamma}{-(\alpha+\beta)} + \epsilon})$ is an absorbing set of the discrete system. In other words, the radius of the absorbing set of the discrete system approaches to that of the underlying system as the stepsize approaches to zero.

Competing interests

The author declares that he has no competing interests.

Acknowledgements

This work is supported by the National Natural Science Foundation of China (No. 11171352).

References

1. Hill, AT: Global dissipativity for A-stable methods. SIAM J. Numer. Anal. **34**, 119-142 (1997)
2. Hill, AT: Dissipativity of Runge-Kutta methods in Hilbert spaces. BIT Numer. Math. **37**, 37-42 (1997)
3. Humphries, AR, Stuart, AM: Model problems in numerical stability theory for initial value problems. SIAM Rev. **36**, 226-257 (1994)
4. Humphries, AR, Stuart, AM: Runge-Kutta methods for dissipative and gradient dynamical systems. SIAM J. Numer. Anal. **31**, 1452-1485 (1994)
5. Xiao, A: Dissipativity of general linear methods for dissipative dynamical systems in Hilbert spaces. Math. Numer. Sin. **22**, 429-436 (2000) (in Chinese)
6. Huang, C: Dissipativity of Runge-Kutta methods for dissipative systems with delays. IMA J. Numer. Anal. **20**, 153-166 (2000)
7. Huang, C: Dissipativity of one-leg methods for dissipative systems with delays. Appl. Numer. Math. **35**, 11-22 (2000)
8. Huang, C, Chang, Q: Dissipativity of multistep Runge-Kutta methods for dynamical systems with delays. Math. Comput. Model. **40**(11-12), 1285-1296 (2004)
9. Gan, S: Exact and discretized dissipativity of the pantograph equation. J. Comput. Math. **25**(1), 81-88 (2007)
10. Tian, H: Numerical and analytic dissipativity of the θ-method for delay differential equations with a bounded lag. Int. J. Bifurc. Chaos **14**, 1839-1845 (2004)
11. Gan, S: Dissipativity of linear θ-methods for integro-differential equations. Comput. Math. Appl. **52**, 449-458 (2006)
12. Gan, S: Dissipativity of θ-methods for nonlinear Volterra delay-integro-differential equations. J. Comput. Appl. Math. **206**, 898-907 (2007)
13. Tian, H, Guo, N: Asymptotic stability, contractivity and dissipativity of one-leg θ-method for non-autonomous delay functional differential equations. Appl. Math. Comput. **203**, 333-342 (2008)
14. Wen, L: Numerical stability analysis for nonlinear Volterra functional differential equations in abstract spaces. Ph.D. thesis, Xiangtan University (2005) (in Chinese)
15. Wen, L, Li, S: Dissipativity of Volterra functional differential equations. J. Math. Anal. Appl. **324**(1), 696-706 (2006)
16. Wen, L, Yu, Y, Wang, W: Generalized Halanay inequalities for dissipativity of Volterra functional differential equations. J. Math. Anal. Appl. **347**, 169-178 (2008)
17. Li, S: Stability analysis of solutions to nonlinear stiff Volterra functional differential equations in Banach space. Sci. China Ser. A **48**, 372-387 (2005)
18. Li, S: Theory of Computational Methods for Stiff Differential Equations. Hunan Science and Technology Publisher, Changsha (1997) (in Chinese)

Oscillation criteria for a class of higher odd order neutral difference equations with continuous variable

Shuhui Wu[1*], Pargat Singh Calay[2] and Zhanyuan Hou[2]

*Correspondence: s.wu@zust.edu.cn
[1]School of Sciences, Zhejiang University of Science and Technology, 318 Liuhe Road, Hangzhou, 310023, China
Full list of author information is available at the end of the article

Abstract

In this paper, we are mainly concerned with oscillatory behavior of solutions for a class of higher odd order nonlinear neutral difference equations with continuous variable. By converting the above difference equations to the corresponding differential equations and inequalities, the oscillatory criteria are obtained. In addition, examples are given to illustrate the obtained criteria, respectively.

Keywords: neutral difference equations; oscillation; continuous variable

1 Introduction

Difference equations have attracted a great deal of attention of researchers in mathematics, biology, physics, and economy. This is specially due to the applications in various problems of biology, physics, economy. Among the topics studied for oscillation of the solutions has been investigated intensively. Please see [1–18].

In this paper, we deal with the nonlinear neutral difference equation with continuous variable of the form

$$\Delta_\tau^m\big(x(t) - px(t-r)\big) + f\big(t, x(g(t))\big) = 0, \tag{1.1}$$

where $m \geq 3$, $p \geq 0$, τ and r are positive constants, $\Delta_\tau x(t) = x(t+\tau) - x(t)$, $0 < g(t) < t$, $g \in C^1([t_0,\infty), R_+)$, $g'(t) > 0$, and $f \in C([t_0,\infty) \times R, R)$. Throughout this paper we assume that

$$g(t+\tau) \geq g(t) + \tau \quad \text{for } t \geq t_0 \tag{1.2}$$

and

$$f(t,u)/u \geq q(t) > 0 \quad \text{for } u \neq 0 \text{ and some } q \in C(R, R_+). \tag{1.3}$$

Let $t_0' = \min\{g(t_0), t_0 - r\}$ and $I_0 = [t_0', t_0]$. A function x is called the *solution* of (1.1) with $x(t) = \varphi(t)$ for $t \in I_0$ and $\varphi \in C(I_0, R)$ if it satisfies (1.1) for $t \geq t_0$.

A solution x is said to be *oscillatory* if it is neither eventually positive nor eventually negative; it is called *nonoscillatory* if it is not oscillatory.

The organization of this paper is as follows. We will give the main results in Section 2 and leave the proofs to Section 5. Three demonstrated examples will be presented in Section 3. In Section 4, some lemmas will be given to prove the main results.

2 Statement of the main results

For later convenience, let

$$\bar{q}(t) = \alpha \min_{t\le s\le t+m\tau} \{q(s)\} \Big(\min_{g(t)\le s\le g(t)+m\tau} \{ (g^{-1}(s))' \} \Big)^{m}, \tag{2.1}$$

where $0 < \alpha < 1$. Throughout this paper, the function $\bar{q}$ will play an important role in the oscillatory criteria for (1.1). Let

$$\beta_1 = \inf_{t\ge T} \left\{ \frac{(g^{-1}(t) - t)^{m-1} \bar{q}(g^{-1}(t))}{(m-1)!\tau^m} \right\} \tag{2.2}$$

and

$$\beta_2 = \inf_{t\ge T} \left\{ \frac{(g^{-1}(t) - t)^{m-1} \bar{q}(t)}{(m-1)!\tau^m} \right\}, \tag{2.3}$$

where $T \ge t_0$ is sufficiently large.

Theorem 2.1 *Assume that* (1.1) *with* $0 < p < 1$ *satisfies*

$$r\beta_1 \sum_{i=1}^{n} ip^i \ge 1 \tag{2.4}$$

and

$$0 \le \liminf_{t\to\infty} \int_{t-r}^{t} (g^{-1}(s) - s)^{m-1} \bar{q}(g^{-1}(s))\, ds \le \frac{(m-1)!\tau^m (1-p) e^{-1}}{p - p^{n+1}} \tag{2.5}$$

for some integer $n \ge 1$. *Also assume that* $\bar{q}(t)$ *given by* (2.1) *is nonincreasing. Then, for every bounded solution* $x(t)$ *of* (1.1), *either* $x(t)$ *is oscillatory or* $\liminf_{t\to\infty}(|x(t)| - p|x(t-r)|) < 0$.

Corrollary 2.2 *The conclusion of Theorem* 2.1 *still holds if* (2.5) *is replaced by*

$$0 \le \liminf_{t\to\infty} \int_{t-r}^{t} (g^{-1}(s) - s)^{m-1} \bar{q}(g^{-1}(s))\, ds \le \frac{(m-1)!\tau^m}{ep}. \tag{2.6}$$

Corrollary 2.3 *Assume that* (1.1) *with* $0 < p < 1$ *satisfies*

$$r\beta_2 \sum_{i=1}^{n} ip^i \ge 1 \tag{2.7}$$

and

$$0 \le \liminf_{t\to\infty} \int_{t-r}^{t} (g^{-1}(s) - s)^{m-1} \bar{q}(s)\, ds \le \frac{(m-1)!\tau^m (1-p) e^{-1}}{p - p^{n+1}} \tag{2.8}$$

for some integer $n \geq 1$. *Also assume that* $\bar{q}(t)$ *given by* (2.1) *is nondecreasing. Then the conclusion of Theorem* 2.1 *holds.*

Corrollary 2.4 *The conclusion of Corollary* 2.3 *still holds if* (2.8) *is replaced by*

$$0 \leq \liminf_{t\to\infty} \int_{t-r}^{t} \left(g^{-1}(s) - s\right)^{m-1} \bar{q}(s)\, ds \leq \frac{(m-1)!\tau^m}{ep}. \tag{2.9}$$

Corrollary 2.5 *Assume* $0 < p < 1$ *and* $r = k\tau$. *Under the assumptions of either Theorem* 2.1 *or Corollary* 2.2 *or Corollary* 2.3 *or Corollary* 2.4, *every bounded solution* $x(t)$ *of* (1.1) *is oscillatory.*

The following results are for the bounded solutions of (1.1) with $p > 1$.

Theorem 2.6 *Assume that* $p > 1$, $r = k\tau$, $k \in N$, $r \geq t + m\tau - g(t)$, *and*

$$r\beta_2 \sum_{i=1}^{n} \frac{(i-1)}{p^i} \geq 1 \tag{2.10}$$

for some integer $n \geq 2$. *Also assume that* $\bar{q}(t)$ *given by* (2.1) *is nondecreasing. Then every bounded solution* $x(t)$ *of* (1.1) *is oscillatory.*

Corollary 2.7 *Assume that* $p > 1$, $r = k\tau$, $k \in N$, $r \geq t + m\tau - g(t)$, *and*

$$r\beta_1 \sum_{i=1}^{n} \frac{(i-1)}{p^i} \geq 1 \tag{2.11}$$

for some integer $n \geq 2$. *Also assume that* $\bar{q}(t)$ *given by* (2.1) *is nonincreasing. Then every bounded solution* $x(t)$ *of* (1.1) *is oscillatory.*

3 Examples

Three examples will be given in this section to demonstrate the applications of the obtained results. From (2.2) and (2.3) it is clear that both β_1 and β_2 are nondecreasing functions of T. The following examples show that β_1 and β_2 may be independent of T or increasing functions of T.

Example 1 Consider the difference equation

$$\Delta_1^m\left(x(t) - \frac{1}{2}x(t-1)\right) + \left((m-1)! + \frac{1}{t}\right)x(t-1) = 0 \tag{3.1}$$

for $t > 0$, where m is an odd positive integer $m \geq 3$. Viewing (3.1) as (1.1), we have $\tau = 1$, $0 < p = 1/2 < 1$, $r = 1$, $q(t) = (m-1)! + 1/t$ and $g(t) = t - 1$. Then, according to (2.1),

$$\bar{q}(t) = \alpha\left((m-1)! + \frac{1}{t+m}\right).$$

So

$$\beta_1 = \inf_{t\geq T}\left\{\frac{(t+1-t)^{m-1}\cdot\alpha((m-1)! + \frac{1}{t+m+1})}{(m-1)!\cdot 1^m}\right\} = \alpha$$

with $T \geq 3$. Since

$$\beta_1 \sum_{i=1}^{3} irp^i = \alpha \cdot \left(\frac{1}{2} + 2 \times \frac{1}{4} + 3 \times \frac{1}{8}\right) = \frac{11\alpha}{8} \geq 1$$

holds for $\alpha \in [8/11, 1)$ and

$$0 \leq \liminf_{t\to\infty} \int_{t-1}^{t} (s+1-s)^{m-1} \cdot \alpha\left((m-1)! + \frac{1}{s+m+1}\right) ds = \alpha \cdot (m-1)! \leq \frac{2 \cdot (m-1)!}{e}$$

holds for any $\alpha \in (0, 2/e]$, (2.4) and (2.6) are satisfied for $n = 3$ and $\alpha \in [8/11, 2/e]$. Since $r = 1 = \tau$, by Corollaries 2.2 and 2.5, every bounded solution $x(t)$ of (3.1) is oscillatory.

Example 2 Consider the difference equation

$$\Delta^m_{\frac{\pi}{m}} \left(x(t) - 2x(t-4\pi)\right) + 8x(t-\pi) + \frac{8\sigma}{1+t^2} x^3(t-\pi) = 0, \tag{3.2}$$

for $t > 0$, where m is an odd positive integer with $m \geq 3$ and σ is a positive real number. Regarding (3.2) as (1.1), we have $\tau = \pi/m$, $p = 2$, $r = 4\pi$, $g(t) = t - \pi$ and $q(t) = 8$. Then, for some $\alpha \in (0,1)$, $\bar{q} = 8\alpha$ by (2.1). Moreover, $r \geq t + m\tau - g(t)$ and $r = k\tau$ are satisfied. In addition,

$$\beta_2 = \inf_{t \geq T} \left\{ \frac{8\alpha \cdot (t+\pi-t)^{m-1}}{(m-1)!(\frac{\pi}{m})^m} \right\} = \frac{8m^m\alpha}{\pi(m-1)!},$$

where $T \geq 12\pi$. So (2.10) is satisfied since

$$\beta_2 \sum_{i=1}^{3} \frac{4\pi(i-1)}{p^i} = \frac{8m^m\alpha}{\pi(m-1)!} \times 4\pi \times \left(\frac{1}{2^2} + \frac{2}{2^3}\right) = \alpha \frac{16m^m}{(m-1)!} \geq 1$$

holds for $\alpha \in [(m-1)!/(16m^m), 1)$. By Theorem 2.6, every bounded solution $x(t)$ of (3.2) is oscillatory.

Example 3 Consider the difference equation

$$\Delta^m_{\frac{\pi}{m}} \left(x(t) - 2x(t-2\pi)\right) + e^{-\frac{\sigma}{t}} x(t-\pi) = 0, \tag{3.3}$$

for $t > 0$, where m is an odd positive integer with $m \geq 3$ and σ is a positive constant. Regarding (3.3) as (1.1), we have $\tau = \pi/m$, $p = 2$, $r = 2\pi$, $g(t) = t - \pi$, and $q(t) = e^{-\frac{\sigma}{t}}$. Then, for some $\alpha \in (0,1)$, $\bar{q} = \alpha e^{-\frac{\sigma}{t}}$ by (2.1). Moreover, $r \geq t + m\tau - g(t)$ and $r = k\tau$ are satisfied. In addition,

$$\beta_2 = \inf_{t \geq T} \left\{ \frac{\alpha e^{-\frac{\sigma}{t}} \cdot (t+\pi-t)^{m-1}}{(m-1)!(\frac{\pi}{m})^m} \right\} = \frac{m^m\alpha}{e^{\sigma/T}\pi(m-1)!} \to \frac{m^m\alpha}{\pi(m-1)!},$$

as $T \to \infty$. So (2.10) is satisfied when T is large enough since

$$\beta_2 \sum_{i=1}^{3} \frac{2\pi(i-1)}{p^i} \to \frac{\alpha m^m}{\pi(m-1)!} \times 2\pi \times \left(\frac{1}{2^2} + \frac{2}{2^3}\right) = \alpha \frac{m^m}{(m-1)!} > 1$$

as $T \to \infty$ for $\alpha \in ((m-1)!/(m^m), 1)$. By Theorem 2.6, every bounded solution $x(t)$ of (3.3) is oscillatory.

4 Related lemmas

To prove the main results, we need to prove the following lemmas first. The first lemma is about a function $x(t)$ satisfying the differential inequality

$$x'(t) + q(t)x(\tau(t)) \le 0, \tag{4.1}$$

where $q, \tau \in C([t_0, \infty), R_+)$, $\tau(t) \le t$, and $\lim_{t\to\infty} \tau(t) = \infty$. Let

$$\eta = \liminf_{t\to\infty} \int_{\tau(t)}^{t} q(s)\, ds.$$

Lemma 4.1 *Assume that τ is nondecreasing, $0 \le \eta \le e^{-1}$, and $x(t)$ is an eventually positive function satisfying* (4.1). *Set*

$$r = \liminf_{t\to\infty} \frac{x(t)}{x(\tau(t))}.$$

Then r satisfies

$$\frac{1-\eta-\sqrt{1-2\eta-\eta^2}}{2} \le r \le 1.$$

The above lemma can be found in [6], p.18.

Lemma 4.2 *Let $0 \le p < 1$. Assume that $x(t)$ is a bounded and eventually positive (negative) solution of* (1.1) *with $z(t) = x(t) - px(t-r)$ and $\liminf_{t\to\infty} z(t) \ge 0$ ($\limsup_{t\to\infty} z(t) \le 0$). Let*

$$y(t) = \int_{t}^{t+\tau} dt_1 \int_{t_1}^{t_1+\tau} dt_2 \cdots \int_{t_{m-1}}^{t_{m-1}+\tau} z(\theta)\, d\theta.$$

Then $y(t) > 0$ (< 0), $(-1)^k y^{(k)}(t) > 0$ (< 0) for $1 \le k \le m$ eventually. Moreover,

$$\Delta_\tau^m y(t) + \bar{q}(t) \sum_{i=0}^{n} p^i y(g(t) - ir) < 0 \ (> 0) \tag{4.2}$$

holds for any fixed natural number n and for all large enough t.

Proof Suppose $x(t)$ is a bounded and eventually positive solution. Notice that $g(t) < t$ and $g'(t) > 0$ for all $t \ge t_0$. So there exists a $t_1 > t_0$ such that $x(g(t)) > 0$ for all $t \ge t_1$. From (1.1) it follows that

$$\Delta_\tau^m z(t) + f(t, x(g(t))) = 0.$$

By (1.3), we have $f(t, x(g(t))) \ge q(t)x(g(t)) > 0$ for $t \ge t_1$. Therefore,

$$y^{(m)}(t) + q(t)x(g(t)) \le 0 \tag{4.3}$$

for $t \geq t_1$. According to $q(t)x(g(t)) > 0$, $y^{(m)}(t) < 0$ for all $t \geq t_1$. Thus, $y^{(m-1)}(t)$ is decreasing so either $y^{(m-1)}(t) > 0$ for all $t \geq t_1$ or $y^{(m-1)}(t) \leq y^{(m-1)}(t_2) < 0$ for some $t_2 > t_1$ and for all $t \geq t_2$. If the latter holds, then

$$y^{(m-k)}(t) \to -\infty, \quad k = 2, 3, \ldots, m,$$

as $t \to \infty$, a contradiction to the boundedness of x and z. Therefore we have $y^{(m-1)}(t) > 0$ for all $t \geq t_1$. Thus, $y^{(m-2)}(t)$ is increasing so either $y^{(m-2)}(t) < 0$ for all $t \geq t_1$ or $y^{(m-2)}(t) \geq y^{(m-2)}(t_3) > 0$ for some $t_3 \geq t_1$ and all $t \geq t_3$. If the latter holds, then

$$y^{(m-k)}(t) \to \infty, \quad k = 3, 4, \ldots, m,$$

as $t \to \infty$, a contradiction again to the boundedness of x and z. Hence, we must have $y^{(m-2)}(t) < 0$ for all $t \geq t_1$. Repeating the above process, we obtain $(-1)^k y^{(k)}(t) > 0$ for $1 \leq k \leq m$ and all $t \geq t_1$. Therefore, $y(t)$ is decreasing so either $y(t) > 0$ for all $t \geq t_1$ or there is a $t_4 \geq t_1$ such that $y(t) \leq y(t_4) < 0$ for $t \geq t_4$. Suppose the latter case holds. Then

$$\begin{aligned}
&\int_t^{t+\tau} dt_1 \int_{t_1}^{t_1+\tau} dt_2 \cdots \int_{t_{m-1}}^{t_{m-1}+\tau} x(\theta)\, d\theta \\
&= y(t) + p \int_t^{t+\tau} dt_1 \int_{t_1}^{t_1+\tau} dt_2 \cdots \int_{t_{m-1}}^{t_{m-1}+\tau} x(\theta - r)\, d\theta \\
&\leq y(t_4) + p \int_t^{t+\tau} dt_1 \int_{t_1}^{t_1+\tau} dt_2 \cdots \int_{t_{m-1}}^{t_{m-1}+\tau} x(\theta - r)\, d\theta \\
&\ldots \\
&\leq y(t_4) \sum_{i=0}^{s-1} p^i + p^s \int_t^{t+\tau} dt_1 \int_{t_1}^{t_1+\tau} dt_2 \cdots \int_{t_{m-1}}^{t_{m-1}+\tau} x(\theta - sr)\, d\theta \\
&\leq \frac{y(t_4)(1-p^s)}{1-p} + p^s M \tau^m
\end{aligned}$$

for $t \geq t_4 + sr$, where $M = \sup_{t \geq t_0} x(t)$ and s is any positive integer. Let $s \to \infty$ so $t \to \infty$ as well, $p^s M \tau^m$ then is arbitrarily small due to $0 \leq p < 1$. Thus,

$$\int_t^{t+\tau} dt_1 \int_{t_1}^{t_1+\tau} dt_2 \cdots \int_{t_{m-1}}^{t_{m-1}+\tau} x(\theta)\, d\theta < 0,$$

which contradicts the assumption that $x(t)$ is eventually positive. Therefore, we must have $y(t) > 0$ for all $t \geq t_1$.

From (4.3) it follows that

$$\Delta_\tau^m z(t) + q(t) z(g(t)) + p q(t) x(g(t) - r) \leq 0.$$

According to the definition of $z(t)$, the above inequality becomes

$$\Delta_\tau^m z(t) + q(t) z(g(t)) + p q(t) z(g(t) - r) + p^2 q(t) x(g(t) - 2r) \leq 0.$$

Proceeding in the same way as the above, we have

$$\Delta_\tau^m z(t) + q(t)\sum_{i=0}^{n} p^i z\big(g(t) - ir\big) + p^{n+1}q(t)x\big(g(t) - (n+1)r\big) \le 0.$$

Since $q(t)p^{n+1}x(g(t) - (n+1)r) > 0$ when t is large enough, the above inequality implies that

$$\Delta_\tau^m z(t) + q(t)\sum_{i=0}^{n} p^i z\big(g(t) - ir\big) < 0.$$

In order to integrate the above inequality, we need to show that $z(t)$ is positive. If $p = 0$, then $z(t) = x(t) > 0$ holds eventually. Now suppose $0 < p < 1$. Since $y^{(m)}(t) = \Delta_\tau^m z(t) < 0$ for $t \ge t_1$,

$$\Delta_\tau^{m-1} z\big(t + (h+1)\tau\big) - \Delta_\tau^{m-1} z(t + h\tau) = \Delta_\tau^m z(t + h\tau) < 0$$

so $\Delta_\tau^{m-1} z(t + h\tau)$ is decreasing as h increases. By the boundedness of $x(t)$ we know that $\lim_{h\to\infty} \Delta_\tau^{m-1} z(t + h\tau)$ exists. If $\lim_{h\to\infty} \Delta_\tau^{m-1} z(t + h\tau) = S(t) \neq 0$, then

$$\Delta_\tau^{m-2} z\big(t + (h+1)\tau\big) \to -\infty \text{ or } \infty$$

as $h \to \infty$, a contradiction to the boundedness of $\Delta_\tau^{m-2} z(t)$. Thus, for each $t \ge t_1$, $\Delta_\tau^{m-1} z(t + h\tau)$ is decreasing and tends to 0 as $h \to \infty$. Repeating the same procedure, we see that $\Delta_\tau z(t + h\tau)$ is increasing as h increases and $\Delta_\tau z(t + h\tau) \to 0$ as $h \to \infty$; $z(t + h\tau)$ is decreasing as h increases so $\lim_{h\to\infty} z(t + h\tau)$ exists for each $t \ge t_1$. By assumption, $\liminf_{t\to\infty} z(t) \ge 0$. Then $z(t + h\tau)$ is decreasing and $\lim_{h\to\infty} z(t + h\tau) \ge 0$ so $z(t + h\tau) > 0$ for all $t \ge t_1$ and $h \ge 1$. Integrating $q(t)z(g(t) - ir)$, by the assumptions on g and q, we obtain

$$\begin{aligned}
&\int_t^{t+\tau} ds_1 \int_{s_1}^{s_1+\tau} ds_2 \cdots \int_{s_{m-1}}^{s_{m-1}+\tau} z\big(g(\theta) - ir\big)q(\theta)\,d\theta \\
&\ge \min_{t\le s\le t+m\tau}\{q(s)\} \int_t^{t+\tau} ds_1 \int_{s_1}^{s_1+\tau} ds_2 \cdots \int_{s_{m-1}}^{s_{m-1}+\tau} z\big(g(\theta) - ir\big)\,d\theta \\
&\ge \min_{t\le s\le t+m\tau}\{q(s)\} \int_{g(t)}^{g(t+\tau)} \big(g^{-1}(s_1)\big)'\,ds_1 \int_{s_1}^{g(g^{-1}(s_1)+\tau)} \big(g^{-1}(s_2)\big)'\,ds_2 \cdots \\
&\quad \times \int_{s_{m-1}}^{g(g^{-1}(s_{m-1})+\tau)} z(\theta - ir)\big(g^{-1}(\theta)\big)'\,d\theta \\
&\ge \min_{t\le s\le t+m\tau}\{q(s)\}\Big(\min_{g(t)\le s\le g(t)+m\tau} \big(g^{-1}(s)\big)'\Big)^m \int_{g(t)}^{g(t)+\tau} ds_1 \int_{s_1}^{s_1+\tau} ds_2 \cdots \\
&\quad \times \int_{s_{m-1}}^{s_{m-1}+\tau} z(\theta - ir)\,d\theta \\
&\ge \min_{t\le s\le t+m\tau}\{q(s)\}\Big(\min_{g(t)\le s\le g(t)+m\tau} \big(g^{-1}(s)\big)'\Big)^m y\big(g(t) - ir\big) \\
&\ge \bar{q}(t)y\big(g(t) - ir\big).
\end{aligned}$$

Therefore,

$$\Delta_\tau^m y(t) + \bar{q}(t)\sum_{i=0}^{n} p^i y\big(g(t) - ir\big) < 0$$

holds for any fixed natural number n and for all large enough t. If $x(t)$ is a bounded and eventually negative solution, then the above proof with obvious changes shows the conclusion within brackets. □

Lemma 4.3 *Let $0 \le p < 1$ and $r = k\tau$. Assume that $x(t)$ is a bounded and eventually positive (negative) solution of* (1.1). *Let*

$$z(t) = x(t) - px(t - r),$$

$$y(t) = \int_t^{t+\tau} dt_1 \int_{t_1}^{t_1+\tau} dt_2 \cdots \int_{t_{m-1}}^{t_{m-1}+\tau} z(\theta)\, d\theta.$$

Then the conclusion of Lemma 4.2 *holds.*

Proof The proof is the same as that of Lemma 4.2 until $\lim_{h\to\infty} z(t + h\tau)$ exists for each $t \ge t_1$. Suppose there is a $t' > t_1$ such that $\lim_{h\to\infty} z(t' + h\tau) = \delta < 0$. Then $z(t' + h\tau) \le \delta/2 < 0$ for $h \ge h_1 > 0$ so $z(t' + hr + h_1\tau) = z(t' + (kh + h_1)\tau) \le \delta/2$ for $h \ge 0$. Thus, for $h > 0$,

$$\begin{aligned} x\big(t' + hr + h_1\tau\big) &\le \delta/2 + px\big(t' + (h-1)r + h_1\tau\big) \\ &\le \delta\big(1 + p + \cdots + p^{h-1}\big)/2 + p^h x\big(t' + h_1\tau\big). \end{aligned}$$

This implies $x(t' + k(h + h_1)\tau) < 0$ for large h, a contradiction to the assumption that x is eventually positive. Therefore $\lim_{h\to\infty} z(t + h\tau) \ge 0$ for $t \ge t_1$. Since $z(t + h\tau)$ is decreasing as h increases, $z(t) > 0$ for all $t \ge t_1$. The rest of the proof of Lemma 4.2 is still valid here. □

Lemma 4.4 *Under the assumptions of Lemma* 4.2 *or Lemma* 4.3, *let*

$$v(t) = \int_t^{t+\tau} dt_1 \int_{t_1}^{t_1+\tau} dt_2 \cdots \int_{t_{m-1}}^{t_{m-1}+\tau} y(\theta)\, d\theta.$$

Then $v(t) > 0$ (< 0), $(-1)^k v^{(k)}(t) > 0$ (< 0) for $1 \le k \le m$ eventually. Moreover,

$$v^{(m)}(t) + \frac{1}{\tau^m}\bar{q}(t)\sum_{i=0}^{n} p^i v\big(g(t) - ir\big) < 0 \ (> 0) \tag{4.4}$$

holds for any fixed natural number n and for all large enough t.

Proof By the definition of $v(t)$, $v(t)$ has the same sign as $y(t)$ for all $t \ge t_1$. Furthermore, we have

$$v'(t) = \int_t^{t+\tau} dt_1 \int_{t_1}^{t_1+\tau} dt_2 \cdots \int_{t_{m-1}}^{t_{m-1}+\tau} y'(\theta)\, d\theta.$$

Then $v'(t)$ has the same sign as $y'(t)$. Similarly, $v^{(j)}(t)$ has the same sign as $y^{(j)}(t)$ for all $j = 1, 2, \ldots, m$. Notice also that $v^{(m)}(t) = \Delta_\tau^m y(t)$. If $y'(t) < 0$, then

$$\begin{aligned}
v\big(g(t) - ir\big) &= \int_{g(t)}^{g(t)+\tau} dt_1 \int_{t_1}^{t_1+\tau} dt_2 \cdots \int_{t_{m-1}}^{t_{m-1}+\tau} y(\theta - ir)\, d\theta \\
&\le \int_{g(t)}^{g(t)+\tau} dt_1 \int_{t_1}^{t_1+\tau} dt_2 \cdots \int_{t_{m-1}}^{t_{m-1}+\tau} y(t_{m-1} - ir)\, d\theta \\
&\le \tau \int_{g(t)}^{g(t)+\tau} dt_1 \int_{t_1}^{t_1+\tau} dt_2 \cdots \int_{t_{m-2}}^{t_{m-2}+\tau} y(t_{m-1} - ir)\, dt_{m-1} \\
&\cdots \\
&\le \tau^{m-1} \int_{g(t)}^{g(t)+\tau} y(t_1 - ir)\, dt_1 \\
&\le \tau^m y\big(g(t) - ir\big).
\end{aligned}$$

Hence, from (4.2) it follows that

$$v^{(m)}(t) + \frac{1}{\tau^m} \bar{q}(t) \sum_{i=0}^{n} p^i v\big(g(t) - ir\big) < 0$$

holds for any fixed natural number n and for all large enough t. If $y'(t) > 0$, then $v(g(t) - ir) \ge \tau^m y(g(t) - ir)$ so

$$v^{(m)}(t) + \frac{1}{\tau^m} \bar{q}(t) \sum_{i=0}^{n} p^i v\big(g(t) - ir\big) > 0.$$

□

Lemma 4.5 *Under the assumptions of Lemma* 4.4, *for each* $t \ge t_1$ *there is a* $\theta \in (g(t), t)$ *such that*

$$\big|v'\big(g(t)\big)\big| > \frac{(t - g(t))^{m-1}}{(m-1)!} \big|v^{(m)}(\theta)\big|. \tag{4.5}$$

Proof Under the assumptions of Lemma 4.4, we know that $(-1)^j v^{(j)}(t)$ for $j = 1, 2, \ldots, m$ have the same sign. According to Taylor's formula, we have

$$\begin{aligned}
v'\big(g(t)\big) &= v'(t) + v''(t)\big(g(t) - t\big) + \frac{1}{2} v^{(3)}(t)\big(g(t) - t\big)^2 + \cdots \\
&\quad + \frac{1}{(m-1)!} v^{(m)}(\theta)\big(g(t) - t\big)^{m-1}
\end{aligned}$$

for some $\theta \in (g(t), t)$ and (4.5) follows immediately. □

The next lemmas are for the bounded solutions of (1.1) with $p > 1$.

Lemma 4.6 *Let* $p > 1$ *and* $r = k\tau$, $k \in N$. *Assume that* $x(t)$ *is a bounded and eventually positive (negative) solution of* (1.1). *Let*

$$z(t) = x(t) - px(t - r),$$

$$y(t) = \int_t^{t+\tau} dt_1 \int_{t_1}^{t_1+\tau} dt_2 \cdots \int_{t_{m-1}}^{t_{m-1}+\tau} z(\theta)\, d\theta.$$

Then $y(t) < 0$ (> 0), $(-1)^k y^{(k)}(t) > 0$ (< 0) *for* $1 \le k \le m$ *eventually. Moreover,*

$$\Delta_\tau^m y(t) - \bar{q}(t)\sum_{i=1}^{n}\frac{1}{p^i}y\big(g(t)+ir\big) < 0 \ (> 0) \tag{4.6}$$

holds for any fixed integer $n \ge 1$ *and for all large enough* t.

Proof Suppose $x(t)$ is a bounded and eventually positive solution. Since $g(t) < t$ and $g'(t) > 0$, from the assumptions, there exists a $t_1 > t_0$ such that $x(g(t)) > 0$ for all $t \ge t_1$. Notice also that

$$\Delta_\tau^m z(t) + f\big(t, x(g(t))\big) = 0.$$

According to (1.3), we have $f(t, x(g(t))) \ge q(t)x(g(t)) > 0$ for $t \ge t_1$. Therefore

$$\Delta_\tau^m z(t) + q(t)x\big(g(t)\big) \le 0 \tag{4.7}$$

for $t \ge t_1$. By the definition of $y(t)$, $y^{(m)}(t) = \Delta_\tau^m z(t)$. Thus, from (4.7) it follows that

$$y^{(m)}(t) + q(t)x\big(g(t)\big) \le 0 \tag{4.8}$$

for $t \ge t_1$. Due to $q(t)x(g(t)) > 0$, $y^{(m)}(t) < 0$ for all $t \ge t_1$. From the proof of Lemma 4.2 we know that $(-1)^k y^{(k)}(t) > 0$ holds for $1 \le k \le m$ and all $t \ge t_1$. Thus, $y(t)$ is decreasing. We now prove that $y(t) < 0$ for all $t \ge t_1$. Since $y^{(m)}(t) = \Delta_\tau^m z(t)$ for all $t \ge t_1$, from the proof of Lemma 4.2 we know that $z(t + h\tau)$ is decreasing for each fixed $t \ge t_1$ as h increases. Next we show that $z(t) < 0$, so that $y(t) < 0$ for some $t_2 \ge t_1$ and all $t \ge t_2$. Suppose there is a $t' > t_1$ such that $z(t' + h\tau) > 0$ for all $h \ge 1$. Under $r = k\tau$, we then have $z(t' + hr) > 0$ for all $h \ge 1$ so $x(t' + hr) > p^h x(t')$ for all $h \ge 1$. So $x(t' + hr) \to \infty$ as $h \to \infty$, a contradiction to the boundedness of x. Therefore, for each $t \in [t_1, t_1 + \tau]$, $z(t + h\tau)$ is decreasing as h increases and there is an integer $H(t) > 0$ such that $z(t + h\tau) < z(t + H(t)\tau) < 0$ for all $h > H(t)$. Since $z(t)$ is continuous for each $t' \in [t_1, t_1 + \tau]$, there is an open interval $I(t')$ such that $z(t + h\tau) < z(t + H(t')\tau) < 0$ hold for all $t \in I(t')$ and $h > H(t')$. Since $[t_1, t_1 + \tau]$ is compact and $\{I(t') : t' \in [t_1, t_1 + \tau]\}$ is an open cover of $[t_1, t_1 + \tau]$, there is a finite subset of $\{I(t') : t' \in [t_1, t_1 + \tau]\}$ covering $[t_1, t_1 + \tau]$. Therefore, there is a $K > 0$ such that

$$z(t + h\tau) \le z(t + K\tau) < 0$$

for all $t \in [t_1, t_1 + \tau]$ and all $h \ge K$. Hence, there is a $t_3 > t_1$ such that $z(t) < 0$, so that $y(t) < 0$ for all $t \ge t_3$.

From (4.7), we have

$$\Delta_\tau^m z(t) - \frac{q(t)}{p}z\big(g(t)+r\big) + \frac{q(t)}{p}x\big(g(t)+r\big) \le 0.$$

According to the definition of $z(t)$, it follows from the above inequality that

$$\Delta_\tau^m z(t) - \frac{q(t)}{p}z\big(g(t)+r\big) + \frac{q(t)}{p}\left(-\frac{1}{p}z\big(g(t)+2r\big) + \frac{1}{p}x\big(g(t)+2r\big)\right) \le 0.$$

Repeating the above procedure, we obtain

$$\Delta_\tau^m z(t) - q(t)\sum_{i=1}^{n}\frac{1}{p^i}z\big(g(t)+ir\big) + q(t)\frac{1}{p^n}x\big(g(t)+nr\big) \le 0.$$

Since $q(t)x(g(t)+nr) > 0$ for sufficiently large t, we have

$$\Delta_\tau^m z(t) - q(t)\sum_{i=1}^{n}\frac{1}{p^i}z\big(g(t)+ir\big) < 0.$$

Integrating $q(t)z(g(t)+ir)$, by the assumptions on p and g, we obtain

$$\begin{aligned}
&\int_t^{t+\tau} ds_1 \int_{s_1}^{s_1+\tau} ds_2 \cdots \int_{s_{m-1}}^{s_{m-1}+\tau} z\big(g(\theta)+ir\big)q(\theta)\,d\theta \\
&\le \min_{t\le l\le t+m\tau}\{q(l)\} \int_t^{t+\tau} ds_1 \int_{s_1}^{s_1+\tau} ds_2 \cdots \int_{s_{m-1}}^{s_{m-1}+\tau} z\big(g(\theta)+ir\big)\,d\theta \\
&\le \min_{t\le l\le t+m\tau}\{q(l)\} \int_{g(t)}^{g(t+\tau)} \big(g^{-1}(s_1)\big)'\,ds_1 \int_{s_1}^{g(g^{-1}(s_1)+\tau)} \big(g^{-1}(s_2)\big)'\,ds_2 \cdots \\
&\quad \times \int_{s_{m-1}}^{g(g^{-1}(s_{m-1})+\tau)} z(\theta+ir)\big(g^{-1}(\theta)\big)'\,d\theta \\
&\le \min_{t\le l\le t+m\tau}\{q(l)\}\Big(\min_{g(t)\le s\le g(t)+m\tau}\big(g^{-1}(s)\big)'\Big)^m \int_{g(t)}^{g(t)+\tau} ds_1 \int_{s_1}^{s_1+\tau} ds_2 \cdots \\
&\quad \times \int_{s_{m-1}}^{s_{m-1}+\tau} z(\theta+ir)\,d\theta \\
&\le \min_{t\le l\le t+m\tau}\{q(l)\}\Big(\min_{g(t)\le s\le g(t)+m\tau}\big(g^{-1}(s)\big)'\Big)^m y\big(g(t)+ir\big) \\
&\le \bar{q}(t)y\big(g(t)+ir\big).
\end{aligned}$$

Therefore,

$$\Delta_\tau^m y(t) - \bar{q}(t)\sum_{i=1}^{n}\frac{1}{p^i}y\big(g(t)+ir\big) < 0$$

holds for any fixed integer $n \ge 1$ and for all large enough t. If $x(t)$ is a bounded and eventually negative solution, then the conclusion within brackets follows from the above proof with minor modification. □

Lemma 4.7 *Under the assumptions of Lemma* 4.6, *let*

$$v(t) = \int_t^{t+\tau} dt_1 \int_{t_1}^{t_1+\tau} dt_2 \cdots \int_{t_{m-1}}^{t_{m-1}+\tau} y(\theta)\,d\theta.$$

Then $v(t) < 0$ (> 0), $(-1)^k v^{(k)}(t) > 0$ (< 0) *for* $1 \le k \le m$ *eventually. Moreover,*

$$v^{(m)}(t) - \frac{1}{\tau^m}\bar{q}(t)\sum_{i=1}^{n}\frac{1}{p^i}v\big(g(t)-m\tau+ir\big) < 0 \ (> 0) \tag{4.9}$$

holds for any fixed integer $n \ge 1$ *and for all large enough* t.

Proof By the definition of $v(t)$, $v(t)$ has the same sign as $y(t)$. Further, we have

$$v'(t) = \int_{t}^{t+\tau} dt_1 \int_{t_1}^{t_1+\tau} dt_2 \cdots \int_{t_{m-1}}^{t_{m-1}+\tau} y'(\theta)\, d\theta.$$

Then $v'(t)$ has the same sign as $y'(t)$. Similarly, $(-1)^k v^{(k)}(t)$ for $1 \le k \le m$ and $(-1)^j y^{(j)}(t)$ for $1 \le j \le m$ all have the same sign. Note also that $v^{(m)}(t) = \Delta_\tau^m y(t)$. If $y'(t) < 0$, then

$$\begin{aligned}
v(g(t) + ir) &= \int_{g(t)}^{g(t)+\tau} dt_1 \int_{t_1}^{t_1+\tau} dt_2 \cdots \int_{t_{m-1}}^{t_{m-1}+\tau} y(\theta + ir)\, d\theta \\
&\ge \int_{g(t)}^{g(t)+\tau} dt_1 \int_{t_1}^{t_1+\tau} dt_2 \cdots \int_{t_{m-1}}^{t_{m-1}+\tau} y(t_{m-1} + \tau + ir)\, d\theta \\
&\ge \tau \int_{g(t)}^{g(t)+\tau} dt_1 \int_{t_1}^{t_1+\tau} dt_2 \cdots \int_{t_{m-2}}^{t_{m-2}+\tau} y(t_{m-1} + \tau + ir)\, dt_{m-1} \\
&\cdots \\
&\ge \tau^{m-1} \int_{g(t)}^{g(t)+\tau} y(t_1 + (m-1)\tau + ir)\, dt_1 \\
&\ge \tau^m y(g(t) + m\tau + ir).
\end{aligned}$$

Hence, from (4.6) we have

$$v^{(m)}(t) - \frac{1}{\tau^m}\bar{q}(t) \sum_{i=1}^{n} \frac{1}{p^i} v(g(t) - m\tau + ir) < 0$$

for any fixed integer $n \ge 1$ and for all large enough t. If $y'(t) > 0$, then $0 < v(g(t) + ir) \le \tau^m y(g(t) + m\tau + ir)$ so

$$v^{(m)}(t) - \frac{1}{\tau^m}\bar{q}(t) \sum_{i=1}^{n} \frac{1}{p^i} v(g(t) - m\tau + ir) > 0.$$

□

Lemma 4.8 *Assume that $x(t)$ is an eventually positive (negative) and bounded solution of* (1.1). *Let $z(t)$ and $v(t)$ be defined as in Lemma* 4.6 *and Lemma* 4.7. *Then, under the assumptions of Lemma* 4.6, *for any given $t \ge t_1$, there is a $\theta \in (g(t), t)$ such that*

$$|v'(g(t))| > \frac{(t - g(t))^{m-1}}{(m-1)!} |v^{(m)}(\theta)|. \tag{4.10}$$

Proof The proof of Lemma 4.5 is still valid for Lemma 4.8. □

5 Proofs of the main results

Here, the proofs of the main results will be presented.

Proof of Theorem 2.1 Suppose the conclusion is not true. Let $x(t)$ be an eventually positive and bounded solution of (1.1) with $\liminf_{t\to\infty}(x(t) - px(t-r)) \ge 0$. Let $y(t)$ be defined as in Lemma 4.2 and $v(t)$ be defined as in Lemma 4.4. By Lemma 4.4, we know that $v(t) > 0$,

$(-1)^k v^{(k)}(t) > 0$ for $1 \le k \le m$ and (4.4), *i.e.*,

$$v^{(m)}(t) + \frac{1}{\tau^m}\bar{q}(t)\sum_{i=0}^{n} p^i v\big(g(t) - ir\big) < 0$$

holds for any fixed natural number n and for all large enough t. By Lemma 4.5, we know that

$$v'\big(g(t)\big)\frac{(m-1)!}{(t-g(t))^{m-1}} < v^{(m)}(\theta)$$

for some $\theta \in \big(g(t), t\big)$. Since $\bar{q}(\theta) \ge \bar{q}(t)$ by assumption and

$$v\big(g(\theta) - ir\big) \ge v\big(g(t) - ir\big),$$

from (4.4) with t replaced by θ, it follows that

$$v'\big(g(t)\big)\frac{(m-1)!}{(t-g(t))^{m-1}} + \frac{1}{\tau^m}\bar{q}(t)\sum_{i=0}^{n} p^i v\big(g(t) - ir\big) < 0$$

i.e.,

$$v'\big(g(t)\big) + \frac{(t-g(t))^{m-1}}{(m-1)!\tau^m}\bar{q}(t)\sum_{i=0}^{n} p^i v\big(g(t) - ir\big) < 0. \tag{5.1}$$

With the replacement of t by $g^{-1}(t)$, (5.1) yields

$$v'(t) + \frac{(g^{-1}(t)-t)^{m-1}}{(m-1)!\tau^m}\bar{q}\big(g^{-1}(t)\big)\sum_{i=0}^{n} p^i v(t - ir) < 0. \tag{5.2}$$

Assume that $(-1)^k v^{(k)}(t) > 0$ and (5.2) hold for $0 \le k \le m$ and $t \ge t_1 \ge t_0$. Without loss of generality, we may assume $T \ge t_1 + nr$. Let

$$w(t) = \frac{-v'(t)}{v(t)}.$$

Note that $w(t) > 0$ and $v(t) = v(T)\exp\int_T^t -w(\theta)\,d\theta$ for all $t \ge T \ge t_1 + nr$. From (5.2) it follows that

$$w(t) > \frac{(g^{-1}(t)-t)^{m-1}\bar{q}(g^{-1}(t))}{(m-1)!\tau^m}\sum_{i=0}^{n} p^i \exp\int_{t-ir}^{t} w(s)\,ds, \tag{5.3}$$

i.e.,

$$w(t) > \frac{Q_1(t)}{(m-1)!\tau^m}\sum_{i=0}^{n} p^i \exp\int_{t-ir}^{t} w(s)\,ds \tag{5.4}$$

for all $t \ge Tx$, where $Q_1(t) = (g^{-1}(t)-t)^{m-1}\bar{q}x(g^{-1}(t)) > 0$.

Let $w_0(t) \equiv 0$ for $t \geq Tx - nr$ and let

$$w_{k+1}(t) = \frac{Q_1(t)}{(m-1)!\tau^m} \sum_{i=0}^{n} p^i \exp \int_{t-ir}^{t} w_k(s)\, ds$$

for each $k \in N$ and $t \geq Tx + nkr$. Let

$$\alpha_{1k} = \inf_{t \geq T+(k-1)nr} \{w_k(t)\}, \quad k \in \bar{N}.$$

Then

$$\alpha_{1k+1} \geq \inf_{t \geq T} \left\{ \frac{Q_1(t)}{(m-1)!\tau^m} \sum_{i=0}^{n} p^i e^{ir\alpha_{1k}} \right\} = \beta_1 \sum_{i=0}^{n} p^i e^{ir\alpha_{1k}}.$$

Now, (5.4), (2.4), and the definition of $\{\alpha_{1k}\}$ imply that $\{\alpha_{1k}\}$ is an increasing sequence. Suppose

$$\lim_{k \to \infty} \alpha_{1k} = \rho_1 < \infty.$$

So $\rho_1 \geq \beta_1 \sum_{i=0}^{n} p^i e^{ir\rho_1}$. Let

$$F_1(x) = \beta_1 \sum_{i=0}^{n} p^i e^{irx} - x.$$

Then $F_1'(x) = \beta_1 \sum_{i=1}^{n} irp^i e^{irx} - 1$ and $F_1''(x) > 0$, so $F_1'(x)$ is increasing. Since $F_1'(0) = \beta_1 \sum_{i=0}^{n} irp^i - 1 \geq 0$ by (2.4), then $F_1'(x) > 0$ for $x > 0$. Hence $F_1(x)$ is increasing. Thus, from $F_1(0) = \beta_1 \sum_{i=0}^{n} p^i > 0$ we have $F_1(x) > 0$ for all $x \geq 0$. This shows that no positive number ρ_1 satisfies $\rho_1 \geq \beta_1 \sum_{i=0}^{n} p^i e^{ir\rho_1}$. Therefore, we must have $\alpha_{1k} \to \infty$ as $k \to \infty$. Note that $w(t) \geq w_{k+1}(t) \geq \alpha_{1k+1}$ for $t \geq T_3 + nkr$. Thus $w(t) \to \infty$ as $t \to \infty$. Notice also that

$$w(t) \geq w_{k+1}(t) \geq \alpha_{1k+1} \quad \text{for } t \geq T + nkr.$$

Thus $w(t) \to \infty$ as $t \to \infty$, which implies

$$\frac{v(t)}{v(t+r)} = \exp \int_{t}^{t+r} w(s)\, ds \to \infty \quad \text{as } t \to \infty. \tag{5.5}$$

On the other hand, since $v'(t) < 0$ and $v(t) > 0$, (5.2) yields (by dropping the $i = 0$ term)

$$\begin{aligned} v'(t) &< -\frac{(g^{-1}(t) - t)^{m-1}}{(m-1)!\tau^m} \bar{q}(g^{-1}(t)) \sum_{i=1}^{n} p^i v(t - ir) \\ &< -\frac{(g^{-1}(t) - t)^{m-1}}{(m-1)!\tau^m} \cdot \frac{p - p^{n+1}}{1-p} \cdot \bar{q}(g^{-1}(t)) v(t - r). \end{aligned} \tag{5.6}$$

By (2.5) and Lemma 4.1,

$$\liminf_{t \to \infty} \frac{v(t)}{v(t-r)} \in (0,1].$$

Thus $v(t+r)/v(t)$ has a positive lower bound so $v(t)/v(t+r)$ has a positive upper bound. This contradicts (5.5). Assume that $x(t)$ is an eventually negative and bounded solution of (1.1) with $\limsup_{t\to\infty}(x(t)-px(t-r)) \le 0$. Then the above proof with a minor modification also leads to a contradiction. Therefore, the conclusion of the theorem holds. □

Proof of Corollary 2.2 The proof is the same as that of Theorem 2.1 except (5.6). The conclusion still holds if (5.6) is replaced by

$$v'(t) < -\frac{(g^{-1}(t)-t)^{m-1}}{\tau^m(m-1)!}\bar{q}(g^{-1}(t))pv(t-r).$$ □

Proof of Corollary 2.3 The proof of Theorem 2.1 is still valid after the replacement of $\bar{q}(\theta) \ge \bar{q}(t)$ by $\bar{q}(\theta) \ge \bar{q}(g(t))$. □

The proof of Corollary 2.4 is similar to that of Corollary 2.2.

Proof of Corollary 2.5 The proof of Theorem 2.1 is still valid after the replacement of Lemma 4.2 by Lemma 4.3. □

Proof of Corollary 2.6 Suppose the conclusion is not true. Without loss of generality, assume that (1.1) has an eventually positive and bounded solution $x(t)$. Let $y(t)$ be defined as in Lemma 4.6 and $v(t)$ be defined as in Lemma 4.7. By Lemma 4.7, we know that $v(t) < 0$, $(-1)^k v^{(k)}(t) > 0$ for $1 \le k \le m$, and (4.9), *i.e.*,

$$v^{(m)}(t) - \frac{1}{\tau^m}\bar{q}(t)\sum_{i=1}^{n}\frac{1}{p^i}v\big(g(t) - m\tau + ir\big) < 0$$

holds for any fixed integer $n \ge 1$ and for all large enough t. By Lemma 4.8, we know that

$$v'\big(g(t)\big)\frac{(m-1)!}{(t-g(t))^{m-1}} < v^{(m)}(\theta)$$

for some $\theta \in (g(t), t)$. Since $\bar{q}(\theta) \ge \bar{q}(g(t))$ and $v(g(\theta) - ir) \le v(g(g(t)) - ir)$, with the replacement of t by θ, (4.9) yields

$$v'\big(g(t)\big)\frac{(m-1)!}{(t-g(t))^{m-1}} - \frac{1}{\tau^m}\bar{q}\big(g(t)\big)\sum_{i=1}^{n}\frac{1}{p^i}v\big(g\big(g(t)\big) - m\tau + ir\big) \le 0$$

i.e.,

$$v'\big(g(t)\big) - \frac{(t-g(t))^{m-1}}{(m-1)!\tau^m}\bar{q}\big(g(t)\big)\sum_{i=1}^{n}\frac{1}{p^i}v\big(g\big(g(t)\big) - m\tau + ir\big) \le 0. \tag{5.7}$$

With the replacement of t by $g^{-1}(t)$, (5.7) becomes

$$v'(t) - \frac{(g^{-1}(t)-t)^{m-1}}{(m-1)!\tau^m}\bar{q}(t)\sum_{i=1}^{n}\frac{1}{p^i}v\big(g(t) - m\tau + ir\big) \le 0. \tag{5.8}$$

Assume that $v(t)$, $(-1)^k v^{(k)}(t) > 0$ $(1 \le k \le m)$ and (5.8) hold for $t \ge t_1 \ge t_0$ and, without loss of generality, that $T \ge t_1 + nr$. Let

$$w(t) = \frac{v'(t)}{v(t)}.$$

Note that $w(t) > 0$ and $v(t) = v(t')\exp\int_{t'}^{t} w(\theta)\,d\theta$ for all $t, t' \ge T$. From (5.8), we have

$$w(t) \ge \frac{(g^{-1}(t)-t)^{m-1}\bar{q}(t)}{(m-1)!\tau^m}\sum_{i=1}^{n}\frac{1}{p^i}\exp\int_t^{g(t)-m\tau+ir} w(s)\,ds, \tag{5.9}$$

i.e.,

$$w(t) \ge \frac{Q_{m2}(t)}{(m-1)!\tau^m}\sum_{i=1}^{n}\frac{1}{p^i}\exp\int_t^{g(t)-m\tau+ir} w(s)\,ds \tag{5.10}$$

for all $t \ge T$, where $Q_2(t) = (g^{-1}(t)-t)^{m-1}\bar{q}(t) > 0$.
Let $w_0(t) \equiv 0$ for $t \ge T$. For each $k \in \bar{N}$ and $t \ge T$, let

$$w_{k+1}(t) = \frac{Q_2(t)}{(m-1)!\tau^m}\sum_{i=1}^{n}\frac{1}{p^i}\exp\int_t^{g(t)-m\tau+ir} w_k(s)\,ds$$

and

$$\alpha_{2k} = \inf_{t\ge T}\{w_k(t)\}, \quad k \in \bar{N}.$$

So $w(t) \ge w_{k+1}(t) \ge w_k(t) \ge \alpha_{2k}$ for all $k \in N$ and $t \ge T$. By assumption, we therefore have

$$\begin{aligned}\alpha_{2k+1} &\ge \inf_{t\ge T}\left\{Q_2(t)\cdot\frac{1}{(m-1)!\tau^m}\sum_{i=1}^{n}\frac{1}{p^i}e^{[g(t)-t-m\tau+ir]\alpha_{2k}}\right\}\\ &\ge \beta_2\sum_{i=1}^{n}\frac{1}{p^i}e^{(i-1)r\alpha_{2k}}.\end{aligned}$$

Note that $\{\alpha_{2k}\}$ is a bounded nondecreasing sequence and suppose that

$$\lim_{k\to\infty}\alpha_{2k} = \rho_2 < \infty. \tag{5.11}$$

So $\rho_2 \ge \beta_2\sum_{i=1}^{n}(e^{(i-1)r\rho_2}/p^i)$. Let

$$F_2(x) = \beta_2\sum_{i=1}^{n}\frac{e^{(i-1)rx}}{p^i} - x.$$

Then $F_2'(x) = \beta_2\sum_{i=2}^{n}((i-1)re^{(i-1)rx}/p^i) - 1$ and $F_2''(x) > 0$, so $F_2'(x)$ is increasing. Since $F_2'(0) = \beta_2\sum_{i=1}^{n}((i-1)r/p^i) - 1 \ge 0$ by (2.10), $F_2'(x) > 0$ for $x > 0$. Hence $F_2(x)$ is increasing. Thus, from $F_2(0) = \beta_2\sum_{i=1}^{n}1/p^i > 0$ we have $F_2(x) > 0$ for all $x \ge 0$. This shows that no positive number ρ_2 satisfies $\rho_2 \ge \beta_2\sum_{i=1}^{n}(e^{(i-1)r\rho_2}/p^i)$. This contradiction shows that the conclusion holds.

□

Proof of Corollary 2.7 The proof of Theorem 2.6 is still valid after the replacement of $\bar{q}(\theta) \geq \bar{q}(g(t))$ by $\bar{q}(\theta) \geq \bar{q}(t)$. □

Competing interests
The authors declare that they have no competing interests.

Authors' contributions
All authors carried out the proof. All authors conceived of the study and participated in its design and coordination. All authors read and approved the final manuscript.

Author details
[1]School of Sciences, Zhejiang University of Science and Technology, 318 Liuhe Road, Hangzhou, 310023, China. [2]Faculty of Computing, London Metropolitan University, 166–220 Holloway Road, London, N7 8DB, UK.

Acknowledgements
This research is supported by the NNSF of China via Grant 11171306 and the interdisciplinary research funding from Zhejiang University of Science and Technology via Grant 2012JC09Y.

References
1. Agarwal, R: Difference Equations and Inequalities. Dekker, New York (1992)
2. Agarwal, R, Bohner, M, Li, T, Zhang, C: Oscillation of third-order nonlinear delay differential equations. Taiwan. J. Math. **17**, 545-558 (2013)
3. Agarwal, R, Manuel, M, Thandapani, E: Oscillatory and nonoscillatory behavior of second order neutral delay difference equations II. Appl. Math. Lett. **10**(2), 103-110 (1997)
4. Elabbasy, E, Barsom, M, AL-dheleai, F: New oscillation criteria for third-order nonlinear mixed neutral difference equations. Chin. J. Math. **2014**, Article ID 676470 (2014)
5. El-Morshedy, H: New oscillation criteria for second order linear difference equations with positive and negative coefficients. Comput. Math. Appl. **58**, 1988-1997 (2009)
6. Erbe, L, Kong, Q, Zhang, B: Oscillation Theory for Functional Differential Equations. Dekker, New York (1995)
7. Gyori, I, Ladas, G: Oscillation Theory of Delay Differential Equations with Applications. Clarendon Press, Oxford (1991)
8. Karpuz, B, Öalan, Ö, Yıldız, M: Oscillation of a class of difference equations of second order. Math. Comput. Model. **49**, 912-917 (2009)
9. Li, H, Yeh, C: Oscillation criteria for second order neutral delay difference equations. Comput. Math. Appl. **36**(10–12), 123-132 (1998)
10. Meng, Q, Yan, J: Bounded oscillation for second-order nonlinear neutral difference equations in critical and non-critical states. J. Comput. Appl. Math. **211**, 156-172 (2008)
11. Parhi, N, Panda, A: Nonoscillation and oscillation of solutions of a class of third order difference equations. J. Math. Anal. Appl. **336**, 213-223 (2007)
12. Qin, H, Shang, N, Lu, Y: A note on oscillation criteria of second order nonlinear neutral delay differential equations. Comput. Math. Appl. **56**, 2987-2992 (2008)
13. Stavroulakis, I: Oscillation of delay difference equations. Comput. Math. Appl. **29**(7), 83-88 (1995)
14. Wu, S, Hou, Z: Oscillation criteria for a class of neutral difference equations with continuous variable. J. Math. Anal. Appl. **290**, 316-323 (2004)
15. Zhang, Y, Yan, J: Oscillation criteria for difference equations with continuous arguments. Acta Math. Sin. **38**(3), 406-411 (1995)
16. Zhang, Z, Chen, J, Zhang, C: Oscillation of solutions for second order nonlinear difference equations with nonlinear neutral term. Comput. Math. Appl. **41**(12), 1487-1494 (2001)
17. Zhou, X: Oscillatory and asymptotic properties of higher order nonlinear neutral difference equations with oscillating coefficients. Appl. Math. Lett. **21**, 1142-1148 (2008)
18. Zhou, X, Zhang, W: Oscillatory and asymptotic properties of higher order nonlinear neutral difference equations. Appl. Math. Comput. **203**, 679-689 (2008)

Deviating arguments, impulsive effects, and positive solutions for second order singular p-Laplacian equations

Xuemei Zhang[1*] and Meiqiang Feng[2]

*Correspondence: zxm74@sina.com
[1] Department of Mathematics and Physics, North China Electric Power University, Beijing, 102206, People's Republic of China
Full list of author information is available at the end of the article

Abstract

Using a new method for dealing with second order singular p-Laplacian equations with impulsive effects and deviating arguments, several new and more general results are obtained for the existence of at least single, twin or triple positive solutions by using Krasnosel'skii and Zabreiko's fixed point theorem, the fixed point theorem due to Avery and Henderson, and Leggett-Williams' fixed point theorem. We discuss our problems under two cases when the deviating arguments are delayed or advanced. Our results cover equations without deviating arguments and are compared with some recent results by Kajikiya, Lee, and Sim.

Keywords: deviating argument; impulse effect; second order singular p-Laplacian equation; cone and partial ordering

1 Introduction

Second order differential equations with p-Laplacian arise naturally in non-Newtonian mechanics, nonlinear elasticity, glaciology, population biology, combustion theory, and nonlinear flow laws; see [1, 2]. In recent years many cases of the existence, multiplicity, and uniqueness of positive solution of differential equations with p-Laplacian have attracted considerable attention [3–20].

In [5], Kajikiya *et al.* investigated the following one-dimensional p-Laplacian problem:

$$\begin{cases} (\varphi_p(u'))' + \lambda\omega(t)f(u) = 0, & t \in (0,1), \\ u(0) = u(1) = 0, \end{cases}$$

and by virtue of the global bifurcation theory, they obtained the existence, nonexistence, uniqueness, and multiplicity of positive solutions as well as sign-changing solutions under suitable conditions imposed on the nonlinear term f.

In [6], employing the shooting method, Iturriaga *et al.* obtained both existence and the exact number of positive solutions of the problem

$$\begin{cases} \lambda(\varphi_p(u'))' + f(u) = 0 & \text{in } (0,1), \\ u(0) = u(1) = 0. \end{cases}$$

Recently, using global bifurcation theory, Dai and Ma [8] showed the existence of nodal solutions for the following problem:

$$\begin{cases} (\varphi_p(u'))' + f(t,u) = 0, & t \in (0,1), \\ u(0) = u(1) = 0. \end{cases}$$

In [19], Ding and O'Regan studied a second order p-Laplacian differential equation involving the impulsive effect $\Delta u|_{t=t_k} = I_k(u(t_k))$, $k = 1, 2, \ldots, m$. Using Jensen's inequality, the first eigenvalue of a relevant linear operator and the Krasnoselskii-Zabreiko fixed point theorem, the existence and multiplicity of positive solutions were established.

At the same time, a class of p-Laplacian differential equations with deviating arguments both of an advanced or delayed type have received much attention. For example, in [20], Jankowski considered the following third order p-Laplacian differential equation:

$$\begin{cases} (\varphi_p(u''(t)))' + h(t)f(t, u(t), u(\alpha(t))) = 0, & t \in J', \\ u'(t_k^+) = u'(t_k^-) + Q_k(u(t_k)), & k = 1, 2, \ldots, m, \\ \beta u(0) - \gamma u'(0) = 0, \qquad \delta u(1) + \eta u'(1) = 0, & u''(0) = 0, \end{cases}$$

where $\alpha(t) \not\equiv t$ on J. The author obtained the existence of at least three positive solutions. The main tool is a fixed point theorem due to Avery [21], which is a generalization of the Leggett-Williams fixed point theorem.

Of course, a natural question is the following.

Q1. Can the existence of positive solutions for a second order p-Laplacian differential equation with deviating arguments both of an advanced or delayed type be proved?

Remark 1.1 In [20], by means of the properties of Green's function, Jankowski obtain the inequality

$$\min_{t \in [\xi,1]} u(t) \geq \rho \|u\|,$$

where $\xi \in (0,1)$ and $\rho \in (0,1)$.

In fact, the calculation of ρ is very difficult when $t \in [\xi, 1]$. This is probably the main reason that there is almost no paper to study the existence of positive solutions for a class of second order p-Laplacian impulsive differential equations with two parameters and deviating arguments both of an advanced or delayed type. In [20], Jankowski obtained a constant number ρ by means of the properties of Green's function. However, it is well known that there is not any Green's function whatsoever in one-dimensional p-Laplacian boundary value problems of second order differential equations. This implies the following question.

Q2. Can a similar inequality be obtained if there is no Green's function when $t \in [\xi, 1]$?

For this one needs to use a new technique to deal with second order p-Laplacian equations with deviating arguments, especially for second order p-Laplacian equations with impulsive effects.

Consider the one-dimensional p-Laplacian differential equation with deviating arguments

$$-\left(\phi_p\left(u'(t)\right)\right)' = \omega(t) f\left(t, u\left(\alpha(t)\right)\right), \quad t \neq t_k, t \in (0,1), \tag{1.1}$$

subject to one of the following impulsive and boundary conditions:

$$\begin{cases} -\Delta u|_{t=t_k} = I_k(u(t_k)), & k = 1,2,\ldots,n, \\ u'(0) = 0, & u(1) = \int_0^1 g(t)u(t)\,dt \end{cases} \tag{1.2}$$

or

$$\begin{cases} \Delta u|_{t=t_k} = I_k(u(t_k)), & k = 1,2,\ldots,n, \\ u(0) = \int_0^1 h(t)u(t)\,dt, & u'(1) = 0, \end{cases} \tag{1.3}$$

where $\varphi_p(s) = |s|^{p-2}s, p > 1$, $(\varphi_p)^{-1} = \varphi_q$, $\frac{1}{p} + \frac{1}{q} = 1$, ω may be singular at $t = 0$ and/or $t = 1$, t_k ($k = 1,2,\ldots,n$, where n is fixed positive integer) are fixed points with $0 < t_1 < t_2 < \cdots < t_k < \cdots < t_n < 1$, $\Delta u|_{t=t_k}$ denotes the jump of $u(t)$ at $t = t_k$, *i.e.*,

$$\Delta u|_{t=t_k} = u\left(t_k^+\right) - u\left(t_k^-\right),$$

where $u(t_k^+)$ and $u(t_k^-)$ represent the right-hand limit and left-hand limit of $u(t)$ at $t = t_k$, respectively.

Throughout this paper we assume that $k = 1,2,\ldots,n$ and $\alpha(t) \not\equiv t$ on $J = [0,1]$. In addition, $\omega, f, \alpha, I_k, g$, and h satisfy

(H_1) $\omega \in C((0,1),[0,+\infty))$ with $0 < \int_0^1 \omega(s)\,ds < \infty$ and ω does not vanish on any subinterval of $(0,1)$;
(H_2) $f \in C([0,1] \times [0,+\infty),[0,+\infty))$, $\alpha \in C(J,J)$;
(H_3) $I_k \in C([0,+\infty),[0,+\infty))$;
(H_4) $g, h \in L^1[0,1]$ are nonnegative and $\sigma \in [0,1)$, $\varsigma \in [0,1)$, where

$$\sigma = \int_0^1 g(s)\,ds, \qquad \varsigma = \int_0^1 h(s)\,ds. \tag{1.4}$$

Some special cases of problem (1.1)-(1.2) or problem (1.1)-(1.3) have been investigated. For example, Kajikiya *et al.* [22] considered problem (1.1) in the case that $I_k \equiv 0$, $k = 1,2,\ldots,n$, and $\alpha(t) \equiv t$ on J. By using the properties of eigenfunctions and global bifurcation theory, the authors proved the existence, uniqueness, nonexistence, and multiplicity results of positive solutions as well as sign-changing solutions, specially when the nonlinear term is p-linear near 0 and p-sublinear at ∞.

Remark 1.2 There are almost no papers except [20, 23] studying one-dimensional p-Laplacian differential equations with deviating arguments both of an advanced or delayed type using fixed pointed theory. However, in [20, 23], Jankowski only dealt with third order p-Laplacian equations and fourth order p-Laplacian equations. To our knowledge, it is the first paper where positive solutions have been investigated for the class of second order singular p-Laplacian equations with impulsive effects and deviating arguments both of an advanced or delayed type.

In this paper, we will give positive answers to Q1 and Q2. The organization of this paper is as follows. In Section 2, we first present some definitions and lemmas which are needed throughout this paper and then state several fixed point results: Krasnosel'skii and Zabreiko's fixed point theorem, a new fixed point theorem due to Avery, and Henderson and Leggett-Williams' fixed point theorem. In Section 3, we use a fixed point theorem of cone expansion and compression to obtain the existence of at least one or two positive solutions of problem (1.1)-(1.2) with advanced argument α. Section 4 will further discuss the existence of twin positive solutions of problem (1.1)-(1.2) with advanced argument α. Two new results will be presented by a new fixed point theorem due to Avery and Henderson. Section 5 is for developing existence criteria for (at least) three positive solutions of problem (1.1)-(1.2). Finally, in Section 6, we formulate sufficient conditions under which delayed problem (1.1)-(1.3) has at least one, two, and three positive solutions. In particular, our results in these sections are new when $\alpha(t) \equiv t$ on $t \in J$. To the best of the authors' knowledge, it is the first paper where the fixed point theories are applied to p-Laplacian boundary value problems of second order impulsive differential equations with deviating arguments.

2 Preliminaries

Let $J' = J\backslash\{t_1, t_2, \dots, t_n\}$ and E be the Banach space

$$E = \{u | u : J \to R \text{ is continuous at } t \neq t_k, u(t_k^-) = u(t_k) \text{ and } u(t_k^+) \text{ exist}, k = 1, 2, \dots, n\}$$

with $\|u\| = \max_{0 \le t \le 1} |u(t)|$. We denote

$$\Omega_r := \{u \in E : \|u\| < r\}$$

for all $r > 0$ in the sequel.

In our main results, we will make use of the following definitions and lemmas.

Definition 2.1 (See [24]) Let E be a real Banach space over R. A nonempty closed set $P \subset E$ is said to be a cone provided that

(i) $au + bv \in P$ for all $u, v \in P$ and all $a \ge 0$, $b \ge 0$,

(ii) $u, -u \in P$ implies $u = 0$.

Every cone $P \subset E$ induces an ordering in E given by $x \le y$ if and only if $y - x \in P$.

Definition 2.2 The map β is said to be a nonnegative continuous concave functional on a cone P of a real Banach space E provided that $\beta : P \to [0, \infty)$ is continuous and

$$\beta(tx + (1-t)y) \ge t\beta(x) + (1-t)\beta(y)$$

for all $x, y \in P$ and $0 \le t \le 1$.

Definition 2.3 A function $u \in E \cap C^1(0,1)$ is called a solution of (1.1)-(1.2) if it satisfies (1.1)-(1.2). If $u(t) \ge 0$ and $u(t) \not\equiv 0$ on J, then u is called a positive solution of (1.1)-(1.2).

Lemma 2.1 *Assume that* (H_1)-(H_4) *hold. Then $u \in E \cap C^1(0,1)$ is a solution of problem* (1.1)-(1.2) *if and only if $u \in E$ is a solution of the following equations:*

$$u(t) = \frac{1}{1-\sigma}\left\{\int_0^1 g(t)\left[\int_t^1 \phi_q\left(\int_0^s \omega(r)f(r, u(\alpha(r)))\,dr\right)ds\right]dt + \int_0^1 g(t)\left(\sum_{t \le t_k} I_k(u(t_k))\right)dt\right\}$$

$$+\int_t^1 \phi_q\left(\int_0^s \omega(r)f(r,u(\alpha(r)))\,dr\right)ds$$
$$+\sum_{t\le t_k} I_k(u(t_k)) \tag{2.1}$$

and

$$\min_{t\in[\xi,1]} u(t) \ge \delta\|u\|, \tag{2.2}$$

where $0<\xi<t_1$ *and*

$$\delta = \frac{(1-\xi)\int_\xi^1 (1-t)g(t)\,dt}{1-\xi-(1-\xi)\int_\xi^1 g(t)\,dt + \int_\xi^1 (1-t)g(t)\,dt}. \tag{2.3}$$

Proof First suppose that $u\in E\cap C^1(0,1)$ is a solution of problem (1.1)-(1.2). It is easy to see by integration of (1.1) that

$$-\phi_p(u'(t)) + \phi_p(u'(0)) = \int_0^t \omega(s)f(s,u(\alpha(s)))\,ds.$$

By the boundary condition, we have

$$u'(t) = -\phi_q\left(\int_0^t \omega(s)f(s,u(\alpha(s)))\,ds\right). \tag{2.4}$$

If $t_{n-1}<t<t_n$, then integrating (2.4) from t_n to 1, we have

$$u(1)-u(t_n^+) = -\int_{t_n}^1 \phi_q\left(\int_0^s \omega(r)f(r,u(\alpha(r)))\,dr\right)ds. \tag{2.5}$$

Integrating (2.4) from t to t_n, we have

$$u(t_n^-)-u(t) = -\int_t^{t_n} \phi_q\left(\int_0^s \omega(r)f(r,u(\alpha(r)))\,dr\right)ds. \tag{2.6}$$

So, (2.5) and (2.6) imply

$$u(1)-u(t) = -\int_t^1 \phi_q\left(\int_0^s \omega(r)f(r,u(\alpha(r)))\,dr\right)ds - I_n(u(t_n)), \quad t_{n-1}<t<t_n.$$

Repeating the above procession, for $t\in J$, we have

$$u(t) = u(1) + \int_t^1 \phi_q\left(\int_0^s \omega(r)f(r,u(\alpha(r)))\,dr\right)ds + \sum_{t\le t_k} I_k(u(t_k)). \tag{2.7}$$

Substituting $u(1)=\int_0^1 g(t)u(t)\,dt$ into (2.7), we obtain

$$u(t) = \int_0^1 g(t)u(t)\,dt + \int_t^1 \phi_q\left(\int_0^s \omega(r)f(r,u(\alpha(r)))\,dr\right)ds + \sum_{t\le t_k} I_k(u(t_k)).$$

Then we have

$$\int_0^1 g(t)u(t)\,dt = \int_0^1 g(t)u(t)\,dt \times \sigma + \int_0^1 g(t)\Big(\sum_{t\le t_k} I_k\big(u(t_k)\big)\Big)\,dt$$
$$+\int_0^1 g(t)\Big[\int_t^1 \phi_q\Big(\int_0^s \omega(r)f\big(r,u(\alpha(r))\big)\,dr\Big)\,ds\Big]\,dt$$

and

$$u(t) = \frac{1}{1-\sigma}\Big\{\int_0^1 g(t)\Big[\int_t^1 \phi_q\Big(\int_0^s \omega(r)f\big(r,u(\alpha(r))\big)\,dr\Big)\,ds\Big]\,dt$$
$$+\int_0^1 g(t)\Big(\sum_{t\le t_k} I_k\big(u(t_k)\big)\Big)\,dt\Big\}$$
$$+\int_t^1 \phi_q\Big(\int_0^s \omega(r)f\big(r,u(\alpha(r))\big)\,dr\Big)\,ds + \sum_{t\le t_k} I_k\big(u(t_k)\big).$$

Then the proof of sufficiency is complete.

Conversely, we see if $u \in E$ is a solution of (2.1).

Direct differentiation of (2.1) implies, for $t \neq t_k$,

$$u'(t) = -\phi_q\Big(\int_0^t \omega(s)f\big(s,u(\alpha(s))\big)\,ds\Big).$$

Evidently,

$$\big(\phi_p(u'(t))\big)' = -f\big(t,u(\alpha(t))\big),$$
$$\Delta u|_{t=t_k} = -I_k\big(u(t_k)\big) \quad (k=1,2,\dots,n), \qquad u'(0)=0, \qquad u(1) = \int_0^1 g(t)u(t)\,dt.$$

Finally, we show that (2.2) and (2.3) hold. It is clear that $u'(t) = -\phi_q(\int_0^t f(s,u(\alpha(s)))\,ds) < 0$, which implies that

$$\|u\| = u(0), \qquad \min_{t\in[0,1]} u(t) = u(1).$$

On the other hand, for given $s_1, s_2 \in J'$ with $s_1 \le s_2$, one can prove that $u'(s_2) \le u'(s_1)$. Hence, $u'(t)$ is nonincreasing on J'.

So, for every $t \in [\xi, 1)$, we have

$$\frac{u(\xi)-u(1)}{1-\xi} \le \frac{u(t)-u(1)}{1-t},$$

i.e., $(1-\xi)(u(t)-u(1)) \ge (1-t)[u(\xi)-u(1)]$.

Therefore,

$$(1-\xi)\int_\xi^1 g(t)u(t)\,dt - \int_\xi^1 (1-\xi)g(t)u(1)\,dt$$
$$\ge u(\xi)\int_\xi^1 (1-t)g(t)\,dt - \int_\xi^1 (1-t)g(t)u(1)\,dt.$$

This together with the boundary condition $u(1) = \int_0^1 g(t)u(t)\,dt$ implies that

$$u(1) \geq \frac{\int_\xi^1 (1-t)g(t)\,dt}{1-\xi-(1-\xi)\int_\xi^1 g(t)\,dt + \int_\xi^1 (1-t)g(t)\,dt}u(\xi). \tag{2.8}$$

It is clear that (2.8) holds when $t = 1$.

Noticing that

$$\frac{u(0)-u(1)}{1} \leq \frac{u(\xi)-u(1)}{1-\xi},$$

we have

$$u(\xi) \geq \xi u(1) + (1-\xi)u(0) \geq (1-\xi)u(0). \tag{2.9}$$

It follows from (2.8) and (2.9) that (2.2) and (2.3) hold. □

Define a cone K in E by

$$K = \left\{u \in E : u \geq 0, \min_{t\in[\xi,1]} u(t) \geq \delta\|u\|\right\},$$

where δ is defined in (2.3). It is easy to see that K is a closed convex cone of E.

Define $T : K \to E$ by

$$\begin{aligned}(Tu)(t) = {} & \frac{1}{1-\sigma}\left\{\int_0^1 g(t)\left[\int_t^1 \phi_q\left(\int_0^s \omega(r)f(r,u(\alpha(r)))\,dr\right)ds\right]dt\right.\\ & \left. + \int_0^1 g(t)\left(\sum_{t\leq t_k} I_k(u(t_k))\right)dt\right\}\\ & + \int_t^1 \phi_q\left(\int_0^s \omega(r)f(r,u(\alpha(r)))\,dr\right)ds + \sum_{t\leq t_k} I_k(u(t_k)).\end{aligned} \tag{2.10}$$

Obviously, $(Tu)(t) \geq 0$ for $t \in J$.

From the definition of T and the proof of Lemma 2.1, we claim that for each $u \in K$, $Tu \in K$ and satisfies (1.1), and $(Tu)(0)$ is the maximum value of $(Tu)(t)$ on J.

From (2.10) and Lemma 2.1, it is also easy to obtain the following results.

Lemma 2.2 *Assume that* (H_1)-(H_4) *hold. Then problem* (1.1)-(1.2) *is equivalent to the fixed point problem of* T *in* K.

Lemma 2.3 *Assume that* (H_1)-(H_4) *hold. Then* $T : K \to K$ *is completely continuous.*

Lemma 2.4 (Fixed point theorem of cone expansion and compression; see [24]) *Let* Ω_1 *and* Ω_2 *be two bounded open sets in Banach space* E*, such that* $0 \in \Omega_1$ *and* $\bar{\Omega}_1 \subset \Omega_2$*. Let* P *be a cone in* E *and let operator* $A : P \cap (\bar{\Omega}_2 \backslash \Omega_1) \to P$ *be completely continuous. Suppose that one of the following two conditions is satisfied:*

(i) $Ax \not\geq x, \forall x \in P \cap \partial\Omega_1; Ax \not\leq x, \forall x \in P \cap \partial\Omega_2$

and

(ii) $Ax \nleq x, \forall x \in P \cap \partial\Omega_1; Ax \ngeq x, \forall x \in P \cap \partial\Omega_2$.

Then A has at least one fixed point in $P \cap (\Omega_2 \backslash \bar{\Omega}_1)$.

Remark 2.1 To the best of our knowledge, it is the first paper where the fixed point theorem of cone expansion and compression is applied to p-Laplacian differential equations.

Given a nonnegative continuous functional ψ on a cone P of a real Banach space, we define, for each $r > 0$, the set

$$P(\psi, r) = \{u \in P : \psi(u) < r\}.$$

Lemma 2.5 (See [25]) *Let P be a cone in a real Banach space. Let α and γ be increasing nonnegative continuous functional on P, and let θ be a nonnegative continuous functional on P with $\theta(0) = 0$ such that, for some $c > 0$ and $H > 0$,*

$$\gamma(x) \leq \theta(x) \leq \alpha(x) \quad \text{and} \quad \|x\| \leq H\gamma(x)$$

for all $x \in \overline{P(\gamma, c)}$. Suppose there exist a completely continuous operator $A : \overline{P(\gamma, c)} \to P$ and $0 < a < b < c$ such that

$$\theta(\lambda x) \leq \lambda\theta(x) \quad \text{for } 0 \leq \lambda \leq 1 \text{ and } x \in \partial P(\theta, b),$$

and

(i) *$\gamma(Ax) > c$ for all $x \in \partial P(\gamma, c)$;*

(ii) *$\theta(Ax) < b$ for all $x \in \partial P(\theta, b)$;*

(iii) *$P(\alpha, a) \neq \emptyset$ and $\alpha(Ax) > a$ for $x \in \partial P(\alpha, a)$.*

Then A has at least two positive solutions x_1 and x_2 belonging to $\overline{P(\gamma, c)}$ satisfying

$$a < \alpha(x_1) \quad \text{with } \theta(x_1) < b \quad \text{and} \quad b < \theta(x_2) \quad \text{with } \gamma(x_2) < c.$$

Let $0 < a < b$ be given and let β be a nonnegative continuous concave functional on the cone K. Define the convex sets K_a, $K(\beta, a, b)$ by

$$K_a = \{x \in K : \|x\| < a\},$$
$$K(\beta, a, b) = \{x \in K : a \leq \beta(x), \|x\| \leq b\}.$$

Finally we state the Leggett-Williams fixed point theorem [26].

Lemma 2.6 (See [26]) *Let K be a cone in a real Banach space E, $A : \bar{K}_a \to \bar{K}_a$ be completely continuous and β be a nonnegative continuous concave functional on K with $\beta(x) \leq \|x\|$ for all $x \in K_a$. Suppose there exist $0 < d < a < b \leq c$ such that*

(i) *$\{x \in K(\beta, a, b) : \beta(x) > a\} \neq \emptyset$ and $\beta(Ax) > a$ for $x \in K(\beta, a, b)$;*

(ii) *$\|Ax\| < d$ for $\|x\| \leq d$;*

(iii) *$\beta(Ax) > a$ for $x \in K(\beta, a, c)$ with $\|Ax\| > b$.*

Then A has at least three positive solutions, x_1, x_2, x_3, satisfying

$$\|x_1\| < d, \qquad a < \beta(x_2), \qquad \|x_3\| > d \quad \text{and} \quad \beta(x_3) < a.$$

3 Single or twin solutions of problem (1.1)-(1.2) for the case $\alpha(t) \geq t$ on J

For convenience we introduce the following notations:

$$f_{\delta\rho}^{\rho} = \min\left\{\min_{t\in[\xi,1]} \frac{f(t,u)}{\phi_p(\rho)} : u \in [\delta\rho,\rho]\right\}, \qquad I_{\delta\rho}^{\rho}(k) = \min\left\{\frac{I_k(u)}{\rho} : u \in [\delta\rho,\rho]\right\},$$

$$f_0^{\rho} = \max\left\{\max_{t\in J} \frac{f(t,u)}{\phi_p(\rho)} : u \in [0,\rho]\right\}, \qquad I_0^{\rho}(k) = \max\left\{\frac{I_k(u)}{\rho} : u \in [0,\rho]\right\},$$

$$f^{\varrho} = \limsup_{u\to\varrho} \max_{t\in[0,1]} \frac{f(t,u)}{\phi_p(u)}, \qquad I^{\varrho}(k) = \limsup_{u\to\varrho} \frac{I_k(u)}{u},$$

$$f_{\varrho} = \liminf_{u\to\varrho} \min_{t\in[\xi,1]} \frac{f(t,u)}{\phi_p(u)}, \qquad I_{\varrho}(k) = \liminf_{u\to\varrho} \frac{I_k(u)}{u} \qquad (\varrho := \infty \text{ or } 0^+, k = 1,2,\ldots,n),$$

$$\gamma = \varphi_q\left(\int_0^1 \omega(s)\,ds\right), \qquad \sigma_1 = \int_{\xi}^{t_n} g(s)\,ds,$$

$$\frac{1}{l} = \frac{\gamma+n}{1-\sigma}, \qquad \frac{1}{L} = \frac{\sigma_1}{1-\sigma}.$$

Remark 3.1 It follows from the definitions of l and L that $L > l$.

Theorem 3.1 *Suppose* (H_1)-(H_4) *hold,* $\alpha(t) \geq t$ *on* J *and there exist* $\rho_1, \rho_2 \in (0,\infty)$ *with* $\rho_1 < \delta\rho_2$ *such that the following conditions hold:*

(H_5) $f_0^{\rho_1} < \phi_p(l)$, $I_0^{\rho_1}(k) < l$.

(H_6) $I_{\delta\rho_2}^{\rho_2}(k) \geq L$.

Then problem (1.1)-(1.2) *has at least one positive solution* u *with* $\rho_1 < \|u\| \leq \rho_2$.

Proof Considering (H_5), we have $f(t,x) \leq \phi_p(l\rho_1)$, $I_k(x) \leq l\rho_1$ for $0 \leq x \leq \rho_1$, $t \in J$.

Since $0 \leq t \leq \alpha(t) \leq 1$ on J, it follows from $0 \leq x(t) \leq \rho_1$ on J that

$$0 \leq x(\alpha(t)) \leq \rho_1.$$

Next, we prove that

$$Tu \ngeq u, \quad u \in K, \qquad \|u\| = \rho_1. \tag{3.1}$$

In fact, if there exists $u_1 \in K$, $\|u_1\| = \rho_1$ such that $Tu_1 \geq u_1$, then we have

$$\begin{aligned}
\max_{t\in J} u_1(t) \leq \max_{t\in J}\Bigg\{&\frac{1}{1-\sigma}\Bigg\{\int_0^1 g(t)\left[\int_t^1 \phi_q\left(\int_0^s \omega(r)f(r,u_1(\alpha(r)))\,dr\right)ds\right]dt \\
&+ \int_0^1 g(t)\left(\sum_{t\leq t_k} I_k(u_1(t_k))\right)dt\Bigg\} \\
&+ \int_t^1 \phi_q\left(\int_0^s \omega(r)f(r,u_1(\alpha(r)))\,dr\right)ds + \sum_{t\leq t_k} I_k(u_1(t_k))\Bigg\} \\
\leq \frac{1}{1-\sigma}\Bigg\{&\int_0^1 g(t)\left[\int_0^1 \phi_q\left(\int_0^s \omega(r)f(r,u_1(\alpha(r)))\,dr\right)ds\right]dt
\end{aligned}$$

$$\begin{aligned}
&+\int_0^1 g(t)\left(\sum_{k=1}^{n} I_k(u_1(t_k))\right) dt\Bigg\} \\
&+\int_0^1 \phi_q\left(\int_0^s \omega(r)f(r,u_1(\alpha(r)))\,dr\right) ds+\sum_{k=1}^{n} I_k(u_1(t_k)) \\
&=\frac{1}{1-\sigma}\int_0^1 \phi_q\left(\int_0^s \omega(r)f(r,u_1(\alpha(r)))\,dr\right) ds+\frac{1}{1-\sigma}\sum_{k=1}^{n} I_k(u_1(t_k)) \\
&\leq\frac{1}{1-\sigma}\int_0^1 \phi_q\left(\int_0^1 \omega(r)f(r,u_1(\alpha(r)))\,dr\right) ds+\frac{1}{1-\sigma}\sum_{k=1}^{n} I_k(u_1(t_k)) \\
&<\frac{1}{1-\sigma}\phi_q\left(\int_0^1 \omega(r)\phi_p(l\rho_1)\,dr\right)+\frac{1}{1-\sigma}\sum_{k=1}^{n} l\rho_1 \\
&=\frac{l\rho_1}{1-\sigma}\phi_q\left(\int_0^1 \omega(r)\,dr\right)+\frac{1}{1-\sigma}\sum_{k=1}^{n} l\rho_1 \\
&=\frac{l\rho_1}{1-\sigma}\gamma+\frac{n}{1-\sigma}l\rho_1 \\
&=\rho_1=\|u_1\|.
\end{aligned}$$

This implies $\|u_1\| < \|u_1\|$, which is a contraction. Hence (3.1) holds.

Considering (H_6), we have $I_k(x) \geq L\rho_2$ for $\delta\rho_2 \leq x \leq \rho_2$.

Next, we prove that

$$Tu \nleq u, \quad u \in K, \qquad \|u\| = \rho_2. \tag{3.2}$$

In fact, if there exists $u_0 \in K$, $\|u_0\| = \rho_2$ such that $Tu_0 \leq u_0$, then we have

$$\delta\rho_2 \leq u_0 \leq \rho_2$$

for $t \in [\xi, 1]$, and then

$$\begin{aligned}
0 \leq t \in J \quad \Rightarrow \quad u_0(t) &\geq (Tu_0)(t) \\
&=\frac{1}{1-\sigma}\left\{\int_0^1 g(t)\left[\int_t^1 \phi_q\left(\int_0^s \omega(r)f(r,u_0(\alpha(r)))\,dr\right) ds\right] dt\right. \\
&\quad+\left.\int_0^1 g(t)\left(\sum_{t\leq t_k} I_k(u_0(t_k))\right) dt\right\} \\
&\quad+\int_t^1 \phi_q\left(\int_0^s \omega(r)f(r,u_0(\alpha(r)))\,dr\right) ds+\sum_{t\leq t_k} I_k(u_0(t_k)) \\
&\geq\frac{1}{1-\sigma}\int_0^1 g(t)\left(\sum_{t\leq t_k} I_k(u_0(t_k))\right) dt \\
&\geq\frac{1}{1-\sigma}\left[\int_{t_{n-1}}^{t_n} g(t)I_n(u_0(t_n))\,dt\right. \\
&\quad+\int_{t_{n-2}}^{t_{n-1}} g(t)(I_{n-1}(u_0(t_{n-1}))+I_n(u_0(t_n)))\,dt+\cdots
\end{aligned}$$

$$+\int_0^{t_1} g(t)\left(I_1\left(u_0(t_1)\right)+I_2\left(u_0(t_2)\right)+\cdots+I_n\left(u_0(t_n)\right)\right)dt\Bigg]$$

$$=\frac{1}{1-\sigma}\left[\int_0^{t_n} g(t)I_n\left(u_0(t_n)\right)dt\right.$$

$$\left.+\int_0^{t_{n-1}} g(t)I_{n-1}\left(u_0(t_{n-1})\right)dt+\cdots+\int_0^{t_1} g(t)I_1\left(u_0(t_1)\right)dt\right]$$

$$\geq\frac{1}{1-\sigma}\int_0^{t_n} g(t)I_n\left(u_0(t_n)\right)dt$$

$$\geq\frac{1}{1-\sigma}\int_\xi^{t_n} g(t)I_n\left(u_0(t_n)\right)dt$$

$$=\frac{1}{1-\sigma}L\rho_2\sigma_1$$

$$=\rho_2=\|u_0\|.$$

This is also a contraction. Hence (3.2) holds.

Applying (i) of Lemma 2.4 to (3.1) and (3.2) shows that T has a fixed point $u\in K\cap(\Omega_{\rho_2}\backslash\bar{\Omega}_{\rho_1})$. Thus, it follows that problem (1.1)-(1.2) has one positive solution u satisfying $\rho_1<\|u\|\leq\rho_2$. This gives the proof of the theorem. □

Noticing that $L>l$, it follows from the proof of Theorem 3.1 that:

Corollary 3.1 *Suppose* (H_1)-(H_4) *hold,* $\alpha(t)\geq t$ *on* J *and there exist* $\rho_1,\rho_2\in(0,\infty)$ *with* $L\rho_2<l\rho_1$ *such that* (H_5) *and* (H_6) *hold. Then problem* (1.1)-(1.2) *has at least one positive solution* u *with* $\rho_2<\|u\|\leq\rho_1$.

Corollary 3.2 *Suppose* (H_1)-(H_4) *hold,* $\alpha(t)\geq t$ *on* J *and there exist* $\rho_1,\rho_2\in(0,\infty)$ *with* $\rho_1<\rho_2$ *such that the following conditions hold*:

(H_5^*) $f_0^{\rho_2}<\phi_p(l)$, $I_0^{\rho_2}(k)<l$.
(H_6^*) $I_{\delta\rho_1}^{\rho_1}(k)\geq L$.

Then problem (1.1)-(1.2) *has at least one positive solution* u *with* $\rho_1<\|u\|<\rho_2$.

Corollary 3.3 *Suppose* (H_1)-(H_4) *hold and* $\alpha(t)\geq t$ *on* J. *Assume that one of the following conditions holds*:

(H_7) $f^0<\phi_p(l)$, $I^0(k)<ql$, $I_\infty(k)\geq\frac{L}{\delta}$.
(H_8) $I_0(k)\geq\frac{L}{\delta}$, $f^\infty<\phi_p(l)$, $I^\infty(k)<l$.

Then problem (1.1)-(1.2) *has at least one positive solution.*

Proof We show that (H_7) implies (H_5) and (H_6). Suppose that (H_7) holds. Then there exists $0<\rho_1<\delta\rho_2$ such that

$$\frac{f(t,u)}{\phi_p(u)}<\phi_p(l),\qquad\frac{I_k(u)}{u}<l\quad\text{for } t\in J, 0<u\leq\rho_1$$

and

$$\frac{I_k(u)}{u}\geq\frac{L}{\delta}\quad\text{for } u\geq\delta\rho_2.$$

Hence, we obtain

$$f(t,u) < \phi_p(l)\phi_p(u) \le \phi_p(l)\phi_p(\rho_1) = \phi_p(l\rho_1), \qquad I_k(u) < lu \le l\rho_1 \quad \text{for } t \in J, 0 < u \le \rho_1$$

and

$$I_k(u) \ge \frac{L}{\delta}u \ge \frac{L}{\delta}\delta\rho_2 = L\rho_2 \quad \text{for } u \ge \delta\rho_2.$$

Noticing $\delta\rho_2 < \rho_2$, it follows that

$$I_k(u) \ge L\rho_2 \quad \text{for } \delta\rho_2 \le u \le \rho_2.$$

Therefore, (H_5) and (H_6) hold. Hence, by Theorem 3.1, problem (1.1)-(1.2) has at least one positive solution.

Similarly, (H_8) implies that (H_5^*) and (H_6^*) hold. □

Remark 3.2 It is easy to see that Corollary 3.3 holds if we change (H_7) into the following condition:

(H_7^*) $f^0 = 0$, $I^0(k) = 0$, and $I_\infty(k) = \infty$.

However, we give no information whatsoever on the existence of positive solution for problem (1.1)-(1.2) if we change (H_8) into the following condition:

(H_8^*) $f^\infty = 0$, $I^\infty(k) = 0$, and $I_0(k) = \infty$.

Theorem 3.2 *Suppose* (H_1)-(H_4), $\alpha(t) \ge t$ *on* J *and one of the following conditions holds*:

(H_9) *There exist* $\rho_1, \rho_2, \rho_3 \in (0,\infty)$, *with* $\rho_1 < \delta\rho_2$ *and* $\rho_2 < \rho_3$ *such that*

$$f_0^{\rho_1} < \phi_p(l), \qquad I_0^{\rho_1}(k) < l\rho_1, \qquad I_{\delta\rho_2}^{\rho_2}(k) > \phi_p(L),$$
$$f_0^{\rho_3} < \phi_p(l), \qquad I_0^{\rho_3}(k) < l\rho_3.$$

(H_{10}) *There exist* $\rho_1, \rho_2, \rho_3 \in (0,\infty)$, *with* $\rho_1 < \rho_2 < \rho_3$ *such that*

$$I_{\gamma\rho_1}^{\rho_1}(k) > \phi_p(L), \qquad f_0^{\rho_2} < \phi_p(l), \qquad I_0^{\rho_2}(k) < l\rho_2, \qquad I_{\gamma\rho_3}^{\rho_3}(k) > \phi_p(L).$$

Then problem (1.1)-(1.2) *has at least two positive solutions* u_1, u_2 *with* $u_1 \in K \cap (\Omega_{\rho_2} \backslash \bar{\Omega}_{\rho_1})$, $u_2 \in K \cap (\Omega_{\rho_3} \backslash \bar{\Omega}_{\rho_2})$.

Proof We only consider the condition (H_9). If (H_{10}) holds, then the proof is similar to that of the case when (H_9) holds.

In fact, it follows from the proof of Theorem 3.1 that

$$Tu \not\geq u, \quad u \in K, \qquad \|u\| = \rho_3. \tag{3.3}$$

Applying Lemma 2.4 to (3.1)-(3.3) shows that T has two fixed points u_1 and u_2 such that $u_1 \in K \cap (\Omega_{\rho_2} \backslash \bar{\Omega}_{\rho_1})$, $u_2 \in K \cap (\Omega_{\rho_3} \backslash \bar{\Omega}_{\rho_2})$. These are the desired distinct positive solutions of problem (1.1)-(1.2). Then proof of Theorem 3.2 is complete. □

Corollary 3.4 *Suppose* (H_1)-(H_4), $\alpha(t) \geq t$ *on* J, *and there exists* $\rho \in (0, \infty)$ *such that one of the following conditions holds*:

(H_{11}) $0 \leq f^0 < \phi_p(l)$, $0 \leq I^0(k) < l$, $I^{\rho}_{\delta\rho} > L$, *and* $0 \leq f^{\infty} < \phi_p(l)$, $0 \leq I^{\infty}(k) < l$.
(H_{12}) $\phi_p(L) < f_0 \leq \infty$, $L < I_0(k) \leq \infty$, $f_0 < \phi_p(l)$, $I_0(k) < l$, *and*

$$\phi_p(L) < f_{\infty} \leq \infty, \qquad L < I_{\infty}(k) \leq \infty.$$

Then problem (1.1)-(1.2) *has at least two positive solutions in* K.

Proof We show that (H_{11}) implies (H_9). It is easy to verify that $0 \leq f^0 < \phi_p(l)$, $0 \leq I^0(k) < l$ imply that there exists $\rho_1 \in (0, \delta\rho)$ such that

$$f_0^{\rho_1} < l, \qquad I_0^{\rho_1}(k) < l.$$

Let $\eta \in (f^{\infty}, \phi_p(l))$. Then there exists $r > \eta$ such that $\max_{t\in J} f(t,u) \leq \eta\phi_p(u)$ for $u \in [r, \infty)$ since $0 \leq f^{\infty} < \phi_p(u)$. Let

$$\beta = \max\left\{\max_{t\in J} f(t,u) : 0 \leq u \leq r\right\} \quad \text{and} \quad \rho_3 > \max\left\{\phi_q\left(\frac{\beta}{\phi_p(l) - \eta}\right), \rho\right\}.$$

Then we have

$$\max_{0\leq t\leq 1} f(t,u) \leq \eta\phi_p(u) + \beta \leq \eta\phi_p(\rho_3) + \beta < \phi_p(l)\phi_p(\rho_3), \quad \forall u \in [0, \rho_3].$$

This implies that $f_0^{\rho_3} \leq \phi_p(l)$. Similarly, from $0 \leq I_k^{\infty} < l$, we have $I_0^{\rho_3}(k) \leq l\rho_3$. Hence, (H_{11}) implies (H_9). Similarly (H_{12}) implies (H_{10}), and the corollary is proved. □

Theorem 3.2 can be generalized and we obtain many solutions.

Theorem 3.3 *Suppose that* (H_1)-(H_3) *hold and* $\alpha(t) \geq t$ *on* J. *Then we have the following assertions.*

(1) *If there exists* $\{\rho_i\}_{i=1}^{2m_0} \subset (0, \infty)$ *with* $\rho_1 < \delta\rho_2 < \rho_2 < \rho_3 < \delta\rho_4 < \cdots < \rho_{2m_0}$ *such that*

$$f_0^{\rho_{2m-1}} < \phi_p(l), \qquad I_0^{\rho_{2m-1}}(k) < l\rho_{2m-1}, \qquad I_{\delta\rho_{2m}}^{\rho_{2m}}(k) > L, \quad m = 1, 2, \ldots, m_0,$$

then problem (1.1)-(1.2) *has at least* $2m_0$ *solutions in* K.

(2) *If there exists* $\{\rho_i\}_{i=1}^{2m_0} \subset (0, \infty)$ *with* $\rho_1 < \rho_2$ *and* $\rho_2 < \delta\rho_3 < \rho_3 < \rho_4 < \delta\rho_5 < \cdots < \rho_{2m_0+2}$ *such that*

$$I_{\delta\rho_{2m-1}}^{\rho_{2m-1}}(k) > L, \qquad f_0^{\rho_{2m}} < \phi_p(l), \qquad I_0^{\rho_{2m}} < l\rho_{2m}, \quad m = 1, 2, \ldots, m_0,$$

then problem (1.1)-(1.2) *has at least* $2m_0 - 1$ *solutions in* K.

4 Further results on twin solutions for the case of $\alpha(t) \geq t$ on J

In the previous section, we have obtained some results on the existence of at least two positive solutions for problem (1.1)-(1.2). In this section, we will further discuss the existence of two positive solutions for problem (1.1)-(1.2) under the more general case.

For $u \in K$, we define the nonnegative increasing continuous functionals θ, θ_1, and θ_2 by

$$\theta = \min_{t\in[\xi,1]} u(t) = u(1),$$

$$\theta_1 = \max_{t\in[0,\xi]} u(t) = u(0)$$

and

$$\theta_2 = \max_{t\in[0,1]} u(t) = u(0).$$

It is easy to see that, for each $u \in K$,

$$\theta(u) \le \theta_1(u) = \theta_2(u). \tag{4.1}$$

In addition, for each $u \in K, \theta(u) = u(1) \ge \delta\|u\|$, which implies that

$$\|u\| \le \frac{1}{\delta}\theta(u) \quad \text{for all } u \in K. \tag{4.2}$$

Finally, we also note that

$$\theta_1(\lambda u) = \lambda\theta_1(u), \quad 0 \le \lambda \le 1 \text{ and } u \in \partial K(\theta_1, r) \text{ for } r > 0.$$

For notational convenience, we denote

$$N = \frac{\sigma_1}{1-\sigma}, \qquad N_1 = \frac{n+\gamma}{1-\sigma}.$$

We now present the results in this section.

Theorem 4.1 *Suppose that there exist constants $0 < a < b < c$ such that*

$$0 < a < \frac{N}{N_1}b < \frac{\delta N}{N_1}c.$$

Assume f and I_k satisfy the following conditions:

(C_1) $I_k(u) > \frac{c}{N}$ *for* $u \in [c, \frac{c}{\delta}]$;
(C_2) $f(t,u) \le \phi_p(\frac{b}{N_1})$ *for* $(t,u) \in J \times [0, \frac{b}{\delta}]$, $I_k(u) \le \frac{b}{N_1}$ *for* $u \in [0, \frac{b}{\delta}]$;
(C_3) $I_k(u) > \frac{a}{N}$ *for* $u \in [\delta a, a]$.

Then problem (1.1)-(1.2) *has at least two positive solutions u_1 and u_2 such that*

$$a < \theta_2(u_1) \quad \textit{with } \theta_1(u_1) < b \quad \textit{and} \quad b < \theta_1(u_2) \quad \textit{with } \theta(u_2) < c.$$

Proof By the definition of the operator T and its properties, it suffices to show that the conditions of Lemma 2.5 hold with respect to T.

We first show that if $u \in \partial K(\theta, c)$, then $\theta(Tu) > c$.

In fact, if $u \in \partial K(\theta, c)$, then $\theta(u) = \min_{t\in[\xi,1]} u(t) = u(1) = c$. Since $u \in K$, one can get $u(t) \geq c$ for $t \in [\xi, 1]$. Noticing (4.2), we have

$$c \leq u(t) \leq \frac{1}{\delta} c, \quad t \in [\xi, 1].$$

As a consequence of (C_1)

$$I_k(u(t)) > \frac{c}{N} \quad \text{for } t \in [\xi, 1].$$

Also, $Tu \in K$, and hence we get

$$\begin{aligned}
\theta(Tu) &= \min_{t\in[\xi,1]} (Tu)(t) = (Tu)(1) \\
&= \frac{1}{1-\sigma}\left\{\int_0^1 g(t)\left[\int_1^1 \phi_q\left(\int_0^s \omega(r) f(r, u(\alpha(r)))\, dr\right) ds\right] dt \right. \\
&\quad \left. + \int_0^1 g(t)\left(\sum_{t\leq t_k} I_k(u(t_k))\right) dt\right\} \\
&\quad + \int_1^1 \phi_q\left(\int_0^s \omega(r) f(r, u(\alpha(r)))\, dr\right) ds + \sum_{t\leq t_k} I_k(u(t_k)) \\
&= \frac{1}{1-\sigma}\int_0^1 g(t)\left(\sum_{t\leq t_k} I_k(u(t_k))\right) dt + \sum_{t\leq t_k} I_k(u(t_k)) \\
&\geq \frac{1}{1-\sigma}\int_0^1 g(t)\left(\sum_{t\leq t_k} I_k(u(t_k))\right) dt \\
&\geq \frac{1}{1-\sigma}\int_0^{t_n} g(t) I_n(u(t_n))\, dt \\
&\geq \frac{1}{1-\sigma}\int_\xi^{t_n} g(t) I_n(u(t_n))\, dt \\
&> \frac{1}{1-\sigma}\int_\xi^{t_n} g(t)\frac{c}{N}\, dt \\
&= \frac{1}{1-\sigma}\frac{c}{N}\sigma_1 \\
&= c.
\end{aligned}$$

Next, we verify that $\theta_1(Tu) < b$ for $u \in \partial K(\theta_1, b)$.
So, letting $u \in \partial K(\theta_1, b)$, then

$$\theta_1(u) = \max_{t\in[0,\xi]} u(t) = u(0) = b.$$

This implies that $0 \leq u(t) \leq b$ for $t \in [0, \xi]$, and since $u \in K$, we also have

$$u(t) \leq \|u\| \leq \frac{1}{\delta}\theta(u) \leq \frac{1}{\delta}\theta_1(u) = \frac{1}{\delta}u(0) = \frac{1}{\delta}b,$$

which implies that

$$0 \le u(t) \le \frac{1}{\delta} b, \quad t \in J.$$

It follows from (C_2) that

$$f(t,u) \le \phi_p\left(\frac{b}{N_1}\right) \quad \text{for } (t,u) \in J \times \left[0, \frac{b}{\delta}\right], \qquad I_k(u) \le \frac{b}{N_1} \quad \text{for } u \in \left[0, \frac{b}{\delta}\right].$$

Noticing $Tu \in K$, we obtain

$$\begin{aligned}
(Tu)(t) &= \frac{1}{1-\sigma}\left\{\int_0^1 g(t)\left[\int_t^1 \phi_q\left(\int_0^s \omega(r) f\big(r, u(\alpha(r))\big)\, dr\right) ds\right] dt\right. \\
&\quad \left. + \int_0^1 g(t)\left(\sum_{t \le t_k} I_k\big(u(t_k)\big)\right) dt\right\} \\
&\quad + \int_t^1 \phi_q\left(\int_0^s \omega(r) f\big(r, u(\alpha(r))\big)\, dr\right) ds + \sum_{t \le t_k} I_k\big(u(t_k)\big) \\
&\le \frac{1}{1-\sigma}\left\{\int_0^1 g(t)\left[\int_0^1 \phi_q\left(\int_0^s \omega(r) f\big(r, u(\alpha(r))\big)\, dr\right) ds\right] dt\right. \\
&\quad \left. + \int_0^1 g(t)\left(\sum_{k=1}^{n} I_k\big(u(t_k)\big)\right) dt\right\} \\
&\quad + \int_0^1 \phi_q\left(\int_0^s \omega(r) f\big(r, u(\alpha(r))\big)\, dr\right) ds + \sum_{k=1}^{n} I_k\big(u(t_k)\big) \\
&= \frac{1}{1-\sigma}\int_0^1 \phi_q\left(\int_0^s \omega(r) f\big(r, u(\alpha(r))\big)\, dr\right) ds + \frac{1}{1-\sigma}\sum_{k=1}^{n} I_k\big(u(t_k)\big) \\
&\le \frac{1}{1-\sigma}\int_0^1 \phi_q\left(\int_0^s \omega(r)\phi_p\left(\frac{b}{N_1}\right) dr\right) ds + \frac{1}{1-\sigma}\sum_{k=1}^{n} \frac{b}{N_1} \\
&= \frac{1}{1-\sigma}\frac{b}{N_1}\int_0^1 \phi_q\left(\int_0^s \omega(r)\, dr\right) ds + \frac{1}{1-\sigma}\sum_{k=1}^{n} \frac{b}{N_1} \\
&\le \frac{1}{1-\sigma}\frac{b}{N_1}\phi_q\left(\int_0^1 \omega(r)\, dr\right) + \frac{1}{1-\sigma} n \frac{b}{N_1} \\
&= \frac{1}{1-\sigma}\frac{b}{N_1}\gamma + \frac{1}{1-\sigma} n \frac{b}{N_1} \\
&= \frac{b}{N_1(1-\sigma)}(n+\gamma) \\
&= b.
\end{aligned}$$

Finally, we prove that $K(\theta_2, a) \ne \emptyset$ and $\theta_2(Tu) > a$ for all $u \in \partial K(\theta_2, a)$.
In fact, the constant $\frac{a}{3} \in K(\theta_2, a)$. Moreover, for $u \in \partial K(\theta_2, a)$, we have

$$\theta_2(u) = \max_{t \in J} u(t) = u(0) = a,$$

which implies that

$$\delta a \le u(t) \le a, \quad t \in [\xi, 1].$$

Using assumption (C_3), we have

$$I_k(u(t)) > \frac{a}{N}, \quad t \in [\xi, 1].$$

As before $Tu \in K$, and so

$$\begin{aligned}
\theta_2(Tu) &= \max_{t\in J}(Tu)(t) = (Tu)(0) \\
&= \frac{1}{1-\sigma}\left\{\int_0^1 g(t)\left[\int_0^1 \phi_q\left(\int_0^s \omega(r)f(r,u(\alpha(r)))\,dr\right)ds\right]dt\right. \\
&\quad \left. + \int_0^1 g(t)\left(\sum_{t\le t_k} I_k(u(t_k))\right)dt\right\} \\
&\quad + \int_0^1 \phi_q\left(\int_0^s \omega(r)f(r,u(\alpha(r)))\,dr\right)ds + \sum_{t\le t_k} I_k(u(t_k)) \\
&\ge \frac{1}{1-\sigma}\int_0^1 g(t)\left(\sum_{t\le t_k} I_k(u(t_k))\right)dt + \sum_{t\le t_k} I_k(u(t_k)) \\
&\ge \frac{1}{1-\sigma}\int_0^1 g(t)\left(\sum_{t\le t_k} I_k(u(t_k))\right)dt \\
&\ge \frac{1}{1-\sigma}\int_0^{t_n} g(t)I_n(u(t_n))\,dt \\
&\ge \frac{1}{1-\sigma}\int_\xi^{t_n} g(t)I_n(u(t_n))\,dt \\
&> \frac{1}{1-\sigma}\int_\xi^{t_n} g(t)\frac{a}{N}\,dt \\
&= \frac{1}{1-\sigma}\frac{a}{N}\sigma_1 \\
&= a.
\end{aligned}$$

Thus, by Lemma 2.5, there exist at least two fixed points of T which are positive solutions u_1 and u_2, belonging to $\overline{K(\theta, c)}$, of problem (1.1)-(1.2) such that

$$a < \theta_2(u_1) \quad \text{with } \theta_1(u_1) < b \quad \text{and} \quad b < \theta_1(u_2) \quad \text{with } \theta(u_2) < c. \qquad \square$$

We remark that the condition (C_2) in Theorem 4.1 can be replaced by the following conditions:

$(C_2)'$ $f_0^b \le \phi_p(\frac{1}{N_1})$, $I_0^b(k) \le \frac{1}{N_1}$;
$(C_2)''$ $f^0 \le \phi_p(\frac{1}{N_1})$, $I^0(k) \le \frac{1}{N_1}$.

Corollary 4.1 *If the condition* (C_2) *in Theorem* 4.1 *is replaced by* $(C_2)'$ *and* $(C_2)''$, *respectively, then the conclusion of Theorem* 4.1 *also holds.*

Proof It follows from the proof of Theorem 4.1 that Corollary 4.1 holds. □

5 Triple solutions for the case of $\alpha(t) \geq t$ on J

In this section, we shall study the existence of three positive solutions of problem (1.1)-(1.2). Let the nonnegative continuous concave functional $\psi(u) : K \to [0,\infty)$ be defined by

$$\psi(u) = \min_{t\in[\xi,1]} u(t) = u(1), \quad u \in K.$$

Note that for $u \in K$, $\psi(u) \leq \|u\|$.

Theorem 5.1 *Suppose that there exist constants* $0 < d < a < \frac{a}{\delta} \leq c$ *such that*

(A_1) $f(t,u) \leq \phi_p(\frac{d}{N_1})$ *for* $(t,u) \in J \times [0,d]$, $I_k(u) \leq \frac{d}{N_1}$ *for* $u \in [0,d]$;
(A_2) $I_k(u) \geq \frac{a}{N}$ *for* $u \in [a,\frac{a}{\delta}]$;
(A_3) $f(t,u) \leq \phi_p(\frac{c}{N_1})$ *for* $(t,u) \in J \times [0,c]$, $I_k(u) \leq \frac{c}{N_1}$ *for* $u \in [0,c]$.

Then problem (1.1)-(1.2) *has at least three positive solutions* u_1, u_2, *and* u_3 *such that*

$$\|u_1\| < d, \qquad a < \psi(u_2) \quad \text{and} \quad u_3 > d \quad \text{with } \psi(u_3) < a.$$

Proof By the definition of the operator T and its properties, it suffices to show that the conditions of Lemma 2.6 hold with respect to T.

For convenience, we denote $b = \frac{a}{\delta}$.

We first show that if there exists a positive number r such that $f(t,u) \leq \phi_p(\frac{r}{N_1})$ for $(t,u) \in J \times [0,r]$, $I_k(u) \leq \frac{r}{N_1}$ for $u \in [0,r]$, then $T : \bar{K}_r \to K_r$.

Indeed, if $u \in \bar{K}_r$, then

$$\begin{aligned}
(Tu)(t) &= \frac{1}{1-\sigma}\left\{\int_0^1 g(t)\left[\int_t^1 \phi_q\left(\int_0^s \omega(r)f\big(r,u(\alpha(r))\big)\,dr\right)ds\right]dt\right.\\
&\quad \left.+\int_0^1 g(t)\left(\sum_{t\leq t_k} I_k\big(u(t_k)\big)\right)dt\right\}\\
&\quad +\int_t^1 \phi_q\left(\int_0^s \omega(r)f\big(r,u(\alpha(r))\big)\,dr\right)ds+\sum_{t\leq t_k} I_k\big(u(t_k)\big)\\
&\leq \frac{1}{1-\sigma}\left\{\int_0^1 g(t)\left[\int_0^1 \phi_q\left(\int_0^s \omega(r)f\big(r,u(\alpha(r))\big)\,dr\right)ds\right]dt\right.\\
&\quad \left.+\int_0^1 g(t)\left(\sum_{k=1}^n I_k\big(u(t_k)\big)\right)dt\right\}\\
&\quad +\int_0^1 \phi_q\left(\int_0^s \omega(r)f\big(r,u(\alpha(r))\big)\,dr\right)ds+\sum_{k=1}^n I_k\big(u(t_k)\big)\\
&= \frac{1}{1-\sigma}\int_0^1 \phi_q\left(\int_0^s \omega(r)f\big(r,u(\alpha(r))\big)\,dr\right)ds+\frac{1}{1-\sigma}\sum_{k=1}^n I_k\big(u(t_k)\big)\\
&\leq \frac{1}{1-\sigma}\int_0^1 \phi_q\left(\int_0^s \omega(r)\phi_p\left(\frac{r}{N_1}\right)dr\right)ds+\frac{1}{1-\sigma}\sum_{k=1}^n \frac{r}{N_1}
\end{aligned}$$

$$
\begin{aligned}
&= \frac{1}{1-\sigma}\frac{r}{N_1}\int_0^1 \phi_q\left(\int_0^s \omega(r)\,dr\right)ds + \frac{1}{1-\sigma}\sum_{k=1}^{n}\frac{r}{N_1} \\
&\le \frac{1}{1-\sigma}\frac{r}{N_1}\phi_q\left(\int_0^1 \omega(r)\,dr\right) + \frac{1}{1-\sigma}n\frac{r}{N_1} \\
&= \frac{1}{1-\sigma}\frac{r}{N_1}\gamma + \frac{1}{1-\sigma}n\frac{r}{N_1} \\
&= \frac{r}{N_1(1-\sigma)}(n+\gamma) \\
&= r,
\end{aligned}
$$

which implies that $Tu \in K_r$.

Hence, we have shown that if (A_1) and (A_3) hold, then T maps $\bar{K}_d$ into K_d and $\bar{K}_c$ into K_c.

Next, we verify that $\{u \in K(\psi, a, b) : \psi(u) > a\} \neq \emptyset$ and $\psi(Tu) > a$ for all $u \in K(\psi, a, b)$.

Take $u(t) = \frac{a+b}{2}$, $t \in J$. Then

$$
\|u\| = \frac{a+b}{2} < b, \qquad \psi(u) = \min_{t\in[\xi,1]} u(t) = \frac{a+b}{2} > a.
$$

This shows that

$$
\left\{u \in K(\psi, a, b) : \psi(u) > a\right\} \neq \emptyset.
$$

Therefore, it follows from (A_2) that

$$
\begin{aligned}
\psi(Tu) &= \min_{t\in[\xi,1]} (Tu)(t) = (Tu)(1) \\
&= \frac{1}{1-\sigma}\left\{\int_0^1 g(t)\left[\int_1^1 \phi_q\left(\int_0^s \omega(r) f(r, u(\alpha(r)))\,dr\right)ds\right]dt \right.\\
&\quad \left. + \int_0^1 g(t)\left(\sum_{t\le t_k} I_k(u(t_k))\right)dt\right\} \\
&\quad + \int_1^1 \phi_q\left(\int_0^s \omega(r) f(r, u(\alpha(r)))\,dr\right)ds + \sum_{t\le t_k} I_k(u(t_k)) \\
&= \frac{1}{1-\sigma}\int_0^1 g(t)\left(\sum_{t\le t_k} I_k(u(t_k))\right)dt + \sum_{t\le t_k} I_k(u(t_k)) \\
&\ge \frac{1}{1-\sigma}\int_0^1 g(t)\left(\sum_{t\le t_k} I_k(u(t_k))\right)dt \\
&\ge \frac{1}{1-\sigma}\int_0^{t_n} g(t) I_n(u(t_n))\,dt \\
&\ge \frac{1}{1-\sigma}\int_\xi^{t_n} g(t) I_n(u(t_n))\,dt \\
&\ge \frac{1}{1-\sigma}\int_\xi^{t_n} g(t)\frac{a}{N}\,dt \\
&= \frac{1}{1-\sigma}\frac{a}{N}\sigma_1 \\
&= a.
\end{aligned}
$$

Finally, we assert that if $u \in K(\psi, a, c)$ and $\|Tu\| > b$, then $\psi(Tu) > a$.
Suppose $u \in K(\psi, a, c)$ and $\|Tu\| > b$, then

$$\begin{aligned}\psi(Tu) &= \min_{t\in[\xi,1]}(Tu)(t) = (Tu)(1)\\ &\geq \delta\|Tu\|\\ &> \delta b = a.\end{aligned}$$

To sum up, the hypotheses of Lemma 2.6 hold. Therefore, an application of Lemma 2.6 implies problem (1.1)-(1.2) has at least three positive solutions u_1, u_2, and u_3 such that

$$\|u_1\| < d, \qquad a < \psi(u_2) \quad \text{and} \quad u_3 > d \quad \text{with } \psi(u_3) < a. \qquad \square$$

We remark that the condition (A_3) in Theorem 5.1 can be replaced by the following conditions:

$(A_3)'$ $f_0^c \leq \phi_p(\frac{1}{N_1})$, $I_0^c(k) \leq \frac{1}{N_1}$;
$(A_3)''$ $f^0 \leq \phi_p(\frac{1}{N_1})$, $I^0(k) \leq \frac{1}{N_1}$.

Corollary 5.1 *If the condition* (A_3) *in Theorem* 5.1 *is replaced by* $(A_3)'$ *and* $(A_3)''$, *respectively, then the conclusion of Theorem* 5.1 *also holds.*

Proof It follows from the proof of Theorem 5.1 that Corollary 5.1 holds. $\square$

6 Case of $\alpha(t) \leq t$ on J

Now we deal with problem (1.1)-(1.3) for the case of $\alpha(t) \leq t$ on J. Let E be as defined in Section 2. We define a cone K^* in E by

$$K^* = \left\{u \in E : u \geq 0, \min_{t\in[0,\xi^*]} u(t) \geq \delta^*\|u\|\right\},$$

where $\xi^* \in (t_n, 1)$ and

$$\delta^* = \frac{\xi^* \int_0^{\xi^*} th(t)\,dt}{\xi^* - \xi^*\varsigma + \int_0^{\xi^*} th(t)\,dt}.$$

It is easy to see that K^* is a closed convex cone of E.
Define $T^* : K^* \to E$ by

$$\begin{aligned}(T^*u)(t) = {}& \frac{1}{1-\varsigma}\left\{\int_0^1 g(t)\left[\int_0^t \varphi_q\left(\int_s^1 \omega(r)f\big(r, u(\alpha(r))\big)\,dr\right)ds\right]dt\right.\\ &\left.+ \int_0^1 g(t)\sum_{t_k<t} I_k\big(u(t_k)\big)\,dt\right\}\\ &+ \int_0^t \varphi_q\left(\int_s^1 \omega(r)f\big(r, u(\alpha(r))\big)\,dr\right)ds + \sum_{t_k<t} I_k\big(u(t_k)\big).\end{aligned} \tag{6.1}$$

It is clear that $u(t)$ is a positive solution of problem (1.1)-(1.3) if and only of u is a solution of operator equation (6.1).
By analogous methods, we have the following results.

Lemma 6.1 *Assume that* (H_1)-(H_4) *hold. Then problem* (1.1)-(1.3) *is equivalent to the fixed point problem of* T^* *in* K^*.

Lemma 6.2 *Assume that* (H_1)-(H_4) *hold. Then* $T^* : K^* \to K^*$ *is completely continuous.*

Similar to the proofs in Sections 3-5, we have the following results. For convenience we introduce the following notations:

$$f^{\rho}_{\delta^*\rho} = \min\left\{\min_{t\in[0,\xi^*]}\frac{f(t,u)}{\phi_p(\rho)} : u \in \left[\delta^*\rho, \rho\right]\right\}, \qquad I^{\rho}_{\delta^*\rho}(k) = \min\left\{\frac{I_k(u)}{\rho} : u \in \left[\delta^*\rho, \rho\right]\right\},$$

$$\sigma_1^* = \int_{t_1}^{\xi^*} g(s)\,ds, \qquad N^* = \frac{\sigma_1^*}{1-\sigma}, \qquad \frac{1}{L^*} = \frac{\sigma_1^*}{1-\sigma}.$$

Single or twin solutions of problem (1.1)-(1.3) for the case $\alpha(t) \le t$ on J is treated in the following theorems.

Theorem 6.1 *Suppose* (H_1)-(H_4) *hold,* $\alpha(t) \le t$ *on* J*, and there exist* $\rho_1, \rho_2 \in (0,\infty)$ *with* $\rho_1 < \delta^*\rho_2$ *such that the following conditions hold:*

(H_5) $f_0^{\rho_1} < \phi_p(l), I_0^{\rho_1}(k) < l.$
$(H_6)^{**}$ $I^{\rho_2}_{\delta^*\rho_2}(k) \ge L^*.$

Then problem (1.1)-(1.3) *has at least one positive solution* u *with* $\rho_1 < \|u\| \le \rho_2$.

Proof If (H_5) holds, then similar to the proof of (3.1) we have

$$T^*u \not\geq u, \quad u \in K^*, \qquad \|u\| = \rho_1. \tag{6.2}$$

Considering $(H_6)^{**}$, we have $I_k(x) \ge L^*\rho_2$, $k = 1,2,\ldots,n$, for $\delta^*\rho_2 \le x \le \rho_2$, $t \in [0,\xi^*]$. Since $0 \le \alpha(t) \le \xi^*$ on J, it follows from $\delta^*\rho_2 \le x(t) \le \rho_2$ on J that

$$\delta^*\rho_2 \le x(\alpha(t)) \le \rho_2.$$

Next, we prove that

$$T^*u \not\leq u, \quad u \in K^*, \qquad \|u\| = \rho_2. \tag{6.3}$$

In fact, if there exists $u_0 \in K^*$, $\|u_0\| = \rho_2$ such that $T^*u_0 \le u_0$, then we have

$$\begin{aligned} 0 \le t \in J \quad \Rightarrow \quad u_0(t) &\ge (T^*u_0)(t) \\ &= \frac{1}{1-\varsigma}\left\{\int_0^1 g(t)\left[\int_0^t \varphi_q\left(\int_s^1 \omega(r)f(r,u_0(\alpha(r)))\,dr\right)ds\right]dt \right.\\ &\quad \left. + \int_0^1 g(t)\sum_{t_k<t} I_k(u_0(t_k))\,dt\right\} \\ &\quad + \int_0^t \varphi_q\left(\int_s^1 \omega(r)f(r,u_0(\alpha(r)))\,dr\right)ds + \sum_{t_k<t} I_k(u_0(t_k)) \end{aligned}$$

$$\geq \frac{1}{1-\varsigma}\int_0^1 g(t)\Big(\sum_{t_k<t} I_k\big(u_0(t_k)\big)\Big)\,dt$$

$$\geq \frac{1}{1-\varsigma}\Big[\int_{t_1}^{t_2} g(t)I_1\big(t_1,u(t_1)\big)\,dt$$

$$+\int_{t_2}^{t_3} g(t)\big(I_1\big(t_1,u(t_1)\big)+I_2\big(t_2,u(t_2)\big)\big)\,dt+\cdots$$

$$+\int_{t_n}^{1} g(t)\big(I_1\big(t_1,u(t_1)\big)+I_2\big(t_2,u(t_2)\big)+\cdots+I_n\big(t_n,u(t_n)\big)\big)\,dt\Big]$$

$$=\frac{1}{1-\varsigma}\Big[\int_{t_1}^{1} g(t)I_1\big(t_1,u(t_1)\big)\,dt+\int_{t_2}^{1} g(t)I_2\big(t_2,u(t_2)\big)\,dt$$

$$+\cdots+\int_{t_n}^{1} g(t)I_n\big(t_n,u(t_n)\big)\,dt\Big]$$

$$\geq \frac{1}{1-\varsigma}\int_{t_1}^{1} g(t)I_1\big(u_0(t_1)\big)\,dt$$

$$\geq \frac{1}{1-\varsigma}\int_{t_1}^{\xi^*} g(t)I_1\big(u_0(t_1)\big)\,dt$$

$$=\frac{1}{1-\varsigma}L^*\rho_2\sigma_1^*$$

$$=\rho_2=\|u_0\|.$$

This is also a contraction. Hence (6.3) holds.

Applying (i) of Lemma 2.4 to (6.2) and (6.3) shows that T^* has a fixed point $u\in K^*\cap(\Omega_{\rho_2}\backslash\bar{\Omega}_{\rho_1})$. Thus, it follows that problem (1.1)-(1.3) has one positive solution u satisfying $\rho_1<u(t)\leq\rho_2$, $t\in J$. This gives the proof of the theorem. □

Remark 6.1 The method to deal with the impulsive term in the proof of Theorem 6.1 is different from that in the proof of Theorem 3.1.

Theorem 6.2 *Suppose* (H_1)-(H_4), $\alpha(t)\leq t$ *on* J *and one of the following conditions holds:*

(H_9)* *There exist* $\rho_1,\rho_2,\rho_3\in(0,\infty)$, *with* $\rho_1<\gamma\rho_2$ *and* $\rho_2<\rho_3$ *such that*

$$f_0^{\rho_1}<\phi_p(l),\qquad I_0^{\rho_1}(k)<l\rho_1,\qquad I_{\delta^*\rho_2}^{\rho_2}(k)>\phi_p(L^*),$$
$$f_0^{\rho_3}<\phi_p(l),\qquad I_0^{\rho_3}(k)<l\rho_3.$$

(H_{10})* *There exist* $\rho_1,\rho_2,\rho_3\in(0,\infty)$, *with* $\rho_1<\rho_2<\rho_3$ *such that*

$$I_{\delta^*\rho_1}^{\rho_1}(k)>\phi_p(L^*),\qquad f_0^{\rho_2}<\phi_p(l),\qquad I_0^{\rho_2}(k)<l\rho_2,\qquad I_{\delta^*\rho_3}^{\rho_3}(k)>\phi_p(L^*).$$

Then problem (1.1)-(1.3) *has at least two positive solutions* u_1, u_2 *with* $u_1\in\Omega_{\rho_2}\backslash\bar{K}_{\rho_1}^*$, $u_2\in K_{\rho_3}^*\backslash\bar{\Omega}_{\rho_2}$.

The following theorem deals with further results on twin solutions of problem (1.1)-(1.3) for the case of $\alpha(t)\leq t$ on J.

Theorem 6.3 *Assume that* (H_1)-(H_4) *and* $\alpha(t) \le t$ *on* J. *Moreover, suppose that there exist constants* $0 < a < b < c$ *such that*

$$0 < a < \frac{N^*}{N_1} b < \frac{\delta^* N^*}{N_1} c.$$

Let f *and* I_k *satisfy the following conditions:*

$(C_1)^*$ $I_k(u) > \frac{c}{N^*}$ *for* $u \in [c, \frac{c}{\delta^*}]$;
$(C_2)^*$ $f(t,u) \le \phi_p(\frac{b}{N_1})$ *for* $(t,u) \in J \times [0, \frac{b}{\delta^*}]$, $I_k(u) \le \frac{b}{N_1}$ *for* $u \in [0, \frac{b}{\delta^*}]$;
$(C_3)^*$ $I_k(u) > \frac{a}{N^*}$ *for* $u \in [\delta^* a, a]$.

Then problem (1.1)-(1.3) *has at least two positive solutions* u_1 *and* u_2 *such that*

$$a < \theta_2(u_1) \quad \text{with } \theta_1(u_1) < b \quad \text{and} \quad b < \theta_1(u_2) \quad \text{with } \theta(u_2) < c.$$

Finally we consider triple solutions of problem (1.1)-(1.3) for the case of $\alpha(t) \ge t$ on J.

Theorem 6.4 *Assume that* (H_1)-(H_4) *and* $\alpha(t) \le t$ *on* J. *Moreover, suppose that there exist constants* $0 < d < a < \frac{a}{\delta^*} \le c$ *such that*

(A_1) $f(t,u) \le \phi_p(\frac{d}{N_1})$ *for* $(t,u) \in J \times [0,d]$, $I_k(u) \le \frac{d}{N_1}$ *for* $u \in [0,d]$;
$(A_2)^*$ $I_k(u) \ge \frac{a}{N^*}$ *for* $u \in [a, \frac{a}{\delta^*}]$;
(A_3) $f(t,u) \le \phi_p(\frac{c}{N_1})$ *for* $(t,u) \in J \times [0,c]$, $I_k(u) \le \frac{c}{N_1}$ *for* $u \in [0,c]$.

Then problem (1.1)-(1.3) *has at least three positive solutions* u_1, u_2, *and* u_3 *such that*

$$\|u_1\| < d, \qquad a < \psi(u_2) \quad \text{and} \quad u_3 > d \quad \text{with } \psi(u_3) < a.$$

Remark 6.2 Minor adjustments to the proof of Theorem 6.1 are only needed to prove that Theorems 6.2-6.4 hold, so we leave the details to the reader.

Competing interests
The authors declare that they have no competing interests.

Authors' contributions
XZ completed the main study and carried out the results of this article. MF checked the proofs and verified the calculation. All the authors read and approved the final manuscript.

Author details
[1]Department of Mathematics and Physics, North China Electric Power University, Beijing, 102206, People's Republic of China. [2]School of Applied Science, Beijing Information Science & Technology University, Beijing, 100192, People's Republic of China.

Acknowledgements
This work is sponsored by the project NSFC (11301178), the Fundamental Research Funds for the Central Universities (2014ZZD10, 2014MS58) and the Scientific Research Common Program of Beijing Municipal Commission of Education (KM201511232018). The authors are grateful to anonymous referees for their constructive comments and suggestions, which has greatly improved this paper.

References

1. Bobisud, LE: Steady state turbulent flow with reaction. Rocky Mt. J. Math. **21**, 993-1007 (1991)
2. Glowinski, R, Rappaz, J: Approximation of a nonlinear elliptic problem arising in a non-Newtonian fluid flow model in glaciology. Math. Model. Numer. Anal. **37**, 175-186 (2003)
3. Sánchez, J: Multiple positive solutions of singular eigenvalue type problems involving the one-dimensional *p*-Laplacian. J. Math. Anal. Appl. **292**, 401-414 (2004)

4. Kajikiya, R, Lee, YH, Sim, I: One-dimensional p-Laplacian with a strong singular indefinite weight. I. Eigenvalue. J. Differ. Equ. **244**, 1985-2019 (2008)
5. Kajikiya, R, Lee, YH, Sim, I: Bifurcation of sign-changing solutions for one-dimensional p-Laplacian with a strong singular weight: p-superlinear at ∞. Nonlinear Anal. **74**, 5833-5843 (2011)
6. Iturriaga, L, Sánchez, J: Exact number of solutions of stationary reaction-diffusion equations. Appl. Math. Comput. **216**, 1250-1258 (2010)
7. Dai, G, Ma, R, Lu, Y: Bifurcation from infinity and nodal solutions of quasilinear problems without the signum condition. J. Math. Anal. Appl. **397**, 119-123 (2013)
8. Dai, G, Ma, R: Unilateral global bifurcation phenomena and nodal solutions for p-Laplacian. J. Differ. Equ. **252**, 2448-2468 (2012)
9. Zhang, X, Feng, M: Existence of a positive solution for one-dimensional singular p-Laplacian problems and its parameter dependence. J. Math. Anal. Appl. **413**, 566-582 (2014)
10. Zhang, X, Ge, W: Impulsive boundary value problems involving the one-dimensional p-Laplacian. Nonlinear Anal. **70**, 1692-1701 (2009)
11. Cabada, A, Tomeček, J: Extremal solutions for nonlinear functional p-Laplacian impulsive equations. Nonlinear Anal. **67**, 827-841 (2007)
12. Henderson, J, Wang, H: Nonlinear eigenvalue problems for quasilinear systems. Comput. Math. Appl. **49**, 1941-1949 (2005)
13. Yang, Z, O'Regan, D: Positive solutions for one-dimensional p-Laplacian equations. Math. Comput. Model. **55**, 1942-1950 (2012)
14. Lü, H, O'Regan, D, Zhong, C: Multiple positive solutions for the one-dimensional singular p-Laplacian. Appl. Math. Comput. **133**, 407-422 (2002)
15. Agarwal, RP, Lü, H, O'Regan, D: Eigenvalues and the one-dimensional p-Laplacian. J. Math. Anal. Appl. **266**, 383-400 (2002)
16. Wang, H: On the number of positive solutions of nonlinear systems. J. Math. Anal. Appl. **281**, 287-306 (2003)
17. Bonannao, G, Livrea, R: Multiplicity theorems for the Dirichlet problem involving the p-Laplacian. Nonlinear Anal. **54**, 1-7 (2003)
18. Du, Z, Lin, X, Tisdell, CC: A multiplicity result for p-Laplacian boundary value problems via critical points theorem. Appl. Math. Comput. **205**, 231-237 (2008)
19. Ding, Y, O'Regan, D: Positive solutions for a second-order p-Laplacian impulsive boundary value problem. Adv. Differ. Equ. (2012). doi:10.1186/1687-1847-2012-159
20. Jankowski, T: Positive solutions to third-order impulsive Sturm-Liouville boundary value problems with deviating arguments and one-dimensional p-Laplacian. Dyn. Syst. Appl. **20**, 575-586 (2011)
21. Avery, RI: A generalization of the Leggett-Williams fixed-point theorem. Math. Sci. Res. Hot-Line **2**, 9-14 (1998)
22. Kajikiya, R, Lee, YH, Sim, I: Bifurcation of sign-changing solutions for one-dimensional p-Laplacian with a strong singular weight: p-sublinear at ∞. Nonlinear Anal. **71**, 1235-1249 (2009)
23. Jankowski, T: Positive solutions of one-dimensional p-Laplacian boundary value problems for fourth-order differential equations with deviating arguments. J. Optim. Theory Appl. **149**, 47-60 (2011)
24. Guo, D, Lakshmikantham, V: Nonlinear Problems in Abstract Cones. Academic Press, New York (1988)
25. Avery, RI, Henderson, J: Two positive fixed points of nonlinear operator on ordered Banach spaces. Commun. Appl. Nonlinear Anal. **8**, 27-36 (2001)
26. Leggett, R, Williams, L: Multiple positive fixed points of nonlinear operators on ordered Banach spaces. Indiana Univ. Math. J. **28**, 673-688 (1979)

11

Wavelet multilevel augmentation method for linear boundary value problems

Somlak Utudee[1] and Montri Maleewong[2*]

*Correspondence:
Montri.M@ku.ac.th
[2]Department of Mathematics, Faculty of Science, Kasetsart University, Bangkok, 10900, Thailand
Full list of author information is available at the end of the article

Abstract

This work presents a new approach to numerically solve the general linear two-point boundary value problems with Dirichlet boundary conditions. Multilevel bases from the anti-derivatives of the Daubechies wavelets are constructed in conjunction with the augmentation method. The accuracy of numerical solutions can be improved by increasing the number of basis levels, but the computational cost also increases drastically. The multilevel augmentation method can be applied to reduce the computational time by splitting the coefficient matrix into smaller submatrices. Then the unknown coefficients in the higher level can be solved separately. The convergent rate of this method is 2^s, where $1 \leq s \leq p+1$, when the anti-derivatives of the Daubechies wavelets order p are applied. Some numerical examples are also presented to confirm our theoretical results.

MSC: 65J10; 65L10

Keywords: wavelets; multilevel augmentation method; boundary value problems; Dirichlet boundary conditions

1 Introduction

Boundary value problems can be viewed as mathematical models in science and engineering. For real world applications, exact solutions are not available. Numerical methods are required to solve numerically the models. The efficient methods provide approximate solutions by choosing the appropriate subspaces of solution spaces and their suitable bases. By applying suitable formulations, the linear model equation can be discretized to a linear system. A more accurate approximation can be obtained by increasing the number of basis functions. However, this leads to a larger discretized linear system. To save the computational cost, one can use the multilevel augmentation method. The resulting coefficient matrix corresponding to the finer level of approximate spaces is obtained by augmenting a matrix corresponding to a coarser level. Instead of solving the linear system at the finer level, the coefficient matrix can be separated so that the smaller system at the coarser level can be taken. Thus, the additional computational cost is proportional to the dimension of the different space between the spaces of the finer level and the coarser level, not the dimension of the whole finer level. This method allows us to develop faster and accurate algorithms for solving differential equations (see, *e.g.*, [1–3], and [4]). These previous researches considered the second kind of equations and constructed piecewise polynomials as bases for the subspaces of the Sobolev spaces $H_0^m(0,1)$ consisting of elements satisfying the homogeneous boundary conditions

$$u^{(j)}(0) = u^{(j)}(1) = 0, \quad \text{for } j = 0, 1, \ldots, m-1.$$

On the other hand, wavelets can be applied to discretize differential equations (see, *e.g.*, [5, 6]). Related numerical methods with the applications of Haar and Legendre wavelets for solving boundary value problem are proposed by Siraj-ul-Islam *et al.* [7, 8]. The advantage of wavelet basis is its capability to approximate solutions of differential equations. The wavelet Galerkin method is one of the most powerful methods that can be used to solve ordinary and partial differential equations (see, *e.g.*, [9–11], and [12]). In addition, the accuracy of the approximate solutions can easily be improved by merely increasing the numbers of wavelet basis functions and the orders of wavelets. However, the wavelet basis is not straightforwardly adjusted to satisfy general boundary conditions. In 1992, Xu and Shann introduced a different approach to handle the boundary conditions by using the anti-derivatives of Daubechies wavelets [13]. These anti-derivatives form bases for the finite-dimensional subspaces of Sobolev space H^1 and are used to construct an algorithm for approximating solutions.

In this work, we propose the method that combines the main advantages of wavelet bases and multilevel augmentation together. That is, we apply the multilevel augmentation of operators in conjunction with the anti-derivatives of Daubechies wavelets to approximate linear differential equations in the case of Dirichlet boundary conditions. The originality of this work is that we introduce the anti-derivatives of Daubechies wavelets for solving linear boundary value problems (see [13]) and apply this basis type with the augmentation method proposed by Chen (see, *e.g.*, [1–3] and [4]). By this concept, we obtain a new approach to reduce the computational time for solving the linear system resulting from discretizing a linear differential equation.

Given the interval $\Omega := (0, R)$, we use the notation $L^2(\Omega)$ to denote the space of square integrable functions on Ω with standard inner product $(\cdot, \cdot)$ defined by

$$(u, v) = \int_0^R u(x)v(x)\, dx$$

and the associated norm $\| \cdot \|$.

Let $H^s(\Omega)$ denote the standard Sobolev space with the norm $\| \cdot \|_s$ given by

$$\|v\|_s^2 = \sum_{i=0}^{s} \int_0^R \left|v^{(i)}(x)\right|^2 dx.$$

According to the boundary condition, we work on the solution space

$$H_0^1(\Omega) = \left\{v \in H^1(\Omega) \mid v(0) = v(R) = 0\right\},$$

equipped with the inner product

$$[u, v] = \int_0^R u'(x)v'(x)\, dx, \quad \text{for } u, v \in H_0^1(\Omega),$$

and its associated norm $|\cdot|$. It is well known that the norm $|\cdot|$ is equivalent to the standard norm $\|\cdot\|_1$ in this space.

Let $p \in \mathbb{N}$. We will apply the multilevel augmentation method and anti-derivatives of wavelets of order p to find numerical solutions of two-point boundary value problems with Dirichlet boundary conditions.

Assume that there exists a unique weak solution $u \in H_0^1(\Omega)$. To find numerical solutions, we propose the following steps:

1. The solution space $H_0^1(\Omega)$ is decomposed into orthogonal direct sum of subspaces. The anti-derivatives of the Daubechies wavelets are used to construct finite-dimensional subspaces.
2. For $n \in \mathbb{N}$, the multilevel method is applied to obtain the *nth level* solution by solving a linear system with matrix coefficients related to the anti-derivatives of the Daubechies wavelets.
3. To obtain a solution at a higher level, namely *$(n + i)$th level*, the multilevel augmentation method is applied. By the algorithm to be presented, the computational time for solving the linear system is reduced since the dimension of the matrix coefficient is smaller.

Finally, this work is organized as follows. Section 2 gives an introduction to the anti-derivatives of the Daubechies wavelets and the finite-dimensional subspaces of the solution space $H_0^1(\Omega)$. In Section 3, we describe the algorithm to find approximate solutions using the multilevel augmentation method. The optimal error estimates for the approximate solutions are proven in Section 4, while some numerical examples are demonstrated in Section 5. Conclusions and future work are discussed in Section 6.

2 Bases for subspaces of $H_0^1(\Omega)$

In this section, we will introduce the wavelets of order p and their anti-derivatives. These functions form orthonormal bases for the finite-dimensional subspaces S_n of the solution space $H_0^1(\Omega)$. More details can be found in [5] and [13].

To define the Daubechies wavelets, we consider two functions: the *scaling function* $\phi(x)$ and the *wavelet function* $\psi(x)$. The scaling function is obtained from the dilation equation. The wavelet function is defined from the scaling function. Details are described as follows.

Given a positive integer p, consider a sequence $\{c_k\}_{k\in\mathbb{Z}}$ satisfying

$$c_k = 0, \quad \text{for } k \notin \{0, 1, , \dots, 2p-1\},$$

$$\sum_k c_k = 2,$$

$$\sum_k (-1)^k k^m c_k = 2, \quad \text{for } 0 \le m \le p-1,$$

$$\sum_k c_k c_{k-2m} = 2\delta_{0m}, \quad \text{for } 1-p \le m \le p-1.$$

The scaling function $\phi(x)$ is the unique solution of the dilation equation

$$\phi(x) = \sum_{k=0}^{2p-1} c_k \phi(2x-k),$$

where c_k satisfy the above four properties.

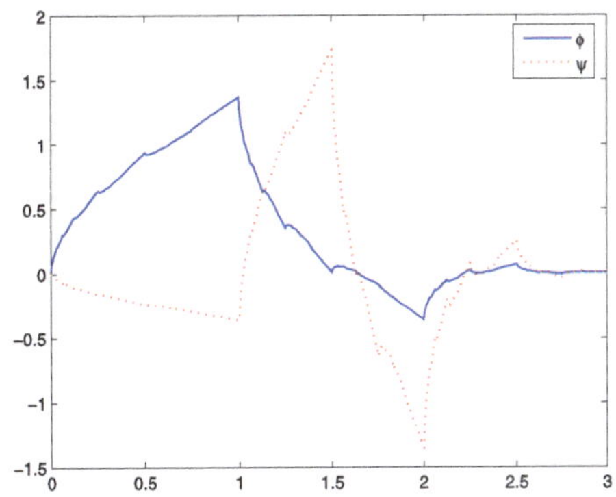

Figure 1 ϕ **and** ψ **for** $p = 2$**.**

Let

$$\widetilde{\psi}(x) = \sum_{k=0}^{2p-1} (-1)^k c_{2p-1-k} \phi(2x - k).$$

The wavelet function $\psi(x)$ is defined by

$$\psi(x) = \widetilde{\psi}(x - p + 1).$$

The graphs of ϕ and ψ for $p = 2$ are shown in Figure 1.

Define

$$\psi_{jk}(x) = \begin{cases} \phi(x-k), & j = -1, \\ \sqrt{2^j}\,\psi(2^j x - k), & j \geq 0. \end{cases}$$

The functions ψ_{jk} with $j \geq -1$ and $(j,k) \in \mathbb{Z} \times \mathbb{Z}$ are called *wavelets of order p*. It is well known that the set of wavelets forms an orthonormal basis for $L^2(\mathbb{R})$.

For $j \geq -1$, define the index set I_j such that

$$k \in I_j \quad \Longleftrightarrow \quad \begin{cases} 2 - 2p \leq k \leq 2p - 2, & j = -1, \\ 2 - 2p \leq k \leq 2^j(2p-1) - 1, & j \geq 0. \end{cases}$$

$\{\psi_{jk}|_\Omega \mid j \geq -1, k \in I_j\}$ is a *frame* of $L^2(\Omega)$. That is, the span$\{\psi_{jk}|_\Omega \mid j \geq -1, k \in I_j\}$ consisting of all linear expansions is equal to $L^2(\Omega)$.

Next, we define the anti-derivatives of wavelets satisfying the Dirichlet boundary condition, namely,

$$\Psi_{jk}(x) = \int_0^x \psi_{jk}\,ds - \frac{x}{R}\int_0^R \psi_{jk}\,ds, \quad \text{for } 0 \leq x \leq R.$$

For $n \in \mathbb{N}_0$, we define a subspace S_n by

$$S_n = \text{span}\{\Psi_{jk} \mid -1 \leq j < n, k \in I_j\}.$$

It is obvious that these subspaces are finite-dimensional subspaces of $H_0^1(\Omega)$ and they are nested, *i.e.*, $S_n \subset S_{n+1}$. Since $\{\Psi_{jk} \mid -1 \leq j < n, k \in I_j\}$ is a frame for S_n, it spans S_n but it

Figure 2 $\Psi_{00}, \Psi_{10}, \Psi_{11}, \Psi_{20}, \Psi_{21}, \Psi_{22}, \Psi_{23}$.

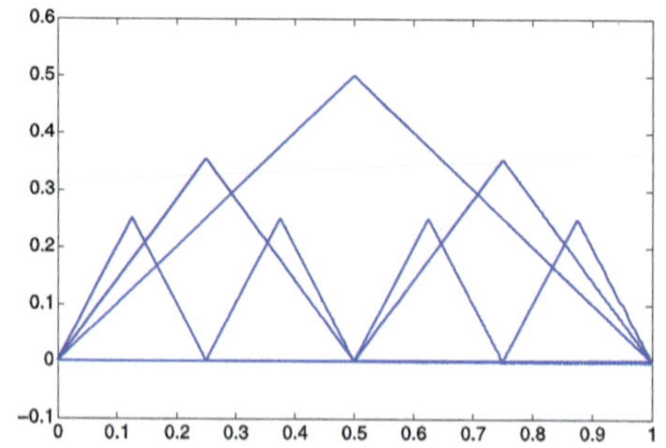

Figure 3 $\Psi_{2,-1}, \ldots, \Psi_{2,10}$.

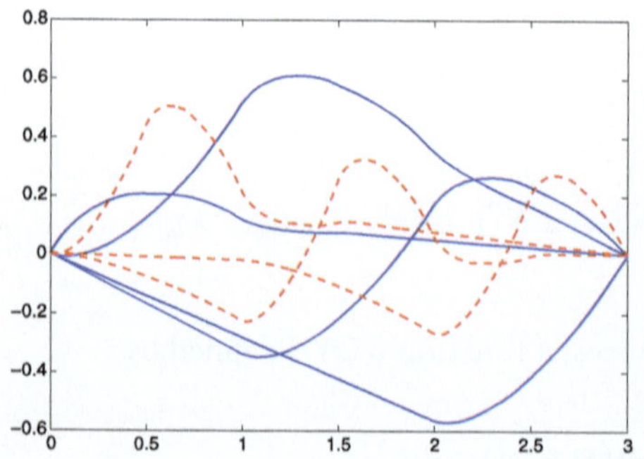

need not be linearly independent. Xu and Shann [13], introduced an index set D_j such that $\{\Psi_{jk} \mid -1 \leq j < n, k \in D_j\}$ is a basis for S_n. The index set D_j is defined by

$$k \in D_j \quad \Longleftrightarrow \quad \begin{cases} 2-2p < k \leq 2p-2, & j=-1, \\ 1-p \leq k \leq 2^j(2p-1)-p, & j \geq 0. \end{cases}$$

For $p = 1$, the basis functions $\{\Psi_{jk} \mid 0 \leq j < n, k \in D_j\}$ for S_n are the piecewise linear hierarchical basis functions with uniform mesh size, $h = 2^{-n}$. The graphs of $\Psi_{00}, \Psi_{10}, \ldots, \Psi_{22}$, and Ψ_{23} are shown in Figure 2, while the graphs of $\Psi_{2,-1}, \ldots, \Psi_{2,10}$ are shown in Figure 3.

It should be noted that the set

$$\{\Psi_{jk} \mid 0 \leq j < n, k \in D_j\}$$

is an orthonormal basis for S_n with the inner product $[\cdot,\cdot]$ of the Sobolev space $H_0^1(\Omega)$ but it is not orthonormal in $L^2(\Omega)$ when equipped with the standard inner product $(\cdot,\cdot)$. Since we require orthonormality in $L^2(\Omega)$ for our augmentation method, we apply the Gram-Schmidt process to obtain an orthonormal basis for S_n with the standard inner product $(\cdot,\cdot)$ in $L^2(\Omega)$. Thus, the set

$$\{\overline{\Psi}_{jk} \mid -1 \leq j < n, k \in D_j\}$$

is defined as orthonormal basis for S_n in the present method.

3 Multilevel augmentation method algorithm

In this section, we describe the multilevel augmentation method for solving boundary value problems.

Let $u \in H_0^1(\Omega)$ be the weak solution of a given differential equation. Suppose that the variational form of the differential equation is

$$a(u,v) = (f,v), \quad \text{for all } v \in H_0^1(\Omega). \tag{1}$$

Let the approximate solution u_n of the given equation be

$$u_n = \sum_{j=-1}^{n} \sum_{k \in D_j} \alpha_{jk} \overline{\Psi}_{jk}.$$

The main idea is to determine the coefficients α_{jk} in such a way that u_n behaves as if it were a weak solution on S_n, that is, u_n satisfies the linear system of equations,

$$a(u_n, \overline{\Psi}_{lm}) = (f, \overline{\Psi}_{lm}), \quad \text{for all } -1 \le l < n, m \in D_j,$$

$$\sum_{-1 \le j < n} \sum_{k \in D_j} \alpha_{jk} a(\overline{\Psi}_{jk}, \overline{\Psi}_{lm}) = (f, \overline{\Psi}_{lm}), \quad \text{for all } -1 \le l < n, m \in D_j.$$

Let $\iota : (j,k) \mapsto i$ be the lexicographically enumerating function. That is,

$$\iota(j,k) \le \iota(l,m) \quad \text{if } j \le l \text{ or } j = l \text{ and } k \le m.$$

We then obtain a linear system of the form $\mathbf{A}_n \mathbf{u}_n = \mathbf{f}_n$, where $\mathbf{A}_n$ is the coefficient matrix, $\mathbf{u}_n$ is the unknown column vector, and $\mathbf{f}_n$ is the column vector defined by

$$\mathbf{A}_n = \left[a(\overline{\Psi}_i, \overline{\Psi}_j) : i,j = 1, \ldots, \dim S_n\right],$$

$$\mathbf{u}_n = [\alpha_j : j = 1, \ldots, \dim S_n],$$

$$\mathbf{f}_n = \left[(f, \overline{\Psi}_j) : j = 1, \ldots, \dim S_n\right].$$

The approximate solution u_n obtained in this way is called the *nth multilevel solution* of (1).

Next, we apply the augmentation method to approximate the next level solution, u_{n+1}. Suppose that u_n is already solved. That is, the matrix representation $\mathbf{u}_n$ of u_n satisfies the equation $\mathbf{A}_n \mathbf{u}_n = \mathbf{f}_n$. We augment the matrix $\mathbf{A}_n$ with submatrices $\mathbf{B}_n$, $\mathbf{C}_n$, and $\mathbf{D}_n$ where

$$\mathbf{B}_n = \left[a(\overline{\Psi}_i, \overline{\Psi}_j) : i = 1, \ldots, \dim S_n, j = \dim S_n + 1, \ldots, \dim S_{n+1}\right],$$

$$\mathbf{C}_n = \left[a(\overline{\Psi}_i, \overline{\Psi}_j) : i = \dim S_n + 1, \ldots, \dim S_{n+1}, j = 1, \ldots, \dim S_n\right],$$

$$\mathbf{D}_n = \left[a(\overline{\Psi}_i, \overline{\Psi}_j) : i,j = \dim S_n + 1, \ldots, \dim S_{n+1}\right].$$

The coefficient matrix $\mathbf{A}_{n+1}$ corresponding to the $(n+1)$st level is identified as

$$\mathbf{A}_{n+1} = \begin{bmatrix} \mathbf{A}_n & \mathbf{B}_n \\ \mathbf{C}_n & \mathbf{D}_n \end{bmatrix}.$$

Instead of finding the full $(n+1)$th level solution, u_{n+1}, from $\mathbf{A}_{n+1} \mathbf{u}_{n+1} = \mathbf{f}_{n+1}$ of size $\dim S_{n+1}$, we will approximate u_{n+1} by decomposition the coefficient matrix $\mathbf{A}_{n+1}$ into the sum of an upper triangular matrix and a lower triangular matrix:

$$\mathbf{A}_{n+1} = \begin{bmatrix} \mathbf{A}_n & \mathbf{B}_n \\ \mathbf{C}_n & \mathbf{D}_n \end{bmatrix} = \begin{bmatrix} \mathbf{A}_n & \mathbf{B}_n \\ 0 & \mathbf{D}_n \end{bmatrix} + \begin{bmatrix} 0 & 0 \\ \mathbf{C}_n & 0 \end{bmatrix}.$$

If the matrices $\mathbf{A}_n$ and $\mathbf{D}_n$ are nonsingular, there exists a (unique) vector $\mathbf{u}_{n,1}$ satisfying

$$\begin{bmatrix} \mathbf{A}_n & \mathbf{B}_n \\ 0 & \mathbf{D}_n \end{bmatrix} \mathbf{u}_{n,1} + \begin{bmatrix} 0 & 0 \\ \mathbf{C}_n & 0 \end{bmatrix} \begin{bmatrix} \mathbf{u}_n \\ 0 \end{bmatrix} = \mathbf{f}_{n+1}. \tag{2}$$

Let $u_{n,1} \in S_{n+1}$ be the corresponding element of $\mathbf{u}_{n,1}$. We call $u_{n,1}$ the $(n+1)$*st multilevel augmentation solution* of (1). The index n refers to the *initial level* n, and 1 refer to a *one step* method to compute u_{n+1} approximately. The linear system of $\dim S_{n+1}$ can be solved by considering two linear systems. One is the size of $\dim S_n$, and another is the size of $\dim S_{n+1} - \dim S_n = 2^n(2p-1)$.

In general, an approximation $u_{n,i+1}$ for $(n+i+1)$st multilevel solution is defined by setting

$$u_{n,0} = u_n.$$

Suppose that $u_{n,i}$ is already solved. We augment the matrix $\mathbf{A}_{n+i}$ with submatrices $\mathbf{B}_{n+i}$, $\mathbf{C}_{n+i}$, and $\mathbf{D}_{n+i}$ where

$$\mathbf{B}_{n+i} = \left[a(\overline{\Psi}_i, \overline{\Psi}_j) : i = 1, \ldots, \dim S_{n+i}, j = (\dim S_{n+i}) + 1, \ldots, \dim S_{n+i+1}\right],$$

$$\mathbf{C}_{n+i} = \left[a(\overline{\Psi}_i, \overline{\Psi}_j) : i = (\dim S_{n+i}) + 1, \ldots, \dim S_{n+i+1}, j = 1, \ldots, \dim S_{n+i}\right],$$

$$\mathbf{D}_{n+i} = \left[a(\overline{\Psi}_i, \overline{\Psi}_j) : i, j = (\dim S_{n+i}) + 1, \ldots, \dim S_{n+i+1}\right].$$

The coefficient matrix $\mathbf{A}_{n+i+1}$ corresponding to the $(n+i+1)$st level is identified as

$$\mathbf{A}_{n+i+1} = \begin{bmatrix} \mathbf{A}_{n+i} & \mathbf{B}_{n+i} \\ \mathbf{C}_{n+i} & \mathbf{D}_{n+i} \end{bmatrix}.$$

We split the coefficient matrix $\mathbf{A}_{n+i+1}$ and solve for $u_{n,i+1} \in S_{n+i+1}$ from the equation

$$\begin{bmatrix} \mathbf{A}_{n+i} & \mathbf{B}_{n+i} \\ 0 & \mathbf{D}_{n+i} \end{bmatrix} \mathbf{u}_{n,i+1} + \begin{bmatrix} 0 & 0 \\ \mathbf{C}_{n+i} & 0 \end{bmatrix} \begin{bmatrix} \mathbf{u}_{n,i} \\ 0 \end{bmatrix} = \mathbf{f}_{n+i+1}. \tag{3}$$

This completes the multilevel augmentation algorithm.

4 Error analysis

In this section, we analyze the convergence of the multilevel augmentation method conjunction with the anti-derivatives of the Daubechies wavelets. First, we consider the distance between the weak solution u of (1) and the nth multilevel solution $u_n \in S_n$ obtained by the wavelets of order p. Theorem 4.1 in [13] stated that, for $u \in H_0^1(\Omega) \cap H^s(\Omega)$, there exists a positive constant C such that

$$\|u - u_n\| + 2^{-n}\|u - u_n\|_1 \le C(2^{-n})^s \|u\|_s, \quad 1 \le s \le p+1.$$

In particular, if we consider separately the distance between u and u_n with the standard L^2 norm $\|\cdot\|$, and the Sobolev norm $\|\cdot\|_1$, we obtain

$$\|u - u_n\| \le C(2^{-n})^s \|u\|_s, \quad 1 \le s \le p+1,$$
$$\|u - u_n\|_1 \le C(2^{-n})^{s-1} \|u\|_s, \quad 1 \le s \le p+1.$$

The above estimations suggest that if we apply the wavelet of order p and $u \in H_0^1(\Omega) \cap H^s(\Omega)$, then the errors measured by the standard norm, and the Sobolev norm in L^2, decrease by the factors of 2^{p+1}, and 2^p, respectively, from n to $n+1$ level.

Next, we consider the distance between the solution u and the $(n+i)$th multilevel augmentation solution, $u_{n,i}$, of (1). In the remaining section, we denote by $\mathcal{A}$ the operator corresponding to the matrix $\mathbf{A}$, and we denote by $\mathbf{u}$ the column matrix representing element u.

For $n \in \mathbb{N}$, if the inverse operators $\mathcal{A}_n^{-1}$ and $\mathcal{D}_n^{-1}$ exist, then the $(n+1)$th multilevel augmentation solution $u_{n,i}$ exists. If there also exist an $N \in \mathbb{N}$, $\alpha > 0$, and $\delta > 0$ such that

$$\|\mathcal{A}_n^{-1}\| \le \alpha, \qquad \|\mathcal{D}_n^{-1}\| \le \delta, \quad \text{for } n \ge N,$$

and

$$\lim_{n\to\infty} \|\mathcal{B}_n\| = \lim_{n\to\infty} \|\mathcal{C}_n\| = 0,$$

the error for our method can be estimated as in the following theorem.

Theorem 1 (Error for multilevel augmentation method) *Let $u \in H_0^1(\Omega)$ be the solution of* (1)*. Suppose that there exist an $N \in \mathbb{N}$, and positive constants α and δ such that for $n \ge N$ the inverse operators $\mathcal{A}_n^{-1}$, $\mathcal{D}_n^{-1}$ exist and*

$$\|\mathcal{A}_n^{-1}\| \le \alpha, \qquad \|\mathcal{D}_n^{-1}\| \le \delta,$$

and

$$\lim_{n\to\infty} \|\mathcal{B}_n\| = \lim_{n\to\infty} \|\mathcal{C}_n\| = 0.$$

If $u \in H_0^1(\Omega) \cap H^s(\Omega)$, then there exist an $M \in \mathbb{N}$ and positive constant c_0 such that, for $n \ge M$ and $i \in \mathbb{N} \cup \{0\}$, we have the estimate

$$\|u - u_{n,i}\| \le c_0 2^{-2(n+i)s} \|u\|_s, \quad 1 \le s \le p+1.$$

Proof By the hypotheses on $\mathcal{A}_n^{-1}$ and $\mathcal{D}_n^{-1}$, we see that the $(n+i)$th multilevel augmentation solution u_{n+i} and the $(n+i)$th multilevel augmentation solution $u_{n,i}$ exist for all $n \ge N$ and

$i \in \mathbb{N} \cup \{0\}$. By Theorem 4.1 in [13], there exists a positive constant C such that

$$\|u - u_{n+i}\| \leq C\left(2^{-(n+i)s}\right)\|u\|_s, \quad 1 \leq s \leq p+1. \tag{4}$$

For $n \geq N$ and $i \in \mathbb{N}$, let W_{n+i-1} be the orthogonal complement of S_{n+i-1} in S_{n+i}. The element $u_{n,i} \in S_{n+i}$ can be uniquely written in the form

$$u_{n,i} = u^1_{n,i-1} + v^1_{n,i-1},$$

where $u^1_{n,i-1} \in S_{n+i-1}$, and $v^1_{n,i-1} \in W_{n+i-1}$. Since the column matrix $\mathbf{u}_{n,i}$ of $u_{n,i}$ satisfies the equation

$$\begin{bmatrix} \mathbf{A}_{n+i-1} & \mathbf{B}_{n+i-1} \\ 0 & \mathbf{D}_{n+i-1} \end{bmatrix} \mathbf{u}_{n,i} + \begin{bmatrix} 0 & 0 \\ \mathbf{C}_{n+i-1} & 0 \end{bmatrix} \begin{bmatrix} \mathbf{u}_{n,i-1} \\ 0 \end{bmatrix} = \mathbf{f}_{n+i},$$

it also satisfies

$$\mathbf{A}_{n+i}\mathbf{u}_{n,i} = \mathbf{f}_{n+i} + \begin{bmatrix} 0 & 0 \\ \mathbf{C}_{n+i-1} & 0 \end{bmatrix} \begin{bmatrix} \mathbf{u}^1_{n,i-1} - \mathbf{u}_{n,i-1} \\ 0 \end{bmatrix}. \tag{5}$$

The $(n+i)$th multilevel solution $u_{n+i} \in S_{n+i}$ satisfies the equation

$$\mathbf{A}_{n+i}\mathbf{u}_{n+i} = \mathbf{f}_{n+i}. \tag{6}$$

Subtracting (6) from (5), we have

$$\mathbf{A}_{n+i}(\mathbf{u}_{n,i} - \mathbf{u}_{n+i}) = \begin{bmatrix} 0 & 0 \\ \mathbf{C}_{n+i-1} & 0 \end{bmatrix} \begin{bmatrix} \mathbf{u}^1_{n,i-1} - \mathbf{u}_{n,i-1} \\ 0 \end{bmatrix}.$$

Since $\mathbf{A}_{n+i}$ is nonsingular, we have the equation

$$\mathbf{u}_{n,i} - \mathbf{u}_{n+i} = \mathbf{A}^{-1}_{n+i} \begin{bmatrix} 0 & 0 \\ \mathbf{C}_{n+i-1} & 0 \end{bmatrix} \begin{bmatrix} \mathbf{u}^1_{n,i-1} - \mathbf{u}_{n,i-1} \\ 0 \end{bmatrix}.$$

Since $\lim_{n\to\infty} \|\mathcal{B}_n\| = \lim_{n\to\infty} \|\mathcal{C}_n\| = 0$, there exist an $M \in \mathbb{N}$ and $c > 0$ such that for $n \geq M$ and $i \in \mathbb{N}$,

$$\|\mathcal{C}_{n+i-1}\| \leq \frac{2^{-s}}{3\alpha}$$

and

$$\left\|u^1_{n,i-1} - u_{n,i-1}\right\| \leq c\left(2^{-(n+i-1)}\right)^s \|u\|_s, \quad 1 \leq s \leq p+1.$$

We have, for $n \geq M$, $i \in \mathbb{N}$, and $1 \leq s \leq p+1$,

$$\|u_{n,i} - u_{n+i}\| \leq \left\|\mathcal{A}^{-1}_{n+i}\right\| \|\mathcal{C}_{n+i-1}\| \left\|u^1_{n,i-1} - u_{n,i-1}\right\| \leq \frac{c}{3}\left(2^{-(n+i)}\right)^s \|u\|_s. \tag{7}$$

From (4) and (7), we conclude that

$$\begin{aligned}\|u-u_{n,i}\| &\le \|u-u_{n+i}\| + \|u_{n,i}-u_{n+i}\| \\ &\le C\big(2^{-(n+i)s}\big)\|u\|_s + \frac{c}{3}\big(2^{-(n+i)s}\big)\|u\|_s \\ &\le \left(C+\frac{c}{3}\right)\big(2^{-(n+i)s}\big)\|u\|_s. \qquad \square\end{aligned}$$

The above theorem suggests that, if the solution $u \in H_0^1(\Omega) \cap H^s(\Omega)$, and we apply the multilevel augmentation method from level $n+i-1$ to $n+i$ by using the anti-derivatives wavelets of order p, the errors measured in $\|\cdot\|$ decrease by a factor of 2^{p+1}. Thus the behaviors of the decreasing error obtained by the multilevel, and the multilevel augmentation methods, are in the same order.

5 Examples

In this section, we illustrate the efficiency of the multilevel augmentation method in conjunction with the anti-derivatives of Daubechies wavelets of order p to find the numerical solutions of the following boundary value problem:

$$-\big(q(x)u'\big)' + r(x)u = f(x), \quad \text{for } x \in (0,R), \tag{8}$$

with the Dirichlet conditions

$$u(0) = u(R) = 0,$$

where $R = 2p-1$. It should be noted that the general interval (α,β) can be changed to $(0,R)$ by the method of changing variables.

We assume that $f \in L^2(\Omega)$, the coefficients q and r are smooth in the closed interval $[0,R]$ with $q > 0$ and $r \ge 0$.

The variational form of (8) is

$$a(u,v) = (f,v), \quad \text{for all } v \in H_0^1(\Omega), \tag{9}$$

where $a(\cdot,\cdot)$ is the bilinear form defined by

$$a(u,v) = \int_0^R qu'v' + ruv\,dx. \tag{10}$$

Since $A(\cdot,\cdot)$ is continuous and coercive on $H_0^1(\Omega)$, by the Lax-Milgram lemma, there exists a unique weak solution $u \in H_0^1(\Omega)$ for (9).

Note that we can also apply the multilevel augmentation method in conjunction with the anti-derivatives of the Daubechies wavelets order p for the cases with nonzero boundary conditions. For example, we consider the boundary value problem

$$-\big(q(x)u'\big)' + r(x)u = f(x), \quad \text{for } x \in (\alpha,\beta), \tag{11}$$

with the boundary conditions

$$u(\alpha) = c, \qquad u(\beta) = d.$$

We notice that the linear function

$$y(x) = \frac{\beta c - \alpha d}{\beta - \alpha} + \frac{d - c}{\beta - \alpha} x$$

satisfies the boundary conditions, that is, $y(\alpha) = c$ and $y(\beta) = d$. Let $u = y + w$ be the weak solution of (11). Since $u(\alpha) = c$ and $u(\beta) = d$, $w(\alpha) = w(\beta) = 0$. The variational form of the boundary value problem in this case is

$$a(w, v) = \int_\alpha^\beta fv\, dx - a(y, v), \quad \text{for all } v \in H_0^1(\alpha, \beta), \tag{12}$$

where $a(\cdot, \cdot)$ is the bilinear form

$$a(w, v) = \int_\alpha^\beta qw'v' + rwv\, dx. \tag{13}$$

Since $a(\cdot, \cdot)$ is continuous and coercive on $H_0^1(\alpha, \beta)$, by the Lax-Milgram lemma, there exists a unique weak solution $w \in H_0^1(\alpha, \beta)$ for (12).

Example 1 For the first example, we consider the two-point boundary value problem

$$-(xu')' + x^2 u = x^4 - x^3 - 4x + 1, \quad \text{for } x \in (0, 1), \tag{14}$$

with boundary conditions $u(0) = u(1) = 0$.

The exact solution is $u(x) = x^2 - x$. Here, we apply the Daubechies wavelets of order $p = 1$ to solve the problem. Numerical results for each level (n) are shown in Table 1. The column of $\|u - u_n\|$ presents the numerical results obtained by the standard multilevel method. When increasing the level of approximations, the norm of the L^2 error decreases by the factor of 2^s where $1 \leq s \leq 2$. The numerical results by the multilevel augmentation method starting from the second and the third levels are shown in the $\|u - u_{1,n-1}\|$ and $\|u - u_{2,n-2}\|$ columns, respectively. At the same level n, the L^2 errors are in the same order as those of the standard multilevel method, except that its values are slightly greater. These additional errors come from the augmentation part which can be seen from the proof of Theorem 1. It can be seen further that the error from the augmentation method is getting closer to the error from the standard multilevel method as n becomes large.

Example 2 Consider the boundary value problem

$$-(e^{-x}u')' = \pi e^{-x}(\cos(\pi x) + \pi \sin(\pi x)), \quad \text{for } x \in (0, 1), \tag{15}$$

with boundary conditions, $u(0) = u(1) = 0$.

The exact solution is $u = \sin(\pi x)$. Numerical results for $p = 1$ are shown in Table 2. The column of $\|u - u_n\|$ presents the numerical results obtained by the multilevel method.

Table 1 Example 1: Numerical results for $p = 1$

n	$\dim S_n$	$\|u - u_n\|$	$\|u - u_{1,n-1}\|$	$\|u - u_{2,n-2}\|$
2	3	8.0368e–001		
3	7	5.6412e–001	5.7409e–001	
4	15	3.7341e–001	3.9796e–001	4.0020e–001
5	31	2.2508e–001	2.7097e–001	2.6985e–001
6	63	1.4568e–001	1.7275e–001	1.7293e–001
7	127	8.8603e–002	1.1148e–001	1.1143e–001

Table 2 Example 2: Numerical results for $p = 1$

n	$\dim S_n$	$\|u - u_n\|$	$\|u - u_{1,n-1}\|$	$\|u - u_{2,n-2}\|$	$\|u - u_{3,n-3}\|$
2	3	1.6427			
3	7	0.6045	0.6458		
4	15	0.2424	0.2488	0.2494	
5	31	0.1124	0.1162	0.1162	0.1170
6	63	0.0589	0.0607	0.0607	0.0606
7	127	0.0309	0.0325	0.0325	0.0325
8	255	0.0154	0.0161	0.0161	0.0161
9	511	0.0085	0.0087	0.0087	0.0087

Table 3 Example 2: Numerical results for $p = 2$

n	$\dim S_n$	$\|u - u_n\|$	$\|u - u_{1,n-1}\|$
0	4	1.6268	
1	7	0.2097	
2	13	0.0250	0.0341
3	25	0.0030	0.0041
4	49	0.0004	0.0005

The L^2 error decreases by the factor of 2^s where $1 \leq s \leq 2$ as expected, and this agrees well with the theoretical results. The L^2 errors obtained by the multilevel augmentation method starting from levels 2, 3, and 4 are shown in the $\|u - u_{1,n-1}\|$, $\|u - u_{2,n-2}\|$, and $\|u - u_{3,n-3}\|$ columns, respectively. Their values are in the same order as the multilevel method.

To apply the Daubechies wavelets of order $p = 2$, we change the interval $[0,1]$ to $[0,3]$. So we obtain the following boundary value problem:

$$-\left(e^{-x}u'\right)' = \frac{\pi}{3}e^{-x}\left(\cos\left(\frac{\pi}{3}x\right) + \frac{\pi}{3}\sin\left(\frac{\pi}{3}x\right)\right), \quad \text{for } x \in (0,3), \tag{16}$$

with conditions $u(0) = u(3) = 0$.

The exact solution of this problem is $u(x) = \sin(\frac{\pi}{3}x)$. Numerical results from the wavelet basis of order $p = 2$ are shown in Table 3. The rate of L^2 error convergence is faster than that of the case $p = 1$. Here, the L^2 error is four times smaller than that of the previous error level. This agrees with the theoretical result that the L^2 error should decrease by a factor of 2^s, where $1 \leq s \leq 3$.

Example 3 Consider the boundary value problem with nonzero boundary conditions

$$u'' - u' = x^2 - 4x + 1, \quad \text{for } x \in (0,1), \tag{17}$$

with boundary conditions, $u(0) = -1$, $u(1) = 1$.

Table 4 Example 3: Numerical results for $p = 1$

n	$\dim S_n$	$\|u - u_n\|$	$\|u - u_{1,n-1}\|$	$\|u - u_{2,n-2}\|$	$\|u - u_{3,n-3}\|$
2	3	0.4221			
3	7	0.1993	0.2123		
4	15	0.0994	0.1051	0.1049	
5	31	0.0502	0.0561	0.0555	0.0555
6	63	0.0278	0.0303	0.0303	0.0302

Table 5 Example 3: Numerical results for $p = 2$

n	$\dim S_n$	$\|u - u_n\|$	$\|u - u_{1,n-1}\|$
0	4	3.3770	
1	7	0.2888	
2	13	0.0351	0.0378
3	25	0.0044	0.0047
4	49	0.0007	0.0008

The exact solution is $u = -\frac{x^3}{3} + x^2 + x + \frac{e^x - 3e + 2}{3(e-1)}$. Numerical results for $p = 1$ are shown in Table 4. Column of $\|u - u_n\|$ presents the numerical results obtained by the multilevel method. The L^2 error is twice smaller than that of the previous error level, and agrees well with the theoretical results. The L^2 errors obtained by the multilevel augmentation method starting from levels 3, 4, 5, and 6 are shown in the $\|u - u_{1,n-1}\|$, $\|u - u_{2,n-2}\|$, and $\|u - u_{3,n-3}\|$ columns, respectively. Their values are in the same order as the multilevel method.

Numerical results from the wavelet basis of order $p = 2$ are shown in Table 5. The rate of L^2 error convergence is faster than that of the case $p = 1$. Here, the L^2 errors is eight times smaller than that of the previous error level.

6 Conclusions

This present work is our attempt to apply the multilevel augmentation method using the anti-derivatives of Daubechies wavelets for approximating two-point boundary value problems with Dirichlet boundary conditions. This method is extended from the multilevel augmentation method that uses polynomial wavelet basis. An error analysis has also been presented. The rate of convergence is by a factor of 2^s, $1 \leq s \leq p + 1$, where p is the Daubechies wavelet order. At the same level, the L^2 error of the multilevel augmentation method is greater than that of the multilevel method, but they are in the same order.

The difficulty of this approach is that the anti-derivatives of Daubechies wavelet cannot be expressed in explicit form. One is required to solve the dilation equation to obtain a wavelet basis in an implicit formula. Here, we have done this using a numerical approximation to obtain the basis function point by point. Also, it is not easy to extend this approach to problems in higher dimension.

The advantage of this method is that we need not solve the full linear system. The unknown coefficients from the previous level can be used to approximate additional unknowns in the next level. Thus, this method can reduce computational time and memory for storing matrix coefficients. Furthermore, by applying the general anti-derivatives of Daubechies wavelets, this method can be used to solve the boundary value problems with Neumann and mixed boundary conditions. Numerical experiments on these problems are in progress and will be reported elsewhere further.

Competing interests
The authors declare that they have no competing interests.

Authors' contributions
All authors contributed equally to the writing of this paper. All authors read and approved the final manuscript.

Author details
[1]Department of Mathematics, Faculty of Science, Chiang Mai University, Chiang Mai, 50200, Thailand. [2]Department of Mathematics, Faculty of Science, Kasetsart University, Bangkok, 10900, Thailand.

Acknowledgements
This research was supported by Chiang Mai University.

References
1. Chen, Z, Micchelli, CA, Xu, Y: A multilevel method for solving operator equations. J. Math. Anal. Appl. **262**, 688-699 (2001)
2. Chen, Z, Wu, B, Xu, Y: Multilevel augmentation methods for differential equations. Adv. Comput. Math. **24**(1-4), 213-238 (2006)
3. Chen, Z, Wu, B, Xu, Y: Multilevel augmentation methods for solving operator equations. Numer. Math. J. Chin. Univ. **14**, 31-55 (2006)
4. Chen, Z, Xu, Y, Yang, H: Multilevel augmentation methods for solving ill-posed operator equations. Inverse Probl. **22**, 155-174 (2006)
5. Daubechies, I: Orthonormal bases of compactly supported wavelets. Commun. Pure Appl. Math. **41**, 909-996 (1988)
6. Glowinski, R, Lawton, W, Ravachol, M, Tenenbaum, E: Wavelet Solutions of Linear and Nonlinear Elliptic, Parabolic and Hyperbolic Problems in One Space Dimension. Computing Methods in Applied Sciences and Engineering, pp. 55-120. SIAM, Philadelphia (1990)
7. Siraj-ul-Islam, Aziz, I, Al-Fahid, AS, Shah, A: A numerical assessment of parabolic partial differential equations using Haar and Legendre wavelets. Appl. Math. Model. **37**, 9455-9481 (2013)
8. Siraj-ul-Islam, Aziz, I, Sarler, B: The numerical solution of second-order boundary value problems by collocation with Haar wavelets. Math. Comput. Model. **52**, 1577-1590 (2010)
9. Amaratunga, K, Williams, JR, Qian, S, Weiss, J: Wavelet-Galerkin solutions for one dimensional partial differential equations, IESL Technical report, 9205, 2703-2716 (1994)
10. Jianhua, S, Xuming, Y, Biquan, Y, Yuantong, S: Wavelet-Galerkin solutions for differential equations. Wuhan Univ. J. Nat. Sci. **3**(4), 403-406 (1998)
11. Mishra, V, Sabina: Wavelet Galerkin solutions of ordinary differential equations. Int. J. Math. Anal. **5**(9), 407-424 (2011)
12. Qian, S, Weiss, J: Wavelets and the numerical solution of boundary value problems. Appl. Math. Lett. **6**(1), 47-52 (1993)
13. Xu, JC, Shann, WC: Galerkin-wavelet methods for two point boundary value problems. Numer. Math. **63**, 123-144 (1992)

12

Successive iteration and positive extremal solutions for nonlinear impulsive q_k-difference equations

Ravi P Agarwal[1,2*], Guotao Wang[3], Bashir Ahmad[2], Lihong Zhang[3] and Aatef Hobiny[2]

*Correspondence: Ravi.Agarwal@tamuk.edu
[1]Department of Mathematics, Texas A&M University, Kingsville, TX 78363-8202, USA
[2]Department of Mathematics, Faculty of Science, King Abdulaziz University, P.O. Box 80203, Jeddah, 21589, Saudi Arabia
Full list of author information is available at the end of the article

Abstract

This paper investigates the existence of positive extremal solutions for nonlinear impulsive q_k-difference equations via a monotone iterative method. The main result is well illustrated with the aid of an example.

MSC: 34B37; 39A13

Keywords: successive iteration; q_k-difference equation; impulse; positive; extremal solutions

1 Introduction

In recent years, there has been put focus on developing the existence theory for initial and boundary value problems of q-difference equations and inclusions. The pioneer work on the topic dates back to the first quarter of the 20th century. In contrast to the classical definition of the derivative, the concept of q-calculus does not involve the idea of limit. The importance of q-difference equations lies in the fact that these equations are always completely controllable and appear in the q-optimal control problems [1]. The q-analog of continuous variational calculus, known as variational q-calculus, is regarded as a generalization of the continuous variational calculus due to the presence of an extra parameter q, which may be physical or economical in nature. The variational calculus on a q-uniform lattice helps to find the extremum of the functional involved in Lagrange problems of q-Euler equations rather than solving the Euler-Lagrange equation itself [2]. The applications of q-calculus appear in several disciplines such as special functions, supersymmetry, control theory, operator theory, combinatorics, initial and boundary value problems of q-difference equations, *etc.* For details and examples, we refer the reader to the books [3–6] and [7–18]. In a recent paper [17], the authors discussed the existence and uniqueness of solutions for impulsive q_k-difference equations. However, it has been found that the study of q_k-difference equations is still at its initial phase and needs further attention.

Motivated by [17], in this paper, we obtain positive extremal solutions for a new class of nonlinear impulsive q_k-difference equations by the method of successive iterations. Precisely, we investigate the following problem:

$$\begin{cases} D_{q_k}u(t) = f(t, u(t)), & 0 < q_k < 1, t \in J', \\ \Delta u(t_k) = I_k(u(t_k)), & k = 1, 2, \dots, m, \\ u(0) = \lambda u(\eta) + d, & \eta \in J_r, r \in \mathbb{Z}, \end{cases} \tag{1.1}$$

where D_{q_k} are q_k-derivatives $(k = 0, 1, 2, \dots, m), f \in C(J \times \mathbb{R}, \mathbb{R}^+), I_k \in C(\mathbb{R}, \mathbb{R}^+), J = [0, T]$, $T > 0$, $0 = t_0 < t_1 < \cdots < t_k < \cdots < t_m < t_{m+1} = T$, $J' = J\backslash\{t_1, t_2, \dots, t_m\}$, $J_r = (t_r, T]$, $0 \le \lambda < 1$, $d \ge 0$, $0 \le r \le m$ and $\Delta u(t_k) = u(t_k^+) - u(t_k^-)$, $u(t_k^+)$, and $u(t_k^-)$ denote the right and the left limits of $u(t)$ at $t = t_k$ $(k = 1, 2, \dots, m)$, respectively.

The paper is organized as follows. In Section 2, we present some preliminary material and prove an auxiliary lemma, which plays a pivotal role in establishing the main result. Section 3 contains the main result, while an illustrative example is presented in Section 4.

2 Preliminaries

Let us fix $J_0 = [0, t_1], J_1 = (t_1, t_2], \dots, J_{m-1} = (t_{m-1}, t_m], J_m = (t_m, T]$ with $T = t_{m+1}$ and introduce the Banach space

$$PC(J, \mathbb{R}) = \{u : J \to \mathbb{R} \mid u \in C(J_k), k = 0, 1, \dots, m, \text{and } u(t_k^+) \text{ exist}, k = 1, 2, \dots, m\}$$

with the norm $\|u\| = \sup_{t \in J} |u(t)|$.

Next we outline some basic concepts of the q_k-calculus [17].

For $0 < q_k < 1$ and $t \in J_k$, we define the q_k-derivatives of a real valued continuous function f as

$$D_{q_k} f(t) = \frac{f(t) - f(q_k t + (1 - q_k)t_k)}{(1 - q_k)(t - t_k)}, \qquad D_{q_k} f(t_k) = \lim_{t \to t_k} D_{q_k} f(t). \tag{2.1}$$

Higher order q_k-derivatives are given by

$$D_{q_k}^0 f(t) = f(t), \qquad D_{q_k}^n f(t) = D_{q_k} D_{q_k}^{n-1} f(t), \quad n \in \mathbb{N}, t \in J_k.$$

The q_k-integral of a function f is defined by

$${}_{t_k}\mathcal{I}_{q_k} f(t) := \int_{t_k}^t f(s)\, d_{q_k}s = (1 - q_k)(t - t_k) \sum_{n=0}^{\infty} q_k^n f\big(q_k^n t + (1 - q_k^n)t_k\big), \quad t \in J_k, \tag{2.2}$$

provided the series converges. If $a \in (t_k, t)$ and f is defined on the interval (t_k, t), then

$$\int_a^t f(s)\, d_{q_k}s = \int_{t_k}^t f(s)\, d_{q_k}s - \int_{t_k}^a f(s)\, d_{q_k}s.$$

Observe that

$$D_{q_k}\big({}_{t_k}\mathcal{I}_{q_k} f(t)\big) = D_{q_k} \int_{t_k}^t f(s)\, d_{q_k}s = f(t),$$

$${}_{t_k}\mathcal{I}_{q_k}\big(D_{q_k} f(t)\big) = \int_{t_k}^t D_{q_k} f(s)\, d_{q_k}s = f(t),$$

$$_a\mathcal{I}_{q_k}\left(D_{q_k}f(t)\right) = \int_a^t D_{q_k}f(s)\,d_{q_k}s = f(t) - f(a), \quad a \in (t_k, t).$$

In the case $t_k = 0$ and $q_k = q$ in (2.1) and (2.2), then $D_{q_k}f = D_q f$, $_{t_k}\mathcal{I}_{q_k}f = {}_0\mathcal{I}_q f$, where D_q and $_0\mathcal{I}_q$ are the well-known q-derivative and q-integral of the function $f(t)$ defined by

$$D_q f(t) = \frac{f(t) - f(qt)}{(1-q)t}, \qquad {}_0\mathcal{I}_q f(t) = \int_0^t f(s)\,d_q s = \sum_{n=0}^{\infty} t(1-q)q^n f(tq^n).$$

Lemma 2.1 *For a given $\sigma(t) \in C(J, \mathbb{R})$, a function $u \in PC(J, \mathbb{R})$ is a solution of the linear impulsive q_k-difference equations*

$$\begin{cases} D_{q_k}u(t) = \sigma(t), & 0 < q_k < 1, t \in J', \\ \Delta u(t_k) = I_k(u(t_k)), & k = 1, 2, \ldots, m, \\ u(0) = \lambda u(\eta) + d, & \eta \in J_r, \end{cases} \tag{2.3}$$

if and only if u satisfies the following impulsive q_k-integral equations:

$$u(t) = \begin{cases} \int_0^t \sigma(s)\,d_{q_0}s + \frac{\lambda}{1-\lambda}\int_{t_r}^{\eta}\sigma(s)\,d_{q_r}s \\ \quad + \frac{\lambda}{1-\lambda}\sum_{i=0}^{r-1}\int_{t_i}^{t_{i+1}}\sigma(s)\,d_{q_i}s + \frac{\lambda}{1-\lambda}\sum_{i=0}^{r} I_i(u(t_i)) + \frac{d}{1-\lambda}, \quad t \in J_0; \\ \int_{t_k}^t \sigma(s)\,d_{q_k}s + \sum_{i=0}^{k-1}\int_{t_i}^{t_{i+1}}\sigma(s)\,d_{q_i}s + \sum_{i=1}^{k} I_i(u(t_i)) + \frac{\lambda}{1-\lambda}\int_{t_r}^{\eta}\sigma(s)\,d_{q_r}s \\ \quad + \frac{\lambda}{1-\lambda}\sum_{i=0}^{r-1}\int_{t_i}^{t_{i+1}}\sigma(s)\,d_{q_i}s + \frac{\lambda}{1-\lambda}\sum_{i=0}^{r} I_i(u(t_i)) + \frac{d}{1-\lambda}, \quad t \in J_k. \end{cases} \tag{2.4}$$

Proof As argued in [18], the solution of the q_k-difference equations (2.3) can be written as

$$u(t) = u(0) + \int_{t_k}^t \sigma(s)\,d_{q_k}s + \sum_{i=0}^{k-1}\int_{t_i}^{t_{i+1}}\sigma(s)\,d_{q_i}s + \sum_{i=1}^{k} I_i\left(u(t_i)\right), \quad t \in J_k. \tag{2.5}$$

Substituting $t = \eta$ in (2.5), we have

$$u(\eta) = u(0) + \int_{t_r}^{\eta} \sigma(s)\,d_{q_r}s + \sum_{i=0}^{r-1}\int_{t_i}^{t_{i+1}}\sigma(s)\,d_{q_i}s + \sum_{i=1}^{r} I_i\left(u(t_i)\right), \quad \eta \in J_r. \tag{2.6}$$

Using the nonlocal boundary condition (2.3), we obtain

$$\begin{aligned} u(t) = &\int_{t_k}^t \sigma(s)\,d_{q_k}s + \sum_{i=0}^{k-1}\int_{t_i}^{t_{i+1}}\sigma(s)\,d_{q_i}s + \sum_{i=1}^{k} I_i\left(u(t_i)\right) + \frac{\lambda}{1-\lambda}\int_{t_r}^{\eta}\sigma(s)\,d_{q_r}s \\ &+ \frac{\lambda}{1-\lambda}\sum_{i=0}^{r-1}\int_{t_i}^{t_{i+1}}\sigma(s)\,d_{q_i}s + \frac{\lambda}{1-\lambda}\sum_{i=0}^{r} I_i\left(u(t_i)\right) + \frac{d}{1-\lambda}, \quad t \in J_k. \end{aligned} \tag{2.7}$$

Conversely, assume that $u(t)$ satisfies the q_k-integral equation (2.4). Then, by employing the operator D_{q_k} on both sides of (2.4) and applying $t = \eta$, we obtain (2.3). This completes the proof. □

3 Main result

Define a cone $P \subset PC(J,\mathbb{R})$ by

$$P = \{u \in PC(J,\mathbb{R}) : u(t) \geq 0, t \in J\}$$

and an operator $\mathcal{Q} : PC(J,\mathbb{R}) \to PC(J,\mathbb{R})$ by

$$\begin{aligned}(\mathcal{Q}u)(t) &= \int_{t_k}^{t} f(s,u(s))\, d_{q_k}s + \sum_{i=0}^{k-1} \int_{t_i}^{t_{i+1}} f(s,u(s))\, d_{q_i}s + \sum_{i=1}^{k} I_i(u(t_i)) \\ &+ \frac{\lambda}{1-\lambda}\int_{t_r}^{\eta} f(s,u(s))\, d_{q_r}s + \frac{\lambda}{1-\lambda}\sum_{i=0}^{r-1}\int_{t_i}^{t_{i+1}} f(s,u(s))\, d_{q_i}s \\ &+ \frac{\lambda}{1-\lambda}\sum_{i=0}^{r} I_i(u(t_i)) + \frac{d}{1-\lambda}. \end{aligned} \tag{3.1}$$

For the sake of convenience, we introduce some notations and assumptions.

Let $b, c > 0$ be constants and $R = b + c + \frac{d}{1-\lambda}$.

(H_1) $f(t,\cdot)$ is nondecreasing on $J \times [0,R]$, and $f(t,u) \leq \frac{b}{M}$ on $J \times [0,R]$, where $M = T + \frac{\lambda\eta}{1-\lambda}$.
(H_2) $I_k(\cdot)$, $k = 1,2,\ldots,m$, are nondecreasing on $[0,R]$, and $I_k(u) \leq \frac{(1-\lambda)c}{m}$ on $[0,R]$.
(H_3) $f(t,0) \not\equiv 0$ on any subinterval of J.

We construct two explicit monotone iterative sequences, which converge to positive extremal solutions of nonlinear impulsive q_k-difference equations (1.1):

$$\begin{aligned} v_{n+1}(t) &= \int_{t_k}^{t} f(s,v_n(s))\, d_{q_k}s + \sum_{i=0}^{k-1} \int_{t_i}^{t_{i+1}} f(s,v_n(s))\, d_{q_i}s + \sum_{i=1}^{k} I_i(v_n(t_i)) \\ &+ \frac{\lambda}{1-\lambda}\int_{t_r}^{\eta} f(s,v_n(s))\, d_{q_r}s + \frac{\lambda}{1-\lambda}\sum_{i=0}^{r-1}\int_{t_i}^{t_{i+1}} f(s,v_n(s))\, d_{q_i}s \\ &+ \frac{\lambda}{1-\lambda}\sum_{i=0}^{r} I_i(v_n(t_i)) + \frac{d}{1-\lambda}, \quad \text{with initial value } v_0(t) = 0, \\ u_{n+1}(t) &= \int_{t_k}^{t} f(s,u_n(s))\, d_{q_k}s + \sum_{i=0}^{k-1} \int_{t_i}^{t_{i+1}} f(s,u_n(s))\, d_{q_i}s + \sum_{i=1}^{k} I_i(u_n(t_i)) \\ &+ \frac{\lambda}{1-\lambda}\int_{t_r}^{\eta} f(s,u_n(s))\, d_{q_r}s + \frac{\lambda}{1-\lambda}\sum_{i=0}^{r-1}\int_{t_i}^{t_{i+1}} f(s,u_n(s))\, d_{q_i}s \\ &+ \frac{\lambda}{1-\lambda}\sum_{i=0}^{r} I_i(u_n(t_i)) + \frac{d}{1-\lambda}, \quad \text{with initial value } u_0(t) = R. \end{aligned} \tag{3.2}$$

Recall that a solution u^* of problem (1.1) is called maximal (minimal) if $u^* \geq (\leq) u$ holds for any solution u of problem (1.1). The maximal and minimal solutions of problem (1.1) are called its extremal solutions.

Theorem 3.1 *If assumptions* (H_1)-(H_3) *hold, then the nonlinear impulsive q_k-difference equations* (1.1) *has positive extremal solutions v^*, u^* in $(0,R]$, which can be achieved by* *monotone iterative sequences defined by* (3.2). *Moreover, the following relation holds*:

$$v_0 \le v_1 \le \cdots \le v_n \cdots \le v^* \le \cdots \le u^* \cdots \le u_n \le \cdots \le u_1 \le u_0. \tag{3.3}$$

Proof By Lemma 2.1, one can transform the nonlocal boundary value problem (1.1) to an equivalent fixed point problem: $u = \mathscr{Q}u$. That is, a fixed point of the operator equation $u = \mathscr{Q}u$ is a solution of the problem (1.1).

Obviously, $\mathscr{Q} : P \to P$. By a similar process to that employed in [18], it is easy to show that $\mathscr{Q} : P \to P$ is completely continuous.

Denote a ball $B = \{u \in P, \|u\| \le R\}$. Now, we show that $\mathscr{Q}(B) \subset B$. Then, for $u \in B$, by (H_1) and (H_2), we have

$$\begin{aligned}
|(\mathscr{Q}u)(t)| &= \int_{t_k}^{t} f(s,u(s))\, d_{q_k}s + \sum_{i=0}^{k-1}\int_{t_i}^{t_{i+1}} f(s,u(s))\, d_{q_i}s + \sum_{i=1}^{k} I_i(u(t_i)) \\
&\quad + \frac{\lambda}{1-\lambda}\int_{t_r}^{\eta} f(s,u(s))\, d_{q_r}s + \frac{\lambda}{1-\lambda}\sum_{i=0}^{r-1}\int_{t_i}^{t_{i+1}} f(s,u(s))\, d_{q_i}s \\
&\quad + \frac{\lambda}{1-\lambda}\sum_{i=0}^{r} I_i(u(t_i)) + \frac{d}{1-\lambda} \\
&\le \frac{b}{M}\left[\int_{t_k}^{t} d_{q_k}s + \sum_{i=0}^{k-1}\int_{t_i}^{t_{i+1}} d_{q_i}s\right] + \frac{k(1-\lambda)c}{m} \\
&\quad + \frac{\lambda b}{(1-\lambda)M}\left[\int_{t_r}^{\eta} d_{q_r}s + \sum_{i=0}^{r-1}\int_{t_i}^{t_{i+1}} d_{q_i}s\right] + \frac{r\lambda c}{m} + \frac{d}{1-\lambda} \\
&\le \frac{b}{M}\left[t - t_k + \sum_{i=0}^{k-1}(t_{i+1}-t_i)\right] + \frac{k(1-\lambda)c}{m} + \frac{\lambda b}{(1-\lambda)M}\left[\eta - t_r + \sum_{i=0}^{r-1}(t_{i+1}-t_i)\right] \\
&\quad + \frac{r\lambda c}{m} + \frac{d}{1-\lambda} \\
&\le \frac{bT}{M} + (1-\lambda)c + \frac{\eta\lambda b}{(1-\lambda)M} + \lambda c + \frac{d}{1-\lambda} \\
&\le R,
\end{aligned} \tag{3.4}$$

which implies that $\|\mathscr{Q}u\| \le R$. Thus $\mathscr{Q}(B) \subset B$.

Next, let us denote the iterative sequence $v_{n+1}(t) = \mathscr{Q}v_n(t)$ $(n = 0,1,2,\ldots)$ and pick $v_0(t) = 0$. Then $v_1 = \mathscr{Q}v_0 = \mathscr{Q}0$, $\forall t \in J$. In view of $v_0(t) = 0 \in B$ and $\mathscr{Q} : B \to B$, it follows that $v_n \in \mathscr{Q}(B) \subset B$ $(n = 0,1,2,\ldots)$. Thus, we have

$$v_1(t) = (\mathscr{Q}0)(t) \ge 0 = v_0(t), \quad \forall t \in J.$$

Applying the conditions (H_1) and (H_2), it is easy to prove the operator $\mathscr{Q}$ is nondecreasing.

So, we have

$$v_2(t) = (\mathscr{Q}v_1)(t) \ge (\mathscr{Q}v_0)(t) = v_1(t), \quad \forall t \in J.$$

By mathematical induction, one can show that the sequence $\{v_n\}_{n=1}^{\infty}$ satisfies

$$v_{n+1}(t) \geq v_n(t), \quad \forall t \in J, n = 0, 1, 2, \ldots. \tag{3.5}$$

Similarly, we denote the iterative sequence $u_{n+1}(t) = \mathcal{Q}u_n(t)$ $(n = 0, 1, 2, \ldots)$ and pick $u_0(t) = R$. Then $u_1 = \mathcal{Q}u_0$. In view of $u_0(t) = R \in B$ and $\mathcal{Q} : B \to B$, $u_n \in \mathcal{Q}(B) \subset B$ $(n = 0, 1, 2, \ldots)$. Thus, by (H_1) and (H_2), we obtain

$$\begin{aligned}
u_1(t) &= \int_{t_k}^{t} f(s, u_0(s))\, d_{q_k}s + \sum_{i=0}^{k-1} \int_{t_i}^{t_{i+1}} f(s, u_0(s))\, d_{q_i}s + \sum_{i=1}^{k} I_i(u_0(t_i)) \\
&\quad + \frac{\lambda}{1-\lambda} \int_{t_r}^{\eta} f(s, u_0(s))\, d_{q_r}s + \frac{\lambda}{1-\lambda} \sum_{i=0}^{r-1} \int_{t_i}^{t_{i+1}} f(s, u_0(s))\, d_{q_i}s \\
&\quad + \frac{\lambda}{1-\lambda} \sum_{i=0}^{r} I_i(u_0(t_i)) + \frac{d}{1-\lambda} \\
&\leq \frac{b}{M}\left[\int_{t_k}^{t} d_{q_k}s + \sum_{i=0}^{k-1} \int_{t_i}^{t_{i+1}} d_{q_i}s\right] + \frac{k(1-\lambda)c}{m} \\
&\quad + \frac{\lambda b}{(1-\lambda)M}\left[\int_{t_r}^{\eta} d_{q_r}s + \sum_{i=0}^{r-1} \int_{t_i}^{t_{i+1}} d_{q_i}s\right] + \frac{r\lambda c}{m} + \frac{d}{1-\lambda} \\
&\leq \frac{bT}{M} + (1-\lambda)c + \frac{\eta\lambda b}{(1-\lambda)M} + \lambda c + \frac{d}{1-\lambda} \\
&\leq R = u_0(t), \quad \forall t \in J.
\end{aligned}$$

Noting that $\mathcal{Q}$ is nondecreasing, we have

$$u_2(t) = (\mathcal{Q}u_1)(t) \leq (\mathcal{Q}u_0)(t) = u_1(t), \quad \forall t \in J.$$

Again, by mathematical induction, it can be shown that the sequence $\{u_n\}_{n=1}^{\infty}$ satisfies

$$u_{n+1}(t) \leq u_n(t), \quad \forall t \in J, n = 0, 1, 2, \ldots. \tag{3.6}$$

By the complete continuity of the operator $\mathcal{Q}$, the sequences $\{v_n\}_{n=1}^{\infty}$ and $\{u_n\}_{n=1}^{\infty}$ are relative compact. It means that $\{v_n\}_{n=1}^{\infty}$ and $\{u_n\}_{n=1}^{\infty}$ have convergent subsequences $\{v_{n_k}\}_{k=1}^{\infty}$ and $\{u_{n_k}\}_{k=1}^{\infty}$, respectively, and there exist $v^*, u^* \in B$ such that $v_{n_k} \to v^*$, $u_{n_k} \to u^*$ as $k \to \infty$. Using this fact together with (3.5) and (3.6) yields

$$\lim_{n\to\infty} v_n = v^*, \qquad \lim_{n\to\infty} u_n = u^*.$$

In consequence, from the continuity of the operator $\mathcal{Q}$, it follows that $\mathcal{Q}v^* = v^*$, $\mathcal{Q}u^* = u^*$. This means that u^* and v^* are two solutions of problem (1.1).

Finally, we prove that u^* and v^* are positive extremal solutions of problem (1.1) in $(0, R]$.

If $w \in [0, R]$ is any solution of problem (1.1), then $\mathcal{Q}w = w$ and $v_0(t) \leq w(t) \leq u_0(t)$. This implies that $\mathcal{Q}$ is nondecreasing and that

$$v_n(t) \leq w(t) \leq u_n(t), \quad \forall t \in J, n = 0, 1, 2, \ldots. \tag{3.7}$$

Thus, employing (3.5)-(3.7), we obtain

$$v_0 \le v_1 \le \cdots \le v_n \cdots \le v^* \le w \le u^* \cdots \le u_n \le \cdots \le u_1 \le u_0. \tag{3.8}$$

In view of $f(t,0) \not\equiv 0$, $\forall t \in J$, 0 is not a solution of the problem (1.1). Consequently, it follows from (3.8) that u^* and v^* are positive extremal solutions of the nonlinear impulsive q_k-difference equations (1.1) in $(0,R]$, which can be achieved by monotone iterative sequences given in (3.2).

This completes the proof. □

4 Example

Example 4.1 Consider the impulsive nonlocal boundary value problem of the nonlinear q_k-difference equation

$$\begin{cases} D_{\frac{1}{5+k}} u(t) = f(t,u) = t^3 + \frac{t}{10}u^2, & t \in [0,\frac{1}{2}], t \neq \frac{k}{3+k}, \\ \Delta u(\frac{k}{3+k}) = \arctan u(\frac{k}{3+k}), & k = 1,2, \\ u(0) = \frac{1}{8}u(\frac{1}{4}) + \frac{7}{8}, \end{cases} \tag{4.1}$$

where $q_k = \frac{1}{5+k}$ $(k = 0,1,2)$, $t_k = \frac{k}{3+k}$ $(k = 1,2)$, $m = 2$, $\lambda = \frac{1}{8}$, $\eta = \frac{1}{4}$, $d = \frac{7}{8}$, $M = \frac{15}{28}$, $f(t,u) = t^3 + \frac{t}{10}u^2$, and $I_k(u) = \arctan u$. Taking $b = 4 - \pi$, $c = \pi$, $R = 5$, it is easy to verify that all conditions of Theorem 3.1 hold. Hence, by Theorem 3.1, the problem (4.1) has positive extremal solutions in $(0,5]$, which can be achieved by the monotone iterative sequences given by (3.2).

Competing interests
The authors declare that they have no competing interests.

Authors' contributions
Each of the authors, RPA, GW, BA, LZ, and AH, contributed to each part of this work equally and read and approved the final version of the manuscript.

Author details
[1]Department of Mathematics, Texas A&M University, Kingsville, TX 78363-8202, USA. [2]Department of Mathematics, Faculty of Science, King Abdulaziz University, P.O. Box 80203, Jeddah, 21589, Saudi Arabia. [3]School of Mathematics and Computer Science, Shanxi Normal University, Linfen, Shanxi 041004, People's Republic of China.

Acknowledgements
The project was funded by the Deanship of Scientific Research (DSR), King Abdulaziz University, under grant no. 65-130-1435-HiCi. The authors, therefore, acknowledge with thanks DSR technical and financial support. The authors also acknowledge the reviewers for their useful comments.

References
1. Bangerezako, G: q-Difference linear control systems. J. Differ. Equ. Appl. **17**, 1229-1249 (2011)
2. Bangerezako, G: Variational q-calculus. J. Math. Anal. Appl. **289**, 650-665 (2004)
3. Gasper, G, Rahman, M: Basic Hypergeometric Series. Cambridge University Press, Cambridge (1990)
4. Kac, V, Cheung, P: Quantum Calculus. Springer, New York (2002)
5. Annaby, MH, Mansour, ZS: q-Fractional Calculus and Equations. Lecture Notes in Mathematics, vol. 2056. Springer, Berlin (2012)
6. Aral, A, Gupta, V, Agarwal, RP: Applications of q-Calculus in Operator Theory. Springer, Berlin (2013)
7. Mozyrska, D, Bartosiewicz, Z: On observability concepts for nonlinear discrete-time fractional order control systems. In: New Trends in Nanotechnology and Fractional Calculus Applications, vol. 4, pp. 305-312 (2010)
8. Agrawal, O: Some generalized fractional calculus operators and their applications in integral equations. Fract. Calc. Appl. Anal. **15**, 700-711 (2012)
9. Ismail, MEH, Simeonov, P: q-Difference operators for orthogonal polynomials. J. Comput. Appl. Math. **233**, 749-761 (2009)

10. El-Shahed, M, Hassan, HA: Positive solutions of q-difference equation. Proc. Am. Math. Soc. **138**, 1733-1738 (2010)
11. Ahmad, B, Ntouyas, SK: Boundary value problems for q-difference inclusions. Abstr. Appl. Anal. **2011**, Article ID 292860 (2011)
12. Ahmad, B, Alsaedi, A, Ntouyas, SK: A study of second-order q-difference equations with boundary conditions. Adv. Differ. Equ. **2012**, Article ID 35 (2012)
13. Bohner, M, Chieochan, R: Floquet theory for q-difference equations. Sarajevo J. Math. **8**(21), 355-366 (2012)
14. Ahmad, B, Nieto, JJ: Basic theory of nonlinear third-order q-difference equations and inclusions. Math. Model. Anal. **18**, 122-135 (2013)
15. Pongarm, N, Asawasamrit, S, Tariboon, J: Sequential derivatives of nonlinear q-difference equations with three-point q-integral boundary conditions. J. Appl. Math. **2013**, Article ID 605169 (2013)
16. Area, I, Atakishiyev, N, Godoy, E, Rodal, J: Linear partial q-difference equations on q-linear lattices and their bivariate q-orthogonal polynomial solutions. Appl. Math. Comput. **223**, 520-536 (2013)
17. Tariboon, J, Ntouyas, SK: Quantum calculus on finite intervals and applications to impulsive difference equations. Adv. Differ. Equ. **2013**, Article ID 282 (2013)
18. Zhang, L, Ahmad, B, Wang, G: Impulsive anti-periodic boundary value problems for nonlinear q_k-difference equations. Abstr. Appl. Anal. **2014**, Article ID 165129 (2014)

13

Impulsive quantum difference systems with boundary conditions

Jessada Tariboon[1*], Sotiris K Ntouyas[2,3] and Phollakrit Thiramanus[1]

*Correspondence:
jessadat@kmutnb.ac.th
[1]Nonlinear Dynamic Analysis Research Center, Department of Mathematics, Faculty of Applied Science, King Mongkut's University of Technology North Bangkok, Bangkok, 10800, Thailand
Full list of author information is available at the end of the article

Abstract

This article studies the existence and uniqueness of solutions for coupled systems of nonlinear impulsive quantum difference equations with coupled and uncoupled boundary conditions. The existence and uniqueness of solutions is established by Banach's contraction principle, while the existence of solutions is derived by using Leray-Schauder's alternative. Examples illustrating our results are also presented.

MSC: 26A33; 39A13; 34A37

Keywords: quantum calculus; impulsive quantum difference equation; existence; uniqueness; fixed point theorems

1 Introduction and preliminaries

In this paper, we concentrate on the study of the existence and uniqueness of solutions for a coupled system of nonlinear impulsive quantum difference equations,

$$\begin{cases} D_{q_k}x(t) = f(t,x(t),y(t)), & t \in J := [0,T], t \neq t_k, \\ D_{p_k}y(t) = g(t,x(t),y(t)), & t \in J, t \neq t_k, \\ \Delta x(t_k) = I_k(x(t_k)), \quad \Delta y(t_k) = I_k^*(y(t_k)), & k = 1,2,\ldots,m, \\ a_1x(0) + b_1y(T) = \lambda_1, \quad a_2y(0) + b_2x(T) = \lambda_2, \end{cases} \tag{1.1}$$

where $0 = t_0 < t_1 < t_2 < \cdots < t_k < \cdots < t_m < t_{m+1} = T$, $f,g : J \times \mathbb{R}^2 \to \mathbb{R}$ are continuous functions, $I_k, I_k^* \in C(\mathbb{R},\mathbb{R})$, $\Delta u(t_k) = u(t_k^+) - u(t_k)$, $u(t_k^+) = \lim_{h\to 0^+} u(t_k + h)$, $u \in \{x,y\}$, for $k = 1,2,\ldots,m$, and $0 < p_k, q_k < 1$ for $k = 0,1,2,\ldots,m$ are given quantum numbers, a_i, b_i, λ_i, $i = 1,2$ are real constants with $a_1a_2 \neq b_1b_2$.

The notions of quantum calculus on finite intervals, q_k-derivatives, and q_k-integrals were introduced in [1]. For a fixed $k \in \mathbb{N} \cup \{0\}$ let $J_k := [t_k, t_{k+1}] \subset \mathbb{R}$ be an interval and $0 < q_k < 1$, $k = 1,2,\ldots,m$ be a constant. We define the q_k-derivative of a function $f : J_k \to \mathbb{R}$ at a point $t \in J_k$ as follows.

Definition 1.1 Assume $f : J_k \to \mathbb{R}$ is a continuous function and let $t \in J_k$. Then the expression

$$D_{q_k}f(t) = \frac{f(t) - f(q_kt + (1-q_k)t_k)}{(1-q_k)(t-t_k)}, \quad t \neq t_k, \qquad D_{q_k}f(t_k) = \lim_{t\to t_k} D_{q_k}f(t), \tag{1.2}$$

is called the q_k-derivative of function f at t.

We say that f is q_k-differentiable on J_k provided $D_{q_k}f(t)$ exists for all $t \in J_k$. Note that if $t_k = 0$ and $q_k = q$ in (1.2), then $D_{q_k}f = D_q f$, where D_q is the well-known q-derivative of the function $f(t)$, defined by

$$D_q f(t) = \frac{f(t) - f(qt)}{(1-q)t}. \tag{1.3}$$

The q_k-integral is defined as follows.

Definition 1.2 Assume $f : J_k \to \mathbb{R}$ is a continuous function. Then the q_k-integral is defined by

$$\int_{t_k}^{t} f(s)\, d_{q_k}s = (1-q_k)(t-t_k)\sum_{n=0}^{\infty} q_k^n f\big(q_k^n t + (1-q_k^n)t_k\big), \tag{1.4}$$

for $t \in J_k$. Moreover, if $a \in (t_k, t)$, then the definite q_k-integral is defined by

$$\begin{aligned}\int_a^t f(s)\, d_{q_k}s = \int_{t_k}^{t} f(s)\, d_{q_k}s - \int_{t_k}^{a} f(s)\, d_{q_k}s &= (1-q_k)(t-t_k)\sum_{n=0}^{\infty} q_k^n f\big(q_k^n t + (1-q_k^n)t_k\big) \\ &\quad - (1-q_k)(a-t_k)\sum_{n=0}^{\infty} q_k^n f\big(q_k^n a + (1-q_k^n)t_k\big).\end{aligned}$$

Note that if $t_k = 0$ and $q_k = q$, then (1.4) reduces to q-integral of a function $f(t)$, defined by $\int_0^t f(s)\, d_q s = (1-q)t\sum_{n=0}^{\infty} q^n f(q^n t)$ for $t \in [0, \infty)$.

For the basic properties of the q_k-derivative and the q_k-integral we refer to [1].

The book by Kac and Cheung [2] covers many of the fundamental aspects of the quantum calculus. In recent years, the topic of q-calculus has attracted the attention of several researchers and a variety of new results can be found in [3–15] and the references cited therein.

Impulsive differential equations serve as basic models to study the dynamics of processes that are subject to sudden changes in their states. The recent development in this field has been motivated by many applied problems, such as control theory, population dynamics, and medicine. For some recent works on the theory of impulsive differential equations, we refer the interested reader to the monographs [16–18]. Moreover, the interested reader is referred to [19–24] for some recent results on impulsive q_k-difference equations.

In this paper we prove existence and uniqueness results for the impulsive boundary value problem (1.1) by using Banach's contraction mapping principle and Leray-Schauder's nonlinear alternative. The rest of this paper is organized as follows: In Section 2 we present an auxiliary lemma which is used to convert the impulsive boundary value problem (1.1) into an equivalent integral equation. In Section 3, we establish an existence and uniqueness result via Banach's contraction principle, and an existence result by applying Leray-Schauder's alternative. Results on uncoupled integral boundary conditions case are in Section 4. Examples illustrating our results are also presented.

2 An auxiliary lemma

Let $J = [0, T]$, $J_0 = [t_0, t_1]$, $J_k = (t_k, t_{k+1}]$ for $k = 1, 2, \dots, m$. To define the solutions of problem (1.1) we need the following lemma, which deals with a linear variant of problem (1.1) and gives a representation of the solutions.

Lemma 2.1 *Given $\phi, \psi \in C(J, \mathbb{R})$, the unique solution of the problem*

$$\begin{cases} D_{q_k}x(t) = \phi(t), & t \in J, t \neq t_k, \\ D_{p_k}y(t) = \psi(t), & t \in J, t \neq t_k, \\ \Delta x(t_k) = I_k(x(t_k)), & \Delta y(t_k) = I_k^*(y(t_k)), \quad k = 1, 2, \ldots, m, \\ a_1x(0) + b_1y(T) = \lambda_1, & a_2y(0) + b_2x(T) = \lambda_2, \end{cases} \tag{2.1}$$

is

$$\begin{aligned} x(t) = \frac{1}{\Omega}\Bigg[& a_2\lambda_1 - a_2b_1\sum_{k=1}^{m}\left(\int_{t_{k-1}}^{t_k}\psi(s)\,d_{p_{k-1}}s + I_k^*\big(y(t_k)\big)\right) - a_2b_1\int_{t_m}^{T}\psi(s)\,d_{p_m}s \\ & - b_1\lambda_2 + b_1b_2\sum_{k=1}^{m}\left(\int_{t_{k-1}}^{t_k}\phi(s)\,d_{q_{k-1}}s + I_k\big(x(t_k)\big)\right) + b_1b_2\int_{t_m}^{T}\phi(s)\,d_{q_m}s\Bigg] \\ & + \sum_{0<t_k<t}\left(\int_{t_{k-1}}^{t_k}\phi(s)\,d_{q_{k-1}}s + I_k\big(x(t_k)\big)\right) + \int_{t_k}^{t}\phi(s)\,d_{q_k}s \end{aligned} \tag{2.2}$$

and

$$\begin{aligned} y(t) = \frac{1}{\Omega}\Bigg[& a_1\lambda_2 - a_1b_2\sum_{k=1}^{m}\left(\int_{t_{k-1}}^{t_k}\phi(s)\,d_{q_{k-1}}s + I_k\big(x(t_k)\big)\right) - a_1b_2\int_{t_m}^{T}\phi(s)\,d_{q_m}s \\ & - b_2\lambda_1 + b_1b_2\sum_{k=1}^{m}\left(\int_{t_{k-1}}^{t_k}\psi(s)\,d_{p_{k-1}}s + I_k^*\big(y(t_k)\big)\right) + b_1b_2\int_{t_m}^{T}\psi(s)\,d_{p_m}s\Bigg] \\ & + \sum_{0<t_k<t}\left(\int_{t_{k-1}}^{t_k}\psi(s)\,d_{p_{k-1}}s + I_k^*\big(y(t_k)\big)\right) + \int_{t_k}^{t}\psi(s)\,d_{p_k}s, \end{aligned} \tag{2.3}$$

where

$$\Omega = a_1a_2 - b_1b_2 \neq 0. \tag{2.4}$$

Proof For $t \in J_0$, q_0-integrating (2.1), it follows that

$$x(t) = x(0) + \int_0^t \phi(s)\,d_{q_0}s,$$

which leads to

$$x(t_1) = x(0) + \int_0^{t_1} \phi(s)\,d_{q_0}s.$$

For $t \in J_1$, taking the q_1-integral for (2.1), we get

$$x(t) = x\big(t_1^+\big) + \int_{t_1}^t \phi(s)\,d_{q_1}s.$$

Since $x(t_1^+) = x(t_1) + I_1(x(t_1))$, we have

$$x(t) = x(0) + \int_0^{t_1} \phi(s)\,d_{q_0}s + \int_{t_1}^t \phi(s)\,d_{q_1}s + I_1\big(x(t_1)\big).$$

Again q_2-integrating (2.1) from t_2 to t, where $t \in J_2$, then

$$\begin{aligned} x(t) &= x(t_2^+) + \int_{t_2}^{t} \phi(s)\, d_{q_2}s \\ &= x(0) + \int_{0}^{t_1} \phi(s)\, d_{q_0}s + \int_{t_1}^{t_2} \phi(s)\, d_{q_1}s + \int_{t_2}^{t} \phi(s)\, d_{q_2}s + I_1(x(t_1)) + I_2(x(t_2)). \end{aligned}$$

Repeating the above process, for $t \in J$, we obtain

$$x(t) = x(0) + \sum_{0<t_k<t} \left(\int_{t_{k-1}}^{t_k} \phi(s)\, d_{q_{k-1}}s + I_k(x(t_k)) \right) + \int_{t_k}^{t} \phi(s)\, d_{q_k}s. \tag{2.5}$$

In the same way, we can obtain

$$y(t) = y(0) + \sum_{0<t_k<t} \left(\int_{t_{k-1}}^{t_k} \psi(s)\, d_{p_{k-1}}s + I_k^*(y(t_k)) \right) + \int_{t_k}^{t} \psi(s)\, d_{p_k}s. \tag{2.6}$$

In particular, for $t = T$, we have

$$\begin{aligned} x(T) &= x(0) + \sum_{k=1}^{m} \left(\int_{t_{k-1}}^{t_k} \phi(s)\, d_{q_{k-1}}s + I_k(x(t_k)) \right) + \int_{t_m}^{T} \phi(s)\, d_{q_m}s, \\ y(T) &= y(0) + \sum_{k=1}^{m} \left(\int_{t_{k-1}}^{t_k} \psi(s)\, d_{p_{k-1}}s + I_k^*(y(t_k)) \right) + \int_{t_m}^{T} \psi(s)\, d_{p_m}s. \end{aligned}$$

Applying the boundary conditions of (2.1), we get the system

$$\begin{aligned} &a_1x(0) + b_1y(0) + b_1 \sum_{k=1}^{m} \left(\int_{t_{k-1}}^{t_k} \psi(s)\, d_{p_{k-1}}s + I_k^*(y(t_k)) \right) + b_1 \int_{t_m}^{T} \psi(s)\, d_{p_m}s = \lambda_1, \\ &a_2y(0) + b_2x(0) + b_2 \sum_{k=1}^{m} \left(\int_{t_{k-1}}^{t_k} \phi(s)\, d_{q_{k-1}}s + I_k(x(t_k)) \right) + b_2 \int_{t_m}^{T} \phi(s)\, d_{q_m}s = \lambda_2, \end{aligned}$$

from which we have

$$\begin{aligned} x(0) = \frac{1}{\Omega} \Bigg[& a_2\lambda_1 - a_2b_1 \sum_{k=1}^{m} \left(\int_{t_{k-1}}^{t_k} \psi(s)\, d_{p_{k-1}}s + I_k^*(y(t_k)) \right) - a_2b_1 \int_{t_m}^{T} \psi(s)\, d_{p_m}s \\ & - b_1\lambda_2 + b_1b_2 \sum_{k=1}^{m} \left(\int_{t_{k-1}}^{t_k} \phi(s)\, d_{q_{k-1}}s + I_k(x(t_k)) \right) + b_1b_2 \int_{t_m}^{T} \phi(s)\, d_{q_m}s \Bigg] \end{aligned}$$

and

$$\begin{aligned} y(0) = \frac{1}{\Omega} \Bigg[& a_1\lambda_2 - a_1b_2 \sum_{k=1}^{m} \left(\int_{t_{k-1}}^{t_k} \phi(s)\, d_{q_{k-1}}s + I_k(x(t_k)) \right) - a_1b_2 \int_{t_m}^{T} \phi(s)\, d_{q_m}s \\ & - b_2\lambda_1 + b_1b_2 \sum_{k=1}^{m} \left(\int_{t_{k-1}}^{t_k} \psi(s)\, d_{p_{k-1}}s + I_k^*(y(t_k)) \right) + b_1b_2 \int_{t_m}^{T} \psi(s)\, d_{p_m}s \Bigg]. \end{aligned}$$

Substituting the values of $x(0)$ and $y(0)$ in (2.5) and (2.6), we obtain the solutions (2.2) and (2.3). □

3 Main results

Let $PC(J,\mathbb{R}) = \{x : J \to \mathbb{R};\, x(t)$ is continuous everywhere except for some t_k at which $x(t_k^+)$ and $x(t_k^-)$ exist and $x(t_k^-) = x(t_k),\, k = 1,2,\ldots,m\}$. $PC(J,\mathbb{R})$ is a Banach space with the norm $\|x\|_{PC} = \sup\{|x(t)|, t \in J\}$. Let us introduce the space $X = \{x(t); x(t) \in PC([0,T])\}$ endowed with the norm $\|x\| = \sup\{|x(t)|, t \in [0,T]\}$. Obviously $(X, \|\cdot\|)$ is a Banach space. Also let $Y = \{y(t); y(t) \in PC([0,T])\}$ be endowed with the norm $\|y\| = \sup\{|y(t)|, t \in [0,T]\}$. Obviously the product space $(X \times Y, \|(x,y)\|)$ is a Banach space with norm $\|(x,y)\| = \|x\| + \|y\|$.

In view of Lemma 2.1, we define an operator $\mathcal{T} : X \times Y \to X \times Y$ by

$$\mathcal{T}(x,y)(t) = \begin{pmatrix} \mathcal{T}_1(x,y)(t) \\ \mathcal{T}_2(x,y)(t) \end{pmatrix},$$

where

$$\begin{aligned}
\mathcal{T}_1(x,y)(t) = \frac{1}{\Omega}\Bigg[& a_2\lambda_1 - a_2 b_1 \sum_{k=1}^{m}\left(\int_{t_{k-1}}^{t_k} g\big(s,x(s),y(s)\big)\, d_{p_{k-1}}s + I_k^*\big(y(t_k)\big)\right) \\
& - a_2 b_1 \int_{t_m}^{T} g\big(s,x(s),y(s)\big)\, d_{p_m}s - b_1\lambda_2 \\
& + b_1 b_2 \sum_{k=1}^{m}\left(\int_{t_{k-1}}^{t_k} f\big(s,x(s),y(s)\big)\, d_{q_{k-1}}s + I_k\big(x(t_k)\big)\right) \\
& + b_1 b_2 \int_{t_m}^{T} f\big(s,x(s),y(s)\big)\, d_{q_m}s\Bigg] \\
& + \sum_{0<t_k<t}\left(\int_{t_{k-1}}^{t_k} f\big(s,x(s),y(s)\big)\, d_{q_{k-1}}s + I_k\big(x(t_k)\big)\right) + \int_{t_k}^{t} f\big(s,x(s),y(s)\big)\, d_{q_k}s
\end{aligned}$$

and

$$\begin{aligned}
\mathcal{T}_2(x,y)(t) = \frac{1}{\Omega}\Bigg[& a_1\lambda_2 - a_1 b_2 \sum_{k=1}^{m}\left(\int_{t_{k-1}}^{t_k} f\big(s,x(s),y(s)\big)\, d_{q_{k-1}}s + I_k\big(x(t_k)\big)\right) \\
& - a_1 b_2 \int_{t_m}^{T} f\big(s,x(s),y(s)\big)\, d_{q_m}s - b_2\lambda_1 \\
& + b_1 b_2 \sum_{k=1}^{m}\left(\int_{t_{k-1}}^{t_k} g\big(s,x(s),y(s)\big)\, d_{p_{k-1}}s + I_k^*\big(y(t_k)\big)\right) \\
& + b_1 b_2 \int_{t_m}^{T} g\big(s,x(s),y(s)\big)\, d_{p_m}s\Bigg] \\
& + \sum_{0<t_k<t}\left(\int_{t_{k-1}}^{t_k} g\big(s,x(s),y(s)\big)\, d_{p_{k-1}}s + I_k^*\big(y(t_k)\big)\right) + \int_{t_k}^{t} g\big(s,x(s),y(s)\big)\, d_{p_k}s.
\end{aligned}$$

For the sake of convenience, we set

$$M_1 = \frac{1}{|\Omega|}\Big[T\big(L_1|a_2||b_1| + K_1|b_1||b_2| + K_1|\Omega|\big) + mK_3\big(|b_1||b_2| + |\Omega|\big)\Big], \tag{3.1}$$

$$M_2 = \frac{1}{|\Omega|}\big[T\big(L_2|a_2||b_1| + K_2|b_1||b_2| + K_2|\Omega|\big) + mL_3|a_2||b_1|\big], \tag{3.2}$$

$$M_3 = \frac{1}{|\Omega|}\big[T\big(N_2|a_2||b_1| + N_1|b_1||b_2| + N_1|\Omega|\big) + m\big(N_4|a_2||b_1| + N_3|b_1||b_2| + N_3|\Omega|\big) + |a_2||\lambda_1| + |b_1||\lambda_2|\big], \tag{3.3}$$

$$M_4 = \frac{1}{|\Omega|}\big[T\big(K_1|a_1||b_2| + L_1|b_1||b_2| + L_1|\Omega|\big) + mK_3|a_1||b_2|\big], \tag{3.4}$$

$$M_5 = \frac{1}{|\Omega|}\big[T\big(K_2|a_1||b_2| + L_2|b_1||b_2| + L_2|\Omega|\big) + mL_3\big(|b_1||b_2| + |\Omega|\big)\big], \tag{3.5}$$

$$M_6 = \frac{1}{|\Omega|}\big[T\big(N_1|a_1||b_2| + N_2|b_1||b_2| + N_2|\Omega|\big) + m\big(N_3|a_1||b_2| + N_4|b_1||b_2| + N_4|\Omega|\big) + |a_1||\lambda_2| + |b_2||\lambda_1|\big]. \tag{3.6}$$

The first result is concerned with the existence and uniqueness of solutions for the problem (1.1) and is based on Banach's contraction mapping principle.

Theorem 3.1 *Assume that:*

(H_1) *The functions $f, g : [0, T] \times \mathbb{R}^2 \to \mathbb{R}$ are continuous and there exist constants $K_i, L_i > 0$, $i = 1, 2$ such that for all $t \in [0, T]$ and $u_i, v_i \in \mathbb{R}$, $i = 1, 2$,*

$$\big|f(t, u_1, u_2) - f(t, v_1, v_2)\big| \le K_1|u_1 - v_1| + K_2|u_2 - v_2|$$

and

$$\big|g(t, u_1, u_2) - g(t, v_1, v_2)\big| \le L_1|u_1 - v_1| + L_2|u_2 - v_2|.$$

(H_2) *The functions $I_k, I_k^* : \mathbb{R} \to \mathbb{R}$ are continuous and there exist constants $K_3, L_3 > 0$ such that for all $t \in [0, T]$ and $u_3, v_3 \in \mathbb{R}$, $k = 1, 2, \dots, m$,*

$$\big|I_k(u_3) - I_k(v_3)\big| \le K_3|u_3 - v_3|$$

and

$$\big|I_k^*(u_3) - I_k^*(v_3)\big| \le L_3|u_3 - v_3|.$$

In addition, assume that

$$M_1 + M_2 + M_4 + M_5 < 1,$$

where M_i, $i = 1, 2, 4, 5$, are given by (3.1)-(3.2) *and* (3.4)-(3.5). *Then the boundary value problem* (1.1) *has a unique solution.*

Proof Define $\sup_{t\in[0,T]} f(t, 0, 0) = N_1 < \infty$, $\sup_{t\in[0,T]} g(t, 0, 0) = N_2 < \infty$, $\sup\{|I_k(0)| : k = 1, 2, \dots, m\} = N_3 < \infty$ and $\sup\{|I_k^*(0)| : k = 1, 2, \dots, m\} = N_4 < \infty$ such that

$$r \ge \max\left\{\frac{M_3}{1 - (M_1 + M_2)}, \frac{M_6}{1 - (M_4 + M_5)}\right\},$$

where M_3 and M_6 are defined by (3.3) and (3.6), respectively.

We show that $\mathcal{T}B_r \subset B_r$, where $B_r = \{(x,y) \in X \times Y : \|(x,y)\| \le r\}$.

For $(x,y) \in B_r$, we have

$$
\begin{aligned}
&\left|\mathcal{T}_1(x,y)(t)\right| \\
&= \sup_{t\in[0,T]}\left\{\frac{1}{\Omega}\left[a_2\lambda_1 - a_2 b_1 \sum_{k=1}^{m}\left(\int_{t_{k-1}}^{t_k} g\big(s,x(s),y(s)\big)\, d_{p_{k-1}}s + I_k^*\big(y(t_k)\big)\right)\right.\right. \\
&\quad - a_2 b_1 \int_{t_m}^{T} g\big(s,x(s),y(s)\big)\, d_{p_m}s - b_1\lambda_2 \\
&\quad \left. + b_1 b_2 \sum_{k=1}^{m}\left(\int_{t_{k-1}}^{t_k} f\big(s,x(s),y(s)\big)\, d_{q_{k-1}}s + I_k\big(x(t_k)\big)\right) + b_1 b_2 \int_{t_m}^{T} f\big(s,x(s),y(s)\big)\, d_{q_m}s\right] \\
&\quad \left. + \sum_{0<t_k<t}\left(\int_{t_{k-1}}^{t_k} f\big(s,x(s),y(s)\big)\, d_{q_{k-1}}s + I_k\big(x(t_k)\big)\right) + \int_{t_k}^{t} f\big(s,x(s),y(s)\big)\, d_{q_k}s\right\} \\
&\le \frac{1}{|\Omega|}\left[|a_2||\lambda_1| + |a_2||b_1| \sum_{k=1}^{m}\left(\int_{t_{k-1}}^{t_k} \left|g\big(s,x(s),y(s)\big) - g(s,0,0)\right| + \left|g(s,0,0)\right| d_{p_{k-1}}s\right.\right. \\
&\quad \left. + \left|I_k^*\big(y(t_k)\big) - I_k^*(0)\right| + \left|I_k^*(0)\right|\right) \\
&\quad + |a_2||b_1| \int_{t_m}^{T} \left|g\big(s,x(s),y(s)\big) - g(s,0,0)\right| + \left|g(s,0,0)\right| d_{p_m}s + |b_1||\lambda_2| \\
&\quad + |b_1||b_2| \sum_{k=1}^{m}\left(\int_{t_{k-1}}^{t_k} \left|f\big(s,x(s),y(s)\big) - f(s,0,0)\right| + \left|f(s,0,0)\right| d_{q_{k-1}}s\right. \\
&\quad \left. + \left|I_k\big(x(t_k)\big) - I_k(0)\right| + \left|I_k(0)\right|\right) \\
&\quad \left. + |b_1||b_2| \int_{t_m}^{T} \left|f\big(s,x(s),y(s)\big) - f(s,0,0)\right| + \left|f(s,0,0)\right| d_{q_m}s\right] \\
&\quad + \sum_{k=1}^{m}\left(\int_{t_{k-1}}^{t_k} \left|f\big(s,x(s),y(s)\big) - f(s,0,0)\right| + \left|f(s,0,0)\right| d_{q_{k-1}}s\right. \\
&\quad \left. + \left|I_k\big(x(t_k)\big) - I_k(0)\right| + \left|I_k(0)\right|\right) + \int_{t_m}^{t} \left|f\big(s,x(s),y(s)\big) - f(s,0,0)\right| + \left|f(s,0,0)\right| d_{q_m}s \\
&\le \frac{1}{|\Omega|}\left[|a_2||\lambda_1| + |a_2||b_1| \sum_{k=1}^{m}\Big(\big(L_1\|x\| + L_2\|y\| + N_2\big)(t_k - t_{k-1}) + L_3\|y\| + N_4\Big)\right. \\
&\quad + |a_2||b_1|\big(L_1\|x\| + L_2\|y\| + N_2\big)(T - t_m) + |b_1||\lambda_2| \\
&\quad + |b_1||b_2| \sum_{k=1}^{m}\Big(\big(K_1\|x\| + K_2\|y\| + N_1\big)(t_k - t_{k-1}) + K_3\|x\| + N_3\Big) \\
&\quad \left. + |b_1||b_2|\big(K_1\|x\| + K_2\|y\| + N_1\big)(T - t_m)\right] \\
&\quad + \sum_{k=1}^{m}\Big(\big(K_1\|x\| + K_2\|y\| + N_1\big)(t_k - t_{k-1}) + K_3\|x\| + N_3\Big)
\end{aligned}
$$

$$
\begin{aligned}
&+ \left(K_1\|x\| + K_2\|y\| + N_1\right)(T - t_m) \\
&= \frac{1}{|\Omega|}\Bigg[|a_2||\lambda_1| + |a_2||b_1| \sum_{k=1}^{m+1} \left(\left(L_1\|x\| + L_2\|y\| + N_2\right)(t_k - t_{k-1})\right) + |a_2||b_1|mL_3\|y\| \\
&\quad + |a_2||b_1|mN_4 + |b_1||\lambda_2| + |b_1||b_2| \sum_{k=1}^{m+1} \left(\left(K_1\|x\| + K_2\|y\| + N_1\right)(t_k - t_{k-1})\right) \\
&\quad + |b_1||b_2|mK_3\|x\| + |b_1||b_2|mN_3 \Bigg] + \sum_{k=1}^{m+1} \left(\left(K_1\|x\| + K_2\|y\| + N_1\right)(t_k - t_{k-1})\right) \\
&\quad + mK_3\|x\| + mN_3 \\
&= \|x\| \left\{ \frac{1}{|\Omega|} \left[\sum_{k=1}^{m+1} (t_k - t_{k-1})\left(L_1|a_2||b_1| + K_1|b_1||b_2| + K_1|\Omega|\right) + mK_3\left(|b_1||b_2| + |\Omega|\right) \right] \right\} \\
&\quad + \|y\| \left\{ \frac{1}{|\Omega|} \left[\sum_{k=1}^{m+1} (t_k - t_{k-1})\left(L_2|a_2||b_1| + K_2|b_1||b_2| + K_2|\Omega|\right) + mL_3|a_2||b_1| \right] \right\} \\
&\quad + \frac{1}{|\Omega|} \Bigg[\sum_{k=1}^{m+1} (t_k - t_{k-1})\left(N_2|a_2||b_1| + N_1|b_1||b_2| + N_1|\Omega|\right) \\
&\quad + m\left(N_4|a_2||b_1| + N_3|b_1||b_2| + N_3|\Omega|\right) + |a_2||\lambda_1| + |b_1||\lambda_2| \Bigg] \\
&= M_1\|x\| + M_2\|y\| + M_3 \\
&\le (M_1 + M_2)r + M_3 \le r.
\end{aligned}
$$

In the same way, we can obtain

$$
\begin{aligned}
&\left|\mathcal{T}_2(x,y)(t)\right| \\
&\quad \le \|x\| \left\{ \frac{1}{|\Omega|} \left[\sum_{k=1}^{m+1} (t_k - t_{k-1})\left(K_1|a_1||b_2| + L_1|b_1||b_2| + L_1|\Omega|\right) + mK_3|a_1||b_2| \right] \right\} \\
&\quad + \|y\| \Bigg\{ \frac{1}{|\Omega|} \Bigg[\sum_{k=1}^{m+1} (t_k - t_{k-1})\left(K_2|a_1||b_2| + L_2|b_1||b_2| + L_2|\Omega|\right) \\
&\quad + mL_3\left(|b_1||b_2| + |\Omega|\right) \Bigg] \Bigg\} \\
&\quad + \frac{1}{|\Omega|} \Bigg[\sum_{k=1}^{m+1} (t_k - t_{k-1})\left(N_1|a_1||b_2| + N_2|b_1||b_2| + N_2|\Omega|\right) \\
&\quad + m\left(N_3|a_1||b_2| + N_4|b_1||b_2| + N_4|\Omega|\right) + |a_1||\lambda_2| + |b_2||\lambda_1| \Bigg] \\
&\quad = M_4\|x\| + M_5\|y\| + M_6 \\
&\quad \le (M_4 + M_5)r + M_6 \le r.
\end{aligned}
$$

Consequently, $\|\mathcal{T}(x,y)(t)\| \le r$.

Now for $(x_2, y_2), (x_1, y_1) \in X \times Y$ and for any $t \in [0, T]$, we get

$$
\begin{aligned}
&\left|\mathcal{T}_1(x_2, y_2)(t) - \mathcal{T}_1(x_1, y_1)(t)\right| \\
&\leq \frac{1}{|\Omega|}\Bigg[|a_2||b_1|\Bigg(\sum_{k=1}^{m+1}\int_{t_{k-1}}^{t_k}\left|g\big(s, x_2(s), y_2(s)\big) - g\big(s, x_1(s), y_1(s)\big)\right| d_{p_{k-1}}s \\
&\quad + \sum_{k=1}^{m}\left|I_k^*\big(y_2(t_k)\big) - I_k^*\big(y_1(t_k)\big)\right|\Bigg) \\
&\quad + |b_1||b_2|\Bigg(\sum_{k=1}^{m+1}\int_{t_{k-1}}^{t_k}\left|f\big(s, x_2(s), y_2(s)\big) - f\big(s, x_1(s), y_1(s)\big)\right| d_{q_{k-1}}s \\
&\quad + \sum_{k=1}^{m}\left|I_k\big(x_2(t_k)\big) - I_k\big(x_1(t_k)\big)\right|\Bigg)\Bigg] \\
&\quad + \sum_{k=1}^{m+1}\int_{t_{k-1}}^{t_k}\left|f\big(s, x_2(s), y_2(s)\big) - f\big(s, x_1(s), y_1(s)\big)\right| d_{q_{k-1}}s + \sum_{k=1}^{m}\left|I_k\big(x_2(t_k)\big) - I_k\big(x_1(t_k)\big)\right| \\
&\leq \frac{1}{|\Omega|}\Bigg[|a_2||b_1|\Bigg(\sum_{k=1}^{m+1}(t_k - t_{k-1})\big(L_1\|x_2 - x_1\| + L_2\|y_2 - y_1\|\big) + mL_3\|y_2 - y_1\|\Bigg) \\
&\quad + |b_1||b_2|\Bigg(\sum_{k=1}^{m+1}(t_k - t_{k-1})\big(K_1\|x_2 - x_1\| + K_2\|y_2 - y_1\|\big) + mK_3\|x_2 - x_1\|\Bigg)\Bigg] \\
&\quad + \sum_{k=1}^{m+1}(t_k - t_{k-1})\big(K_1\|x_2 - x_1\| + K_2\|y_2 - y_1\|\big) + mK_3\|x_2 - x_1\| \\
&= \|x_2 - x_1\|\Bigg\{\frac{1}{|\Omega|}\Bigg[\sum_{k=1}^{m+1}(t_k - t_{k-1})\big(L_1|a_2||b_1| + K_1|b_1||b_2| + K_1|\Omega|\big) \\
&\quad + mK_3\big(|b_1||b_2| + |\Omega|\big)\Bigg]\Bigg\} \\
&\quad + \|y_2 - y_1\|\Bigg\{\frac{1}{|\Omega|}\Bigg[\sum_{k=1}^{m+1}(t_k - t_{k-1})\big(L_2|a_2||b_1| + K_2|b_1||b_2| + K_2|\Omega|\big) + mL_3|a_2||b_1|\Bigg]\Bigg\} \\
&= M_1\|x_2 - x_1\| + M_2\|y_2 - y_1\|,
\end{aligned}
$$

and consequently we obtain

$$
\left\|\mathcal{T}_1(x_2, y_2)(t) - \mathcal{T}_1(x_1, y_1)\right\| \leq (M_1 + M_2)\big[\|x_2 - x_1\| + \|y_2 - y_1\|\big]. \tag{3.7}
$$

Similarly,

$$
\left\|\mathcal{T}_2(x_2, y_2)(t) - \mathcal{T}_2(x_1, y_1)\right\| \leq (M_4 + M_5)\big[\|x_2 - x_1\| + \|y_2 - y_1\|\big]. \tag{3.8}
$$

It follows from (3.7) and (3.8) that

$$
\left\|\mathcal{T}(x_2, y_2)(t) - \mathcal{T}(x_1, y_1)(t)\right\| \leq (M_1 + M_2 + M_4 + M_5)\big[\|x_2 - x_1\| + \|y_2 - y_1\|\big].
$$

Since $M_1 + M_2 + M_4 + M_5 < 1$, therefore, $\mathcal{T}$ is a contraction operator. So, by Banach's fixed point theorem, the operator $\mathcal{T}$ has a unique fixed point, which is the unique solution of problem (1.1). This completes the proof. □

In the next result, we prove the existence of solutions for problem (1.1) by applying the Leray-Schauder alternative.

For the sake of convenience, we set

$$M_7 = \frac{1}{|\Omega|}\left[T\left(B_1|a_2||b_1| + A_1|b_1||b_2| + A_1|\Omega|\right) + mA_4\left(|b_1||b_2| + |\Omega|\right)\right], \tag{3.9}$$

$$M_8 = \frac{1}{|\Omega|}\left[T\left(B_2|a_2||b_1| + A_2|b_1||b_2| + A_2|\Omega|\right) + mB_4|a_2||b_1|\right], \tag{3.10}$$

$$\begin{aligned} M_9 = \frac{1}{|\Omega|}\Big[&T\left(B_0|a_2||b_1| + A_0|b_1||b_2| + A_0|\Omega|\right) \\ &+ m\left(B_3|a_2||b_1| + A_3|b_1||b_2| + A_3|\Omega|\right) + |a_2||\lambda_1| + |b_1||\lambda_2|\Big], \end{aligned} \tag{3.11}$$

$$M_{10} = \frac{1}{|\Omega|}\left[T\left(A_1|a_1||b_2| + B_1|b_1||b_2| + B_1|\Omega|\right) + mA_4|a_1||b_2|\right], \tag{3.12}$$

$$M_{11} = \frac{1}{|\Omega|}\left[T\left(A_2|a_1||b_2| + B_2|b_1||b_2| + B_2|\Omega|\right) + mB_4\left(|b_1||b_2| + |\Omega|\right)\right], \tag{3.13}$$

$$\begin{aligned} M_{12} = \frac{1}{|\Omega|}\Big[&T\left(A_0|a_1||b_2| + B_0|b_1||b_2| + B_0|\Omega|\right) \\ &+ m\left(A_3|a_1||b_2| + B_3|b_1||b_2| + B_3|\Omega|\right) + |a_1||\lambda_2| + |b_2||\lambda_1|\Big], \end{aligned} \tag{3.14}$$

and

$$M_0 = \min\left\{1 - (M_7 + M_{10}), 1 - (M_8 + M_{11})\right\}. \tag{3.15}$$

Lemma 3.1 (Leray-Schauder alternative) ([25], p.4) *Let $F : E \to E$ be a completely continuous operator (i.e., a map that is restricted to any bounded set in E is compact). Let*

$$\mathcal{E}(F) = \left\{x \in E : x = \lambda F(x)\, \textit{for some } 0 < \lambda < 1\right\}.$$

Then either the set $\mathcal{E}(F)$ is unbounded, or F has at least one fixed point.

Theorem 3.2 *Assume that:*

(H_3) *The functions $f, g : [0, T] \times \mathbb{R}^2 \to \mathbb{R}$ are continuous and there exist constants $A_i, B_i \geq 0$ $(i = 1, 2)$ and $A_0, B_0 > 0$ such that $\forall x_i \in \mathbb{R}$ $(i = 1, 2)$*

$$\left|f(t, x_1, x_2)\right| \leq A_0 + A_1|x_1| + A_2|x_2|$$

and

$$\left|g(t, x_1, x_2)\right| \leq B_0 + B_1|x_1| + B_2|x_2|.$$

(H$_4$) *The functions* $I_k, I_k^* : \mathbb{R} \to \mathbb{R}$ *are continuous and there exist constants* $A_4, B_4 \geq 0$ *and* $A_3, B_3 > 0$ *such that* $\forall x_3 \in \mathbb{R}$, $k = 1, 2, \ldots, m$

$$|I_k(x_3)| \leq A_3 + A_4|x_3|$$

and

$$|I_k^*(x_3)| \leq B_3 + B_4|x_3|.$$

In addition it is assumed that

$$M_7 + M_{10} < 1 \quad \text{and} \quad M_8 + M_{11} < 1,$$

where M_7, M_8, M_{10}, M_{11} *are given by* (3.9)-(3.10) *and* (3.12)-(3.13). *Then there exists at least one solution for the boundary value problem* (1.1).

To prove the theorem we use the following lemma.

Lemma 3.2 *Assume that* (H$_3$) *and* (H$_4$) *hold. Then the operator* $\mathcal{T} : X \times Y \to X \times Y$ *is completely continuous.*

Proof By continuity of functions f and g, the operator $\mathcal{T}$ is continuous.

Let $\Theta \subset X \times Y$ be bounded. Then there exist positive constants P_1, P_2, P_3, and P_4 such that

$$|f(t, x(t), y(t))| \leq P_1, \qquad |g(t, x(t), y(t))| \leq P_2, \quad \forall (x, y) \in \Theta,$$
$$|I_k(x(t))| \leq P_3, \qquad |I_k^*(y(t))| \leq P_4, \quad k = 1, 2, \ldots, m.$$

Then for any $(x, y) \in \Theta$, we have

$$\begin{aligned}
&\|\mathcal{T}_1(x, y)\| \\
&\leq \frac{1}{|\Omega|}\Bigg[|a_2||\lambda_1| + |a_2||b_1|\left(\sum_{k=1}^{m+1}\int_{t_{k-1}}^{t_k}|g(s, x(s), y(s))|\, d_{p_{k-1}}s + \sum_{k=1}^{m}|I_k^*(y(t_k))|\right) \\
&\quad + |b_1||\lambda_2| + |b_1||b_2|\left(\sum_{k=1}^{m+1}\int_{t_{k-1}}^{t_k}|f(s, x(s), y(s))|\, d_{q_{k-1}}s + \sum_{k=1}^{m}|I_k(x(t_k))|\right)\Bigg] \\
&\quad + \sum_{k=1}^{m+1}\int_{t_{k-1}}^{t_k}|f(s, x(s), y(s))|\, d_{q_{k-1}}s + \sum_{k=1}^{m}|I_k(x(t_k))| \\
&\leq \frac{1}{|\Omega|}\left[|a_2||\lambda_1| + |a_2||b_1|(P_2T + mP_4) + |b_1||\lambda_2| + |b_1||b_2|(P_1T + mP_3)\right] \\
&\quad + P_1T + mP_3 \\
&:= D_1.
\end{aligned}$$

Similarly, we get

$$\begin{aligned}\left\|\mathcal{T}_2(x,y)\right\| &\leq \frac{1}{|\Omega|}\left[|a_1||\lambda_2| + |a_1||b_2|(P_1T + mP_3) + |b_2||\lambda_1| + |b_1||b_2|(P_2T + mP_4)\right] \\ &\quad + P_2T + mP_4 \\ &:= D_2.\end{aligned}$$

Thus, it follows from the above inequalities that the operator $\mathcal{T}$ is uniformly bounded.

Next, we show that $\mathcal{T}$ is equicontinuous. Let $\nu_1, \nu_2 \in (t_l, t_{l+1})$ for some $l = 0, 1, \ldots, m$ with $\nu_1 < \nu_2$. Then we have

$$\begin{aligned}&\left|\mathcal{T}_1\big(x(\nu_2), y(\nu_2)\big) - \mathcal{T}_1\big(x(\nu_1), y(\nu_1)\big)\right| \\ &\quad = \left|\int_{t_l}^{\nu_2} f\big(s, x(s), y(s)\big)\, d_{q_l}s - \int_{t_l}^{\nu_1} f\big(s, x(s), y(s)\big)\, d_{q_l}s\right| \\ &\quad \leq P_1|\nu_2 - \nu_1|.\end{aligned}$$

Analogously, we can obtain

$$\begin{aligned}&\left|\mathcal{T}_2\big(x(\nu_2), y(\nu_2)\big) - \mathcal{T}_2\big(x(\nu_1), y(\nu_1)\big)\right| \\ &\quad = \left|\int_{t_l}^{\nu_2} g\big(s, x(s), y(s)\big)\, d_{p_l}s - \int_{t_l}^{\nu_1} g\big(s, x(s), y(s)\big)\, d_{p_l}s\right| \\ &\quad \leq P_2|\nu_2 - \nu_1|.\end{aligned}$$

Therefore, the operator $\mathcal{T}(x,y)$ is equicontinuous, and thus the operator $\mathcal{T}(x,y)$ is completely continuous. □

Proof of Theorem 3.2 By Lemma 3.2 the operator $\mathcal{T}(x,y)$ is completely continuous.

Now, it will be verified that the set $\mathcal{E} = \{(x,y) \in X \times Y | (x,y) = \lambda\mathcal{T}(x,y), 0 \leq \lambda \leq 1\}$ is bounded. Let $(x,y) \in \mathcal{E}$, then $(x,y) = \lambda\mathcal{T}(x,y)$. For any $t \in [0,T]$, we have

$$x(t) = \lambda\mathcal{T}_1(x,y)(t), \qquad y(t) = \lambda\mathcal{T}_2(x,y)(t).$$

Then

$$\begin{aligned}\left|x(t)\right| &\leq \|x\|\left\{\frac{1}{|\Omega|}\left[\sum_{k=1}^{m+1}(t_k - t_{k-1})\big(B_1|a_2||b_1| + A_1|b_1||b_2| + A_1|\Omega|\big)\right.\right. \\ &\quad \left.\left. + mA_4\big(|b_1||b_2| + |\Omega|\big)\right]\right\} \\ &\quad + \|y\|\left\{\frac{1}{|\Omega|}\left[\sum_{k=1}^{m+1}(t_k - t_{k-1})\big(B_2|a_2||b_1| + A_2|b_1||b_2| + A_2|\Omega|\big) + mB_4|a_2||b_1|\right]\right\} \\ &\quad + \frac{1}{|\Omega|}\left[\sum_{k=1}^{m+1}(t_k - t_{k-1})\big(B_0|a_2||b_1| + A_0|b_1||b_2| + A_0|\Omega|\big)\right. \\ &\quad \left. + m\big(B_3|a_2||b_1| + A_3|b_1||b_2| + A_3|\Omega|\big) + |a_2||\lambda_1| + |b_1||\lambda_2|\right]\end{aligned}$$

and

$$
\begin{aligned}
|y(t)| \le \|x\| &\left\{ \frac{1}{|\Omega|} \left[\sum_{k=1}^{m+1} (t_k - t_{k-1})\big(A_1|a_1||b_2| + B_1|b_1||b_2| + B_1|\Omega|\big) + mA_4|a_1||b_2| \right] \right\} \\
&+ \|y\| \left\{ \frac{1}{|\Omega|} \left[\sum_{k=1}^{m+1} (t_k - t_{k-1})\big(A_2|a_1||b_2| + B_2|b_1||b_2| + B_2|\Omega|\big) \right. \right. \\
&\left. \left. + mB_4\big(|b_1||b_2| + |\Omega|\big) \right] \right\} \\
&+ \frac{1}{|\Omega|} \left[\sum_{k=1}^{m+1} (t_k - t_{k-1})\big(A_0|a_1||b_2| + B_0|b_1||b_2| + B_0|\Omega|\big) \right. \\
&\left. + m\big(A_3|a_1||b_2| + B_3|b_1||b_2| + B_3|\Omega|\big) + |a_1||\lambda_2| + |b_2||\lambda_1| \right].
\end{aligned}
$$

Hence we have

$$\|x\| \le M_7\|x\| + M_8\|y\| + M_9$$

and

$$\|y\| \le M_{10}\|x\| + M_{11}\|y\| + M_{12},$$

which imply that

$$\|x\| + \|y\| \le (M_7 + M_{10})\|x\| + (M_8 + M_{11})\|y\| + M_9 + M_{12}.$$

Consequently,

$$\big\|(x,y)\big\| \le \frac{M_9 + M_{12}}{M_0},$$

for any $t \in [0,T]$, where M_0 is defined by (3.15), which proves that $\mathcal{E}$ is bounded. Thus, by Lemma 3.1, the operator $\mathcal{T}$ has at least one fixed point. Hence the boundary value problem (1.1) has at least one solution. The proof is complete. □

3.1 Examples

Example 3.1 Consider the following coupled system of impulsive quantum difference equations with coupled boundary conditions

$$
\begin{cases}
D_{\frac{2k+1}{k^2+k+2}} x(t) = \frac{t\cos^2(\pi t)}{(3e^t+4)^2}\frac{|x(t)|}{|x(t)|+1} + \frac{t+1}{(2t+4)^3}\frac{|y(t)|}{|y(t)|+1} + \frac{3}{2}, & t \in [0,2], t \neq t_k, \\
D_{\frac{\sqrt{k+1}}{e^k+1}} y(t) = \frac{1}{(2^{t+1}+5)^2}\sin x(t) + \frac{e^{-2(t+1)}}{7}\cos y(t) + \frac{t^2+1}{3}, & t \in [0,2], t \neq t_k, \\
\Delta x(t_k) = \frac{|x(t_k)|}{6(k+5)+|x(t_k)|}, \quad \Delta y(t_k) = \frac{|y(t_k)|}{7(k+4)+|y(t_k)|}, & t_k = \frac{k}{2}, k = 1,2,3, \\
2x(0) + 4y(2) = 5, \quad 3y(0) - 2x(2) = -6.
\end{cases}
\tag{3.16}
$$

Here $q_k = (2k+1)/(k^2+k+2)$, $p_k = (\sqrt{k+1})/(e^k+1)$, $k = 0,1,2,3$, $m = 3$, $T = 2$, $a_1 = 2$, $a_2 = 3$, $b_1 = 4$, $b_2 = -2$, $\lambda_1 = 5$, $\lambda_2 = -6$, $f(t,x,y) = (t\cos^2(\pi t)|x|)/(((3e^t+4)^2)(|x|+1)) + ((t+1)|y|)/(((2t+4)^3)(|y|+1)) + 3/2$, $g(t,x,y) = (\sin x)/(2^{t+1}+5)^2 + (e^{-2(t+1)}\cos y)/7 + (t^2+1)/3$, $I_k(x) = |x|/(6(k+5)+|x|)$, and $I_k^*(y) = |y|/(7(k+4)+|y|)$. We have $|f(t,x_1,y_1) - f(t,x_2,y_2)| \leq ((2/49)|x_1-x_2| + (3/64)|y_1-y_2|)$, $|g(t,x_1,y_1) - g(t,x_2,y_2)| \leq ((1/49)|x_1-x_2| + (1/(7e^2))|y_1-y_2|)$, $|I_k(x) - I_k(y)| \leq (1/36)|x-y|$, and $|I_k^*(x) - I_k^*(y)| \leq (1/35)|x-y|$. We can find

$$\Omega = a_1a_2 - b_1b_2 = 14 \neq 0.$$

With the given values, it is found that $K_1 = 2/49$, $K_2 = 3/64$, $K_3 = 1/36$, $L_1 = 1/49$, $L_2 = 1/(7e^2)$, $L_3 = 1/35$, $M_1 \simeq 0.29422$, $M_2 \simeq 0.25393$, $M_4 \simeq 0.11127$, $M_5 \simeq 0.22224$,, and

$$M_1 + M_2 + M_4 + M_5 \simeq 0.88167 < 1.$$

Thus all the conditions of Theorem 3.1 are satisfied. Therefore, by the conclusion of Theorem 3.1, problem (3.16) has a unique solution on $[0,2]$.

Example 3.2 Consider the following coupled system of impulsive quantum difference equations with coupled boundary conditions:

$$\begin{cases} D_{\frac{k+1}{\sqrt{k^2+e^k+1}}}x(t) = \frac{1}{4} + \frac{1}{2(t+5)^2}\sin x(t) + \frac{1}{7\pi^2}\tan^{-1}y(t), & t \in [0,3], t \neq t_k, \\ D_{\frac{1}{3}\sin(\frac{k+1}{10}\pi)}y(t) = \frac{t+2}{e} + \frac{1}{40}x(t)\cos y(t) + \frac{1}{2^t+45}y(t), & t \in [0,3], t \neq t_k, \\ \Delta x(t_k) = \frac{1}{4}\tan^{-1}(\frac{x(t_k)}{8}) + 2, \quad t_k = \frac{k}{3}, k = 1,2,\ldots,8, \\ \Delta y(t_k) = \frac{1}{5}\sin(\frac{y(t_k)}{6}) + 3, \quad t_k = \frac{k}{3}, k = 1,2,\ldots,8, \\ -x(0) + 5y(3) = -2, \qquad 2y(0) + 3x(3) = 5. \end{cases} \tag{3.17}$$

Here $q_k = (k+1)/(\sqrt{k^2+e^k+1})$, $p_k = (\sin(((k+1)\pi)/10))/3$, $k = 0,1,2,\ldots,8$, $m = 8$, $T = 3$, $a_1 = -1$, $a_2 = 2$, $b_1 = 5$, $b_2 = 3$, $\lambda_1 = -2$, $\lambda_2 = 5$, $f(t,x,y) = (1/4) + (\sin x)/(2(t+5)^2) + (\tan^{-1}y)/(7\pi^2)$, $g(t,x,y) = ((t+2)/e) + (x\cos y)/40 + (y)/(2^t+45)$, $I_k(x) = (\tan^{-1}(x/8))/4 + 2$, and $I_k^*(y) = (\sin(y/6))/5 + 3$. We get

$$\Omega = a_1a_2 - b_1b_2 = -17 \neq 0.$$

Since $|f(t,x,y)| \leq A_0 + A_1|x| + A_2|y|$, $|g(t,x,y)| \leq B_0 + B_1|x| + B_2|y|$, where $A_0 = 1/4$, $A_1 = 1/50$, $A_2 = 1/(7\pi^2)$, $B_0 = 5/e$, $B_1 = 1/40$, $B_2 = 1/46$, it is found that $M_7 \simeq 0.62765$, $M_8 \simeq 0.27696$, $M_{10} \simeq 0.19588$, $M_{11} \simeq 0.63239$. Furthermore,

$$M_7 + M_{10} \approx 0.82353 < 1$$

and

$$M_8 + M_{11} \approx 0.90935 < 1.$$

Thus all the conditions of Theorem 3.2 holds true and consequently the conclusion of Theorem 3.2; problem (3.17) has at least one solution on $[0,3]$.

4 Uncoupled boundary conditions case

In this section, we consider again the system

$$\begin{cases} D_{q_k}x(t) = f(t, x(t), y(t)), & t \in J, t \neq t_k, \\ D_{p_k}y(t) = g(t, x(t), y(t)), & t \in J, t \neq t_k, \\ \Delta x(t_k) = I_k(x(t_k)), & \Delta y(t_k) = I_k^*(y(t_k)), \quad k = 1, 2, \ldots, m, \\ a_1 x(0) + b_1 x(T) = \lambda_1, & a_2 y(0) + b_2 y(T) = \lambda_2. \end{cases} \tag{4.1}$$

Lemma 4.1 (Auxiliary lemma) *For $h \in C([0, T], \mathbb{R})$, the unique solution of the problem*

$$\begin{cases} D_{q_k}x(t) = h(t), & t \in J, t \neq t_k, \\ \Delta x(t_k) = I_k(x(t_k)), & k = 1, 2, \ldots, m, \\ a_1 x(0) + b_1 x(T) = \lambda_1, \end{cases} \tag{4.2}$$

is given by

$$\begin{aligned} x(t) &= \frac{\lambda_1}{\Lambda} - \frac{b_1}{\Lambda}\left(\sum_{k=1}^{m+1}\int_{t_{k-1}}^{t_k} h(s)\, d_{q_{k-1}}s + \sum_{k=1}^{m} I_k\big(x(t_k)\big)\right) \\ &\quad + \sum_{0<t_k<t}\left(\int_{t_{k-1}}^{t_k} h(s)\, d_{q_{k-1}}s + I_k\big(x(t_k)\big)\right) + \int_{t_k}^{t} h(s)\, d_{q_k}s, \end{aligned} \tag{4.3}$$

where

$$\Lambda := a_1 + b_1 \neq 0. \tag{4.4}$$

In view of Lemma 4.1, we define an operator $\mathfrak{T} : X \times Y \to X \times Y$ by

$$\mathfrak{T}(u, v)(t) = \begin{pmatrix} \mathfrak{T}_1(u, v)(t) \\ \mathfrak{T}_2(u, v)(t) \end{pmatrix},$$

where

$$\begin{aligned} \mathfrak{T}_1(u, v)(t) &= \frac{\lambda_1}{\Lambda} - \frac{b_1}{\Lambda}\left(\sum_{k=1}^{m+1}\int_{t_{k-1}}^{t_k} f\big(s, u(s), v(s)\big)\, d_{q_{k-1}}s + \sum_{k=1}^{m} I_k\big(u(t_k)\big)\right) \\ &\quad + \sum_{0<t_k<t}\left(\int_{t_{k-1}}^{t_k} f\big(s, u(s), v(s)\big)\, d_{q_{k-1}}s + I_k\big(u(t_k)\big)\right) + \int_{t_k}^{t} f\big(s, u(s), v(s)\big)\, d_{q_k}s \end{aligned}$$

and

$$\begin{aligned} \mathfrak{T}_2(u, v)(t) &= \frac{\lambda_2}{\Phi} - \frac{b_2}{\Phi}\left(\sum_{k=1}^{m+1}\int_{t_{k-1}}^{t_k} g\big(s, u(s), v(s)\big)\, d_{p_{k-1}}s + \sum_{k=1}^{m} I_k^*\big(v(t_k)\big)\right) \\ &\quad + \sum_{0<t_k<t}\left(\int_{t_{k-1}}^{t_k} g\big(s, u(s), v(s)\big)\, d_{p_{k-1}}s + I_k^*\big(v(t_k)\big)\right) \\ &\quad + \int_{t_k}^{t} g\big(s, u(s), v(s)\big)\, d_{p_k}s, \end{aligned}$$

where

$$\Phi := a_2 + b_2 \neq 0.$$

We remark that $\mathfrak{T}_1$ depends only on f and $\mathfrak{T}_2$ only on g. We call the above system, for convenience, a 'coupled system with uncoupled boundary conditions'.

In the sequel, we set the constants

$$\overline{M}_1 = \frac{1}{|\Lambda|}(T\overline{K}_1 + m\overline{K}_3)\big(|b_1| + |\Lambda|\big), \tag{4.5}$$

$$\overline{M}_2 = \frac{1}{|\Lambda|}T\overline{K}_2\big(|b_1| + |\Lambda|\big), \tag{4.6}$$

$$\overline{M}_3 = \frac{1}{|\Lambda|}\big[(T\overline{N}_1 + m\overline{N}_3)\big(|b_1| + |\Lambda|\big) + |\lambda_1|\big], \tag{4.7}$$

$$\overline{M}_4 = \frac{1}{|\Phi|}T\overline{L}_1\big(|b_2| + |\Phi|\big), \tag{4.8}$$

$$\overline{M}_5 = \frac{1}{|\Phi|}(T\overline{L}_2 + m\overline{L}_3)\big(|b_2| + |\Phi|\big), \tag{4.9}$$

$$\overline{M}_6 = \frac{1}{|\Phi|}\big[(T\overline{N}_2 + m\overline{N}_4)\big(|b_2| + |\Phi|\big) + |\lambda_2|\big]. \tag{4.10}$$

Now we present the existence and uniqueness result for problem (4.1). We do not provide the proof of this result as it is similar to the one for Theorem 3.1.

Theorem 4.1 *Assume that:*

(H$_5$) *The functions* $f, g : [0,T] \times \mathbb{R}^2 \to \mathbb{R}$ *are continuous and there exist constants* $\overline{K}_i, \overline{L}_i > 0$, $i = 1,2$ *such that for all* $t \in [0,T]$ *and* $u_i, v_i \in \mathbb{R}$, $i = 1,2$,

$$\big|f(t,u_1,u_2) - f(t,v_1,v_2)\big| \le \overline{K}_1|u_1 - v_1| + \overline{K}_2|u_2 - v_2|$$

and

$$\big|g(t,u_1,u_2) - g(t,v_1,v_2)\big| \le \overline{L}_1|u_1 - v_1| + \overline{L}_2|u_2 - v_2|.$$

(H$_6$) *The functions* $I_k, I_k^* : \mathbb{R} \to \mathbb{R}$ *are continuous and there exist constants* $\overline{K}_3, \overline{L}_3 > 0$ *such that for all* $t \in [0,T]$ *and* $u_3, v_3 \in \mathbb{R}$, $k = 1,2,\ldots,m$

$$\big|I_k(u_3) - I_k(v_3)\big| \le \overline{K}_3|u_3 - v_3|$$

and

$$\big|I_k^*(u_3) - I_k^*(v_3)\big| \le \overline{L}_3|u_3 - v_3|.$$

In addition, assume that

$$\overline{M}_1 + \overline{M}_2 + \overline{M}_4 + \overline{M}_5 < 1,$$

where $\overline{M}_1$, $\overline{M}_2$, $\overline{M}_4$, $\overline{M}_5$ *are given by* (4.5)-(4.6) *and* (4.8)-(4.9), *respectively. Then the boundary value problem* (4.1) *has a unique solution.*

Example 4.1 Consider the following coupled system of impulsive quantum difference equations with uncoupled boundary conditions

$$\begin{cases} D_{(\frac{2}{7})^k}x(t) = \frac{\sin(\pi t)}{(e^t+5)^2}\frac{|x(t)|}{|x(t)|+1} + \frac{\pi^t}{(t+4)^3}\frac{|y(t)|}{|y(t)|+1} + 3, & t \in [0,1], t \neq t_k, \\ D_{(\frac{3+k}{4+2k})^k}y(t) = \frac{1}{10(2^t+4)}\cos x(t) + \frac{1}{6\pi(t+3)}|y(t)| + 2, & t \in [0,1], t \neq t_k, \\ \Delta x(t_k) = \frac{|x(t_k)|}{3(k+9)+|x(t_k)|}, \quad \Delta y(t_k) = \frac{|y(t_k)|}{5(k+6)+|y(t_k)|}, & t_k = \frac{k}{5}, k = 1,2,3,4, \\ 3x(0) - 8x(1) = 7, \quad 4y(0) + 5y(1) = 2. \end{cases} \tag{4.11}$$

Here $q_k = (2/7)^k$, $p_k = ((3+k)/(4+2k))^k$, $k = 0,1,2,3,4$, $m = 4$, $T = 1$, $a_1 = 3$, $a_2 = 4$, $b_1 = -8$, $b_2 = 5$, $\lambda_1 = 7$, $\lambda_2 = 2$, $f(t,x,y) = (\sin(\pi t)|x|)/(((e^t+5)^2)(|x|+1)) + (\pi^t|y|)/(((t+4)^3)(|y|+1)) + 3$, $g(t,x,y) = (\cos x)/(10(2^t+4)) + (|y|)/(6\pi(t+3)) + 2$, $I_k(x) = |x|/(3(k+9)+|x|)$, and $I_k^*(y) = |y|/(5(k+6)+|y|)$. Since $|f(t,x_1,y_1) - f(t,x_2,y_2)| \leq ((1/36)|x_1 - x_2| + (\pi/64)|y_1 - y_2|)$, $|g(t,x_1,y_1) - g(t,x_2,y_2)| \leq ((1/50)|x_1 - x_2| + (1/(18\pi))|y_1 - y_2|)$, $|I_k(x) - I_k(y)| \leq (1/30)|x - y|$, and $|I_k^*(x) - I_k^*(y)| \leq (1/35)|x - y|$. We can find

$$\Lambda = a_1 + b_1 = -5 \neq 0 \quad \text{and} \quad \Phi = a_2 + b_2 = 9 \neq 0.$$

With the given values, it is found that $\overline{K}_1 = 1/36$, $\overline{K}_2 = \pi/64$, $\overline{K}_3 = 1/30$, $\overline{L}_1 = 1/50$, $\overline{L}_2 = 1/(18\pi)$, $\overline{L}_3 = 1/35$, $\overline{M}_1 \simeq 0.41889$, $\overline{M}_2 \simeq 0.12763$, $\overline{M}_4 \simeq 0.03111$, $\overline{M}_5 \simeq 0.20529$, and

$$\overline{M}_1 + \overline{M}_2 + \overline{M}_4 + \overline{M}_5 \simeq 0.78291 < 1.$$

Thus all the conditions of Theorem 4.1 are satisfied. Therefore, by the conclusion of Theorem 4.1, problem (4.11) has a unique solution on $[0,1]$.

The second result dealing with the existence of solutions for the problem (4.1) is analogous to Theorem 3.2 and is given below.

In the sequel, we set constants

$$\overline{M}_7 = \frac{1}{|\Lambda|}(T\overline{A}_1 + m\overline{A}_4)\big(|b_1| + |\Lambda|\big), \tag{4.12}$$

$$\overline{M}_8 = \frac{1}{|\Lambda|}T\overline{A}_2\big(|b_1| + |\Lambda|\big), \tag{4.13}$$

$$\overline{M}_9 = \frac{1}{|\Lambda|}\big[(T\overline{A}_0 + m\overline{A}_3)\big(|b_1| + |\Lambda|\big) + |\lambda_1|\big], \tag{4.14}$$

$$\overline{M}_{10} = \frac{1}{|\Phi|}T\overline{B}_1\big(|b_2| + |\Phi|\big), \tag{4.15}$$

$$\overline{M}_{11} = \frac{1}{|\Phi|}(T\overline{B}_2 + m\overline{B}_4)\big(|b_2| + |\Phi|\big), \tag{4.16}$$

$$\overline{M}_{12} = \frac{1}{|\Phi|}\big[(T\overline{B}_0 + m\overline{B}_3)\big(|b_2| + |\Phi|\big) + |\lambda_2|\big]. \tag{4.17}$$

Theorem 4.2 *Assume that:*

(H_7) *The functions $f, g : [0,T] \times \mathbb{R}^2 \to \mathbb{R}$ are continuous and there exist constants $\overline{A}_i, \overline{B}_i \geq 0$ $(i = 1,2)$ and $\overline{A}_0, \overline{B}_0 > 0$ such that $\forall x_i \in \mathbb{R}$ $(i = 1,2)$*

$$\big|f(t,x_1,x_2)\big| \leq \overline{A}_0 + \overline{A}_1|x_1| + \overline{A}_2|x_2|$$

and

$$\left|g(t,x_1,x_2)\right| \le \overline{B}_0 + \overline{B}_1|x_1| + \overline{B}_2|x_2|.$$

(H$_8$) *The functions* $I_k, I_k^* : \mathbb{R} \to \mathbb{R}$ *are continuous and there exist constants* $\overline{A}_4, \overline{B}_4 \ge 0$ *and* $\overline{A}_3, \overline{B}_3 > 0$ *such that* $\forall x_3 \in \mathbb{R}$, $k = 1,2,\dots,m$,

$$\left|I_k(x_3)\right| \le \overline{A}_3 + \overline{A}_4|x_3|$$

and

$$\left|I_k^*(x_3)\right| \le \overline{B}_3 + \overline{B}_4|x_3|.$$

In addition it is assumed that

$$\overline{M}_7 + \overline{M}_{10} < 1 \quad \text{and} \quad \overline{M}_8 + \overline{M}_{11} < 1,$$

where $\overline{M}_7, \overline{M}_8, \overline{M}_{10}, \overline{M}_{11}$ *are given by* (4.12)-(4.13) *and* (4.15)-(4.16), *respectively. Then the boundary value problem* (4.1) *has at least one solution.*

Proof Setting

$$\overline{M}_0 = \min\left\{1 - (\overline{M}_7 + \overline{M}_{10}), 1 - (\overline{M}_8 + \overline{M}_{11})\right\},$$

the proof is similar to that of Theorem 3.2. So we omit it. □

Example 4.2 Consider the following coupled system of impulsive quantum difference equations with uncoupled boundary conditions:

$$\begin{cases} D_{\frac{2}{3+k}} x(t) = 3 + \frac{1}{20}\sin(\frac{x(t)}{2}) + \frac{1}{(t+3)^2}\tan^{-1}(\frac{y(t)}{4}), & t \in [0,1], t \neq t_k, \\ D_{\frac{3+k}{5+2k+k^2}} y(t) = 4 + \frac{t}{10(2^t+1)}x(t) + \frac{\sin(\pi t)}{2(2t+5)^2}y(t), & t \in [0,1], t \neq t_k, \\ \Delta x(t_k) = \frac{1}{10\pi^2}\sin(\frac{\pi x(t_k)}{2}) + \frac{1}{2}, \quad t_k = \frac{k}{10}, k = 1,2,\dots,9, \\ \Delta y(t_k) = \frac{1}{5}\sin(\frac{y(t_k)}{6}) + 3, \quad t_k = \frac{k}{10}, k = 1,2,\dots,9, \\ 2x(0) - 7x(1) = -3, \qquad 3y(0) - 5y(1) = -10. \end{cases} \tag{4.18}$$

Here $q_k = 2/(3+k)$, $p_k = (3+k)/(5+2k+k^2)$, $k = 0,1,2,\dots,9$, $m = 9$, $T = 1$, $a_1 = 2$, $a_2 = 3$, $b_1 = -7$, $b_2 = -5$, $\lambda_1 = -3$, $\lambda_2 = -10$, $f(t,x,y) = 3 + (\sin(x/2))/20 + (\tan^{-1}(y/4))/(t+3)^2$, $g(t,x,y) = 4 + (tx)/(10(2^t+1)) + (\sin(\pi t)y)/(2(2t+5)^2)$, $I_k(x) = (\sin(\pi x/2))/(10\pi^2) + 1/2$, and $I_k^*(y) = y/(20e + y^2) + \pi/4$. We get

$$\Lambda = a_1 + b_1 = -5 \neq 0 \quad \text{and} \quad \Phi = a_2 + b_2 = -2 \neq 0.$$

Since $|f(t,x,y)| \le \overline{A}_0 + \overline{A}_1|x| + \overline{A}_2|y|$, $|g(t,x,y)| \le \overline{B}_0 + \overline{B}_1|x| + \overline{B}_2|y|$, where $\overline{A}_0 = 3$, $\overline{A}_1 = 1/40$, $\overline{A}_2 = 1/36$, $\overline{B}_0 = 4$, $\overline{B}_1 = 1/20$, $\overline{B}_2 = 1/50$, it is found that $\overline{M}_7 \simeq 0.40377$, $\overline{M}_8 \simeq 0.06667$, $\overline{M}_{10} \simeq 0.175$, $\overline{M}_{11} \simeq 0.64941$. Furthermore,

$$\overline{M}_7 + \overline{M}_{10} \approx 0.57877 < 1$$

and

$$\overline{M}_8 + \overline{M}_{11} \approx 0.71608 < 1.$$

Thus all the conditions of Theorem 4.2 holds true and consequently the conclusion of Theorem 4.2; problem (4.18) has at least one solution on $[0,1]$.

Competing interests

The authors declare that they have no competing interests.

Authors' contributions

All authors contributed equally in this article. They read and approved the final manuscript.

Author details

[1]Nonlinear Dynamic Analysis Research Center, Department of Mathematics, Faculty of Applied Science, King Mongkut's University of Technology North Bangkok, Bangkok, 10800, Thailand. [2]Department of Mathematics, University of Ioannina, Ioannina, 451 10, Greece. [3]Nonlinear Analysis and Applied Mathematics (NAAM)-Research Group, Department of Mathematics, Faculty of Science, King Abdulaziz University, P.O. Box 80203, Jeddah, 21589, Saudi Arabia.

Acknowledgements

We would like to thank the reviewers for their valuable comments and suggestions on the manuscript. This research was funded by King Mongkut's University of Technology North Bangkok. Contract no. KMUTNB-GOV-58-10.

References

1. Tariboon, J, Ntouyas, SK: Quantum calculus on finite intervals and applications to impulsive difference equations. Adv. Differ. Equ. **2013**, Article ID 282 (2013)
2. Kac, V, Cheung, P: Quantum Calculus. Springer, New York (2002)
3. Bangerezako, G: Variational q-calculus. J. Math. Anal. Appl. **289**, 650-665 (2004)
4. Dobrogowska, A, Odzijewicz, A: Second order q-difference equations solvable by factorization method. J. Comput. Appl. Math. **193**, 319-346 (2006)
5. Gasper, G, Rahman, M: Some systems of multivariable orthogonal q-Racah polynomials. Ramanujan J. **13**, 389-405 (2007)
6. Ismail, MEH, Simeonov, P: q-Difference operators for orthogonal polynomials. J. Comput. Appl. Math. **233**, 749-761 (2009)
7. Bohner, M, Guseinov, GS: The h-Laplace and q-Laplace transforms. J. Math. Anal. Appl. **365**, 75-92 (2010)
8. El-Shahed, M, Hassan, HA: Positive solutions of q-difference equation. Proc. Am. Math. Soc. **138**, 1733-1738 (2010)
9. Ahmad, B: Boundary-value problems for nonlinear third-order q-difference equations. Electron. J. Differ. Equ. **2011**, Article ID 94 (2011)
10. Ahmad, B, Alsaedi, A, Ntouyas, SK: A study of second-order q-difference equations with boundary conditions. Adv. Differ. Equ. **2012**, Article ID 35 (2012)
11. Ahmad, B, Ntouyas, SK, Purnaras, IK: Existence results for nonlinear q-difference equations with nonlocal boundary conditions. Commun. Appl. Nonlinear Anal. **19**, 59-72 (2012)
12. Ahmad, B, Nieto, JJ: On nonlocal boundary value problems of nonlinear q-difference equations. Adv. Differ. Equ. **2012**, Article ID 81 (2012)
13. Ahmad, B, Ntouyas, SK: Boundary value problems for q-difference inclusions. Abstr. Appl. Anal. **2011**, Article ID 292860 (2011)
14. Zhou, W, Liu, H: Existence solutions for boundary value problem of nonlinear fractional q-difference equations. Adv. Differ. Equ. **2013**, Article ID 113 (2013)
15. Yu, C, Wang, J: Existence of solutions for nonlinear second-order q-difference equations with first-order q-derivatives. Adv. Differ. Equ. **2013**, Article ID 124 (2013)
16. Lakshmikantham, V, Bainov, DD, Simeonov, PS: Theory of Impulsive Differential Equations. World Scientific, Singapore (1989)
17. Samoilenko, AM, Perestyuk, NA: Impulsive Differential Equations. World Scientific, Singapore (1995)
18. Benchohra, M, Henderson, J, Ntouyas, SK: Impulsive Differential Equations and Inclusions, vol. 2. Hindawi Publishing Corporation, New York (2006)
19. Sudsutad, W, Tariboon, J, Ntouyas, SK: Existence of solutions for second-order impulsive q-difference equations with integral boundary conditions. Appl. Math. Inf. Sci. **9**, 1147-1157 (2015)
20. Tariboon, J, Ntouyas, SK: Three-point boundary value problems for nonlinear second-order impulsive q-difference equations. Adv. Differ. Equ. **2014**, Article ID 31 (2014)
21. Tariboon, J, Ntouyas, SK: Boundary value problems for first-order impulsive functional q-integro-difference equations. Abstr. Appl. Anal. **2014**, Article ID 374565 (2014)
22. Thaiprayoon, C, Tariboon, J, Ntouyas, SK: Separated boundary value problems for second-order impulsive q-integro-difference equations. Adv. Differ. Equ. **2014**, Article ID 88 (2014)
23. Asawasamrit, S, Tariboon, J, Ntouyas, SK: Existence of solutions for fractional q-integrodifference equations with nonlocal fractional q-integral conditions. Abstr. Appl. Anal. **2014**, Article ID 474138 (2014)
24. Thiramanus, P, Ntouyas, SK, Tariboon, J: Nonlinear second-order impulsive q-difference equations. Commun. Appl. Nonlinear Anal. **21**(3), 89-102 (2014)
25. Granas, A, Dugundji, J: Fixed Point Theory. Springer, New York (2003)

14

Entire solutions of certain class of differential-difference equations

Fengrong Zhang[1], Nana Liu[1], Weiran Lü[1*] and Chungchun Yang[2]

*Correspondence: luwr@upc.edu.cn
[1]College of Science, China University of Petroleum (East China), Qingdao, 266580, P.R. China
Full list of author information is available at the end of the article

Abstract

As a continuation of our previous studies Liu *et al.* (J. Inequal. Appl. 2014:63, 2014), we will discuss the transcendental entire solutions of the following type of differential-difference equation: $f^3(z) + P_1(z, \Delta f, \ldots, f', \ldots, f^{(k)}) = \lambda_1 e^{\alpha_1 z} + \lambda_2 e^{\alpha_2 z}$, where P_1 is a linear polynomial in $f, \Delta f, \ldots, f^{(k)}$, with polynomials as its coefficients, and $\lambda_1, \lambda_2, \alpha_1, \alpha_2 \in \mathbb{C}$ are nonzero constants such that $\alpha_1 \neq \alpha_2$.

MSC: 30D35; 30A06

Keywords: transcendental entire solution; differential-difference equation; Nevanlinna theory

1 Introduction

In this paper, we shall adopt the standard notations in Nevanlinna's value distribution theory of meromorphic functions. For example, the characteristic function $T(r,f)$, the counting function of the poles $N(r,f)$, and the proximity function $m(r,f)$ (see, *e.g.*, [1, 2]).

Let f be a meromorphic function. Recall that $\alpha \not\equiv 0, \infty$ is a small function with respect to f, if $T(r,\alpha) = S(r,f)$, where $S(r,f)$ denotes any quantity satisfying $S(r,f) = o\{T(r,f)\}$ as $r \to +\infty$, possibly outside a set of r of finite linear measure.

For completeness, we recall basic notions to this end: Given a meromorphic function f and a constant c, $f(z+c)$ is called a shift of $f(z)$. For the sake of simplicity, we let $\Delta f(z) = f(z+1) - f(z)$, $\Delta^n f(z) = \Delta(\Delta^{n-1} f(z))$ $(n \geq 2)$. As for a difference product, we mean a difference monomial of type $\prod_{j=1}^{k} f(z+c_j)^{n_j}$, where $c_1, \ldots, c_k$ are complex constants, and $n_1, \ldots, n_k$ are natural numbers.

Definition 1.1 A difference polynomial, resp. a differential-difference polynomial, in f is a finite sum of difference products of f and its shifts, resp. of products of f, derivatives of f and of their shifts, with all the coefficients of these monomials being small functions of f.

Consider a transcendental meromorphic function f and let

$$P(z,f) = \sum_{l=1}^{n} a_l(z) \prod_{j=0}^{k} \{f^{(j)}(z + c_{lj})\}^{n_{lj}},$$

where a_l $(l = 1, 2, \ldots, n)$ are small functions of f, and c_{lj} $(l = 1, 2, \ldots, n; j = 0, 1, \ldots, k)$ are complex constants, and n_{lj} $(l = 1, 2, \ldots, n; j = 0, 1, \ldots, k)$ are natural numbers.

Definition 1.2 We define the total degree d of $P(z,f)$ by

$$d := d(P(z,f)) = \max_{1\leq l\leq n}\left\{\sum_{j=0}^{k} n_{lj}\right\}.$$

Moreover, a functional equation involving f and its shifts as well as their derivatives is called a differential-difference equation.

Recently, many papers (see, *e.g.*, [3] and [4]) are focused on complex difference polynomials and difference equations. Lots of results have been obtained by using value distribution theory (see, *e.g.*, [5] and [6, 7]).

Yang and Laine [8] considered the following difference equation and proved the following.

Theorem A *A non-linear difference equation*

$$f^3(z) + q(z)f(z+1) = c\sin bz,$$

where q is a non-constant polynomial and $b, c \in \mathbb{C}$ are nonzero constants, does not admit entire solutions of finite order. If q is a nonzero constant, then the above equation possesses three distinct entire solutions of finite order, provided that $b = 3n\pi$ and $q^3 = (-1)^{n+1}c^2 27/4$ for a nonzero integer n.

Very recently, Liu *et al.* [9] proved the following.

Theorem B *Let $n \geq 4$ be an integer, q be a polynomial, and p_1, p_2, α_1, α_2 be nonzero constants such that $\alpha_1 \neq \alpha_2$. If f is an entire solution of finite order to the following equation:*

$$f^n(z) + q(z)\Delta f(z) = p_1e^{\alpha_1 z} + p_2e^{\alpha_2 z},$$

then q is a constant, and one of the following relations holds:

(1) $f(z) = c_1\exp(\frac{\alpha_1}{n}z)$, *and* $c_1(\exp(\alpha_1/n) - 1)q = p_2$, $\alpha_1 = n\alpha_2$;

(2) $f(z) = c_2\exp(\frac{\alpha_2}{n}z)$, *and* $c_2(\exp(\alpha_2/n) - 1)q = p_1$, $\alpha_2 = n\alpha_1$,

where c_1, c_2 are constants satisfying $c_1^n = p_1$, $c_2^n = p_2$.

In this paper, we are going to discuss the case when $n = 3$ in the above theorem and prove some results in Section 3.

2 Preliminaries

In order to prove our conclusions, we need some lemmas.

Lemma 2.1 ([8]) *Let f be a transcendental meromorphic solution of finite order ρ of a difference equation of the form*

$$H(z,f)P(z,f) = Q(z,f),$$

where $H(z,f)$, $P(z,f)$, $Q(z,f)$ are difference polynomials in f such that the total degree of $H(z,f)$ in f and its shifts is n, and that the corresponding total degree of $Q(z,f)$ is less or *equal to n. If H(z,f) contains just one term of maximal total degree, then for any small enough* $\varepsilon > 0$,

$$m(r, P(z,f)) = O(r^{\rho-1+\varepsilon}) + S(r,f).$$

The following result is a Clunie [10] type lemma for the difference-differential polynomials of a meromorphic function f. It can be proved by applying Lemma 2.1 and stated as follows.

Proposition 2.1 ([8, 11]) *If in the above lemma,* $H(z,f) = f^n$, $P(z,f)$, *and* $Q(z,f)$ *are differential-difference polynomials in* f, *then* $m(r, P(z,f)) = S(r,f)$.

The following result is an analogue of a result due to Mohon'ko and Mohon'ko [12] for differential equations.

Lemma 2.2 ([4]) *Let* f *be a non-constant meromorphic solution with finite order of*

$$P(z,f) = 0,$$

where $P(z,f)$ *is difference polynomial in* f, *and let* $\delta < 1$ *and* $\varepsilon > 0$. *If* $P(z,a) \not\equiv 0$ *for a slowly moving target function* a, *that is,* $T(r,a) = S(r,f)$, *then*

$$m\left(r, \frac{1}{f-a}\right) = o\left(\frac{T(r+|c|,f)^{1+\varepsilon}}{r^{\delta}}\right) + o(T(r,f))$$

for all r outside of a possible exceptional set with finite logarithmic measure.

Remark 2.1 By some further analysis on the proof of Theorem 2.4 in [6], we can obtain the conclusion of Lemma 2.2 when $P(z,f)$ is a differential-difference polynomial in f.

Lemma 2.3 ([13]) *Assume that c is a nonzero constant,* α *is a non-constant meromorphic function. Then the differential equation* $f^2 + (cf^{(n)})^2 = \alpha$ *has no transcendental meromorphic solutions satisfying* $T(r,\alpha) = S(r,f)$.

3 Main results

Now, we shall derive the following result similar to that of Theorems A and B, for extended differential-difference equations. First, we will prove the following result.

Theorem 3.1 *Let q be a polynomial, and* p_1, p_2, α_1, α_2 *be nonzero constants such that* $\alpha_1 \neq \alpha_2$. *If* f *is an entire solution of finite order to the following equation:*

$$f^3(z) + q(z)\Delta f(z) = p_1 e^{\alpha_1 z} + p_2 e^{\alpha_2 z}, \tag{3.1}$$

then q is a constant, and one of the following relations holds:

(1) $T(r,f) = N_{1)}(r, \frac{1}{f}) + S(r,f)$;

(2) $f(z) = c_1 \exp(\frac{\alpha_1}{3} z)$, *and* $c_1(\exp(\alpha_1/3) - 1)q = p_2$, $\alpha_1 = 3\alpha_2$;

(3) $f(z) = c_2 \exp(\frac{\alpha_2}{3} z)$, *and* $c_2(\exp(\alpha_2/3) - 1)q = p_1$, $\alpha_2 = 3\alpha_1$,

where $N_{1)}(r, \frac{1}{f})$ *denotes the counting function corresponding to simple zeros of* f, *and* c_1, c_2 *are constants satisfying* $c_1^3 = p_1$, $c_2^3 = p_2$.

Proof of Theorem 3.1 Suppose that f is a transcendental entire solution of finite order of (3.1). For the simplicity, we replace $f(z), f'(z), \Delta f(z), q(z)$ by $f, f', \Delta f, q$. By differentiating both sides of (3.1), we have

$$3f^2 f' + (q\Delta f)' = \alpha_1 p_1 e^{\alpha_1 z} + \alpha_2 p_2 e^{\alpha_2 z}. \tag{3.2}$$

From (3.1) and (3.2), we obtain

$$\alpha_2 f^3 + \alpha_2 q\Delta f - 3f^2 f' - (q\Delta f)' = (\alpha_2 - \alpha_1) p_1 e^{\alpha_1 z}, \tag{3.3}$$

$$\alpha_1 f^3 + \alpha_1 q\Delta f - 3f^2 f' - (q\Delta f)' = (\alpha_1 - \alpha_2) p_2 e^{\alpha_2 z}. \tag{3.4}$$

Differentiating (3.3) yields

$$3\alpha_2 f^2 f' + \alpha_2 (q\Delta f)' - 6f(f')^2 - 3f^2 f'' - (q\Delta f)'' = \alpha_1(\alpha_2 - \alpha_1) p_1 e^{\alpha_1 z}. \tag{3.5}$$

It follows from (3.3) and (3.5) that

$$f\varphi = T(z,f), \tag{3.6}$$

where

$$\varphi = \alpha_1\alpha_2 f^2 - 3(\alpha_1 + \alpha_2) f' f + 6(f')^2 + 3f'' f,$$

$$T(z,f) = -\alpha_1\alpha_2 q\Delta f + (\alpha_1 + \alpha_2)(q\Delta f)' - (q(z)\Delta f)''.$$

Since $T(z,f)$ is a differential-difference polynomial in f of degree 1, it follows from (3.6) and Proposition 2.1 that $m(r,\varphi) = S(r,f)$, and $T(r,\varphi) = S(r,f)$. If $\varphi \not\equiv 0$, Lemma 2.2 gives $m(r, \frac{1}{f}) = S(r,f)$. This together with the first theorem will result in

$$T(r,f) = N_{1)}\left(r, \frac{1}{f}\right) + S(r,f), \tag{3.7}$$

where $N_{1)}(r, \frac{1}{f})$ denotes the counting function corresponding to simple zeros of f, which is our desired result.

Next, we can assume that φ in (3.6) vanishes identically, dividing with f^2, and recalling $f''/f = (f'/f)' + (f'/f)^2$, we get the Riccati equation

$$t' + 3t^2 - (\alpha_1 + \alpha_2)t + \alpha_1\alpha_2/3 = 0,$$

where $t := f'/f$. This equation has two constant solutions $t = \alpha_1/3$, $t = \alpha_2/3$. By Corollary 5.2 in the paper by Bank *et al.* [14], all other meromorphic solutions are of infinite order. Consider $t = \alpha_1/3$, hence $f(z) = A\exp(\alpha_1 z/3)$ with a constant A and look at (3.6). Denote $s(z) := q(z)\Delta f(z)$ and $u(z) := s'(z)/s(z)$. Then, by (3.6), we get

$$u' = u^2 + (\alpha_1 + \alpha_2)u + \alpha_1\alpha_2.$$

As above, the only solutions of finite order are $u = \alpha_1$, $u = \alpha_2$, hence $s(z) = D_1 \exp(\alpha_1 z)$ or $s(z) = D_2 \exp(\alpha_2 z)$. Now, it is easy to see that the case $s(z) = D_1 \exp(\alpha_1 z)$ results in a contradiction. Therefore, we have $f(z) = A\exp(\alpha_1 z/3)$ and $q(z)\Delta f(z) = D_2 \exp(\alpha_2 z)$. The assertions $A(\exp(\alpha_1/3) - 1)q = p_2$, $A^3 = p_1$ and $\alpha_1 = 3\alpha_2$ now immediately follow, and the claim (2) is done. As to the claim (2), the Riccati equations for t and u are symmetric with respect to α_1, α_2, therefore the claim (3) follows.

This completes the proof of Theorem 3.1. □

Remark 3.1 It is easy to see that $f(z) = e^{\pi iz} + e^{-\pi iz} = 2i\sin(\pi iz)$ is a solution of the following equation:

$$f^3(z) + 3\Delta f(z) = e^{3\pi iz} + e^{-3\pi iz}.$$

Note $f(z) = e^{\pi iz} + e^{-\pi iz} = 2i\sin(\pi iz)$ has infinitely many zeros, and $T(r,f) = N_{1)}(r, \frac{1}{f}) + S(r,f)$. Thus, we would like to pose the following conjecture.

Conjecture 3.1 *If $\alpha_1 \neq \alpha_2$, $\alpha_1 + \alpha_2 \neq 0$, then the conclusion (1) of Theorem 3.1 is impossible, in fact, any entire solution f of (3.1) must have 0 as its Picard exceptional value.*

For $\alpha_1 + \alpha_2 = 0$, we have the following.

Theorem 3.2 *Let q_j $(j = 1, 2, 3)$ and p be polynomials, and $\lambda_1, \lambda_2, \alpha$ be nonzero constants. Then non-linear difference equation*

$$f^3(z) + q_3\Delta^3 f(z) + q_2\Delta^2 f(z) + q_1\Delta f(z) = \lambda_1 e^{\alpha z} + \lambda_2 e^{-\alpha z} + p(z) \tag{3.8}$$

possesses solutions of finite order of the form $f(z) = c_1 e^{\alpha z/3} + c_2 e^{-\alpha z/3}$ with $c_1^3 = \lambda_1$, $c_2^3 = \lambda_2$, and $p \equiv 0$, $v = e^{\alpha/3}$, q_1, q_2, q_3 satisfy the condition:

$$(v-1)^3(v^3+1)q_3 + (v-1)^2(v^3 - v)q_2 + (v-1)(v^3+v^2)q_1 = 0.$$

Moreover, if $\alpha = (6n\pi \pm 3\pi)i$, then $(8q_3 - 4q_2 + 2q_1)^3 = 27\lambda_1\lambda_2$; or if $\alpha = (6n\pi \pm \pi)i$, then $q_2 + q_1 = 0$ and $(\pm\sqrt{3}iq_1 - q_3)^3 = 27\lambda_1\lambda_2$, where n is an integer; if $v \neq -1, \frac{1\pm\sqrt{3}i}{2}$, then q_1, q_2, q_3 satisfy the equation

$$(v-1)^2(v^2 - v + 1)q_3 + v(v-1)^2 q_2 + v^2 q_1 = 0.$$

Proof of Theorem 3.2 For the simplicity the argument z will be omitted in $f(z), f'(z), \Delta f(z)$ etc. Suppose that f is a transcendental entire solution of finite order of (3.8), and by differentiating (3.8) we get

$$\begin{aligned} &3f^2f' + q_3'\Delta^3 f + q_3(\Delta^3 f)' + q_2'\Delta^2 f + q_2(\Delta^2 f)' + q_1'\Delta f + q_1(\Delta f)' \\ &\quad = \lambda_1\alpha e^{\alpha z} - \lambda_2\alpha e^{-\alpha z} + p'. \end{aligned} \tag{3.9}$$

Combining (3.8) and (3.9), we obtain

$$\begin{aligned} &\alpha(f^3 + q_3\Delta^3 f + q_2\Delta^2 f + q_1\Delta f) + 3f^2f' + q_3'\Delta^3 f + q_3(\Delta^3 f)' \\ &\quad + q_2'\Delta^2 f + q_2(\Delta^2 f)' + q_1'\Delta f + q_1(\Delta f)' = 2\lambda_1\alpha e^{\alpha z} + \alpha p + p' \end{aligned}$$

and

$$\alpha\left(f^3 + q_3\Delta^3 f + q_2\Delta^2 f + q_1\Delta f\right) - \left\{3f^2 f' + q_3'\Delta^3 f + q_3\left(\Delta^3 f\right)' + q_2'\Delta^2 f + q_2\left(\Delta^2 f\right)' + q_1'\Delta f + q_1(\Delta f)'\right\} = 2\lambda_2\alpha e^{-\alpha z} + \alpha p - p',$$

thus through the above two equations, we find

$$f^4\left[-9\left(f'\right)^2 + \alpha^2 f^2\right] = T_4(z,f), \tag{3.10}$$

where $T_4(z,f)$ is a differential-difference polynomial of f, and its total degree is at most 4.

If $T_4(z,f) \equiv 0$, it follows from (3.10) that $-9(f')^2 + \alpha^2 f^2 \equiv 0$, then $f' = \pm\frac{\alpha}{3}f$, and there exists a nonzero constant c such that $f = ce^{\pm(\alpha z)/3}$. Substituting the above expression of f into (3.8), we obtain a contradiction. Therefore, $T_4(f) \not\equiv 0$. Set $\beta = -9(f')^2 + \alpha^2 f^2$, then Proposition 2.1 shows that $m(r,\beta) = S(r,f)$, *i.e.* β is a small function of f. Moreover, by Lemma 2.3, it is easy to see that β is a constant. By differentiating both sides of $\beta = -9(f')^2 + \alpha^2 f^2$, we get

$$f'' - \left(\frac{\alpha}{3}\right)^2 f = 0. \tag{3.11}$$

It follows from (3.11) that

$$f(z) = c_1 e^{\alpha z/3} + c_2 e^{-\alpha z/3}, \tag{3.12}$$

where c_1, c_2 are nonzero constants. Substituting (3.12) into (3.8), we have

$$a_1 w^6 + a_2 w^4 + p w^3 + a_3 w^2 + a_4 = 0,$$

where $w = e^{(\alpha z)/3}$, $a_1 = c_1^3 - \lambda_1$, $a_2 = 3c_1^2 c_2 + q_3 c_1 e^{\alpha} - 3q_3 c_1 e^{2\alpha/3} + 3q_3 c_1 e^{\alpha/3} - c_1 q_3 + q_2 c_1 e^{2\alpha/3} - 2q_2 c_1 e^{\alpha/3} + q_2 c_1 + q_1 c_1 e^{\alpha/3} - q_1 c_1$, $a_3 = 3c_1 c_2^2 + c_2 q_3 e^{-\alpha} - 3q_3 c_2 e^{-2\alpha/3} + 3q_3 c_2 e^{-\alpha/3} - c_2 q_3 + c_2 q_2 e^{-2\alpha/3} - 2c_2 q_2 e^{-\alpha/3} + c_2 q_2 + c_2 q_1 e^{-\alpha/3} - c_2 q_1$, $a_4 = c_2^3 - \lambda_2$.

Since w is transcendental, we must have $a_1 = a_4 = 0$, and $a_2 = a_3 = p \equiv 0$. Therefore,

$$q_3\left(e^{\alpha/3} - 1\right)^3 + q_2\left(e^{\alpha/3} - 1\right)^2 + q_1\left(e^{\alpha/3} - 1\right) = q_3\left(e^{-\alpha/3} - 1\right)^3 + q_2\left(e^{-\alpha/3} - 1\right)^2 + q_1\left(e^{-\alpha/3} - 1\right). \tag{3.13}$$

Set $v = e^{\alpha/3}$, it follows by (3.13) that

$$q_3(v-1)^3 + q_2(v-1)^2 + q_1(v-1) = q_3\left(\frac{1}{v} - 1\right)^3 + q_2\left(\frac{1}{v} - 1\right)^2 + q_1\left(\frac{1}{v} - 1\right)$$

or

$$\left(v^2 - 1\right)\left\{(v-1)^2\left(v^2 - v + 1\right)q_3 + v(v-1)^2 q_2 + v^2 q_1\right\} = 0. \tag{3.14}$$

Now we distinguish four cases.

Case 1. Assume that $v = 1$, then $\alpha = 6n\pi i$, substituting $v = 1$ into a_2 gives $c_1c_2 = 0$, this is a contradiction, thus $v \neq 1$.

Case 2. If $v = -1$, then $\alpha = (6n\pi \pm 3\pi)i$. By substituting $v = -1$ into a_2, we deduce $(8q_3 - 4q_2 + 2q_1)^3 = 27\lambda_1\lambda_2$.

Case 3. While $v = \frac{1\pm\sqrt{3}i}{2}$, in a similar way to above, we get $\alpha = (6n\pi \pm \pi)i$ and $q_1 + q_2 = 0$. Moreover, $(\pm\sqrt{3}iq_1 - q_3)^3 = 27\lambda_1\lambda_2$ and $\deg q_1 = \deg q_2 = \deg q_3$.

Case 4. When $v \neq -1, \frac{1\pm\sqrt{3}i}{2}$, by (3.14), we could get the following equation:

$$(v-1)^2(v^2 - v + 1)q_3 + v(v-1)^2q_2 + v^2q_1 = 0.$$

This also completes the proof of Theorem 3.2. □

Theorem 3.3 *Let p_j ($j = 1, 2$) be polynomials, λ_1, λ_2, α be nonzero constants, d_1, d_2 be constants, and k be a positive integer. If f is an entire solution of finite order to the following non-linear differential-difference equation:*

$$f^3(z) + p_1(z)f^{(k)}(z + d_1) + p_2(z)f(z + d_2) = \lambda_1e^{\alpha z} + \lambda_2e^{-\alpha z}, \tag{3.15}$$

then $f(z) = c_1e^{\alpha z/3} + c_2e^{-\alpha z/3}$, $c_1^3 = \lambda_1$, $c_2^3 = \lambda_2$,

$$3c_1c_2 + p_1e^{d_1\alpha/3}(\alpha/3)^k + p_2e^{d_2\alpha/3} \equiv 0$$

and

$$3c_1c_2 + p_1e^{-d_1\alpha/3}(-\alpha/3)^k + p_2e^{-d_2\alpha/3} \equiv 0.$$

Proof of Theorem 3.3 For simplicity the argument z will be omitted in $f(z), f'(z), f'(z + d_1)$ *etc.* Suppose that f is a transcendental entire solution of finite order of (3.15), by differentiating (3.15) we get

$$3f^2f' + Q(z,f) = \lambda_1\alpha e^{\alpha z} - \lambda_2\alpha e^{-\alpha z}, \tag{3.16}$$

where $Q(z,f) = p_1'(z)f^{(k)}(z + d_1) + p_1(z)f^{(k+1)}(z + d_1) + p_2'(z)f(z + d_2) + p_2(z)f'(z + d_2)$. Obviously, $Q(z,f)$ is a differential-difference polynomial of f, and its total degree is at most 1.

Set $Q_1(z,f) = p_1(z)f^{(k)}(z + d_1) + p_2(z)f(z + d_2)$. It follows by (3.15) and (3.16) that

$$\alpha f^3 + 3f^2f' + \alpha Q_1(z,f) + Q(z,f) = 2\lambda_1\alpha e^{\alpha z}$$

and

$$\alpha f^3 - 3f^2f' + \alpha Q_1(z,f) - Q(z,f) = 2\lambda_2\alpha e^{-\alpha z}.$$

Thus, for the above two equations, we can also get (3.10) and by a similar method to the one above, we have (3.12). Consequently

$$f(z + d_2) = c_1e^{d_2\alpha/3}e^{\alpha z/3} + c_2e^{-d_2\alpha/3}e^{-\alpha z/3} \tag{3.17}$$

and

$$f^{(k)}(z+d_1)=c_1\left(\frac{\alpha}{3}\right)^k e^{d_1\alpha/3}e^{\alpha z/3}+c_2\left(-\frac{\alpha}{3}\right)^k e^{-d_1\alpha/3}e^{-\alpha z/3}. \tag{3.18}$$

In view of (3.15), (3.12), (3.17), and (3.18), we obtain $f(z)=c_1e^{\alpha z/3}+c_2e^{-\alpha z/3}$ with $c_1^3=\lambda_1$, $c_2^3=\lambda_2$. Moreover, we have

$$3c_1c_2+p_1e^{d_1\alpha/3}(\alpha/3)^k+p_2e^{d_2\alpha/3}\equiv 0$$

and

$$3c_1c_2+p_1e^{-d_1\alpha/3}(-\alpha/3)^k+p_2e^{-d_2\alpha/3}\equiv 0.$$

Thus, we obtain the desired result and finish the proof of this theorem. □

It is not difficult to see that a similar argument can be used to obtain the following result.

Theorem 3.4 *Let f be an entire solution of finite order to the following non-linear differential-difference equation*:

$$f^3(z)+P_1\big(z,\Delta f(z),\dots,f'(z),\dots,f^{(k)}(z)\big)=\lambda_1e^{\alpha z}+\lambda_2e^{-\alpha z},$$

here P_1 is a linear polynomial in $\Delta f(z),\dots,f'(z),\dots,f^{(k)}(z)$, with polynomials as its coefficients, and $\lambda_1,\lambda_2,\alpha\in\mathbb{C}$ are nonzero constants. Then $f(z)=c_1e^{\alpha z/3}+c_2e^{-\alpha z/3}$ with $c_1^3=\lambda_1$, $c_2^3=\lambda_2$.

4 Conclusion

In studying differential equations in the complex plane $\mathbb{C}$, it is always an interesting and quite difficult problem to prove the existence or uniqueness of the entire or meromorphic solution of a given equation. There have been many studies and results obtained lately that relate to the existence or growth of entire or meromorphic solutions of various types of complex difference equations and difference-differential equations. By our methods, one can discuss the entire solutions for the algebraic difference equation and difference-differential equation of the form

$$f^n(z)+P_l\big(z,\Delta f(z),\dots,f'(z),\dots,f^{(k)}(z)\big)=\lambda_1(z)e^{\alpha_1(z)}+\lambda_2(z)e^{\alpha_2(z)},$$

here P_l is a difference-differential polynomial in $\Delta f(z),\dots,f'(z),\dots,f^{(k)}(z)$, with polynomials as its coefficients, and $\deg(P_l)\le n-2$, λ_1, λ_2, α_1, α_2 are nonzero polynomials.

Competing interests
The authors declare that they have no competing interests.

Authors' contributions
All authors contributed in drafting this manuscript. All authors read and approved the final manuscript.

Author details
[1]College of Science, China University of Petroleum (East China), Qingdao, 266580, P.R. China. [2]Department of Mathematics, Nanjing University, Nanjing, 210093, P.R. China.

Acknowledgements
The authors would like to thank the referee for his/her several important suggestions and for pointing out some errors in our original manuscript. These comments greatly improved the readability of the paper. This work was supported by the research funds of China University of Petroleum (No. 14CX06085A), the Special Funds of the National Natural Science Foundation of China (No. 11426217) and the Fundamental Research Funds for the Central Universities (No. 14CX02012A).

References

1. Hayman, WK: Meromorphic Functions. Clarendon, Oxford (1964)
2. Yi, HX, Yang, CC: Uniqueness Theory of Meromorphic Functions. Science Press, Beijing (1995)
3. Chiang, YM, Feng, SJ: On the Nevanlinna characteristic of $f(z+\eta)$ and difference equations in the complex plane. Ramanujan J. **16**, 105-129 (2008)
4. Halburd, RG, Korhonen, RJ: Difference analogue of the lemma on logarithmic derivative with applications to difference equations. J. Math. Anal. Appl. **314**, 477-487 (2006)
5. Chen, ZX: Growth and zeros of meromorphic solution of some linear difference equations. J. Math. Anal. Appl. **373**, 235-241 (2011)
6. Laine, I, Yang, CC: Clunie theorems for difference and q-difference polynomials. J. Lond. Math. Soc. **76**, 556-566 (2007)
7. Laine, I, Yang, CC: Value distribution of difference polynomials. Proc. Jpn. Acad., Ser. A, Math. Sci. **83**, 148-151 (2007)
8. Yang, CC, Laine, I: On analogies between nonlinear difference and differential equations. Proc. Jpn. Acad., Ser. A, Math. Sci. **86**, 10-14 (2010)
9. Liu, NN, Lü, WR, Shen, TT, Yang, CC: Entire solutions of certain type of difference equations. J. Inequal. Appl. **2014**, 63 (2014)
10. Clunie, J: On integral and meromorphic functions. J. Lond. Math. Soc. **37**, 17-27 (1962)
11. Zhang, RR, Chen, ZX: Value distribution of difference polynomials of meromorphic functions. Sci. Sin., Math. **42**, 1115-1130 (2012) (in Chinese)
12. Mohon'ko, AZ, Mohon'ko, VD: Estimates of the Nevanlinna characteristics of certain classes of meromorphic functions and their applications to differential equations. Sib. Mat. Zh. **15**, 1305-1322 (1974) (in Russian)
13. Yang, CC, Li, P: On the transcendental solutions of a certain type of nonlinear differential equations. Arch. Math. **82**, 442-448 (2004)
14. Bank, BS, Gundersen, GG, Laine, I: Meromorphic solutions of the Riccati differential equation. Ann. Acad. Sci. Fenn., Ser. A 1 Math. **6**, 369-398 (1981)

15

Uniform asymptotic stability implies exponential stability for nonautonomous half-linear differential systems

Masakazu Onitsuka* and Tomomi Soeda

*Correspondence: onitsuka@xmath.ous.ac.jp
Department of Applied Mathematics, Okayama University of Science, Okayama, 700-0005, Japan

Abstract

The present paper is considered a two-dimensional half-linear differential system: $x' = a_{11}(t)x + a_{12}(t)\phi_{p^*}(y)$, $y' = a_{21}(t)\phi_p(x) + a_{22}(t)y$, where all time-varying coefficients are continuous; p and p^* are positive numbers satisfying $1/p + 1/p^* = 1$; and $\phi_q(z) = |z|^{q-2}z$ for $q = p$ or $q = p^*$. In the special case, the half-linear system becomes the linear system $\mathbf{x}' = A(t)\mathbf{x}$ where $A(t)$ is a 2×2 continuous matrix and $\mathbf{x}$ is a two-dimensional vector. It is well known that the zero solution of the linear system is uniformly asymptotically stable if and only if it is exponentially stable. However, in general, uniform asymptotic stability is not equivalent to exponential stability in the case of nonlinear systems. The aim of this paper is to clarify that uniform asymptotic stability is equivalent to exponential stability for the half-linear differential system. Moreover, it is also clarified that exponential stability, global uniform asymptotic stability, and global exponential stability are equivalent for the half-linear differential system. Finally, the converse theorems on exponential stability which guarantee the existence of a strict Lyapunov function are presented.

MSC: 34D05; 34D20; 34D23; 37C75; 93D05; 93D20; 93D30

Keywords: uniform asymptotic stability; exponential stability; half-linear differential system; Lyapunov function; converse theorem

1 Introduction

We consider a system of differential equations of the form

$$\begin{aligned} x' &= a_{11}(t)x + a_{12}(t)\phi_{p^*}(y), \\ y' &= a_{21}(t)\phi_p(x) + a_{22}(t)y, \end{aligned} \tag{1.1}$$

where the prime denotes d/dt; the coefficients $a_{11}(t)$, $a_{12}(t)$, $a_{21}(t)$, and $a_{22}(t)$ are continuous on $I = [0, \infty)$; the numbers p and p^* are positive and satisfy

$$\frac{1}{p} + \frac{1}{p^*} = 1;$$

the real-valued function $\phi_q(z)$ is defined by

$$\phi_q(z) = \begin{cases} |z|^{q-2}z & \text{if } z \neq 0, \\ 0 & \text{if } z = 0, \end{cases} \qquad z \in \mathbb{R}$$

with $q = p$ or $q = p^*$. Note that ϕ_{p^*} is the inverse function of ϕ_p, and the numbers p and p^* are naturally greater than 1. Hence, we also note that the right-hand side of (1.1) does not satisfy Lipschitz condition at the origin since the function ϕ_p satisfies

$$\lim_{z\to 0} \frac{d}{dz}\phi_p(z) = \lim_{z\to 0}(p-1)|z|^{p-2} = \infty,$$

if $1 < p < 2$, and the function ϕ_{p^*} satisfies

$$\lim_{z\to 0} \frac{d}{dz}\phi_{p^*}(z) = \lim_{z\to 0}(p^*-1)|z|^{p^*-2} = \infty,$$

if $p > 2$. Since $\phi_p(0) = 0 = \phi_{p^*}(0)$, system (1.1) has the zero solution $(x(t), y(t)) \equiv (0,0)$. The type of system (1.1) appeared in [1–9]. Let $u = \exp(-\int a_{11}(t)\,dt)x$ and $v = \exp(-\int a_{22}(t)\,dt)y$. Then we can transform system (1.1) into the simpler system

$$u' = a(t)\phi_{p^*}(v), \qquad v' = b(t)\phi_p(u),$$

where

$$a(t) = a_{12}(t)\exp\left(\int \{(p^*-1)a_{22}(t) - a_{11}(t)\}\,dt\right)$$

and

$$b(t) = a_{21}(t)\exp\left(\int \{(p-1)a_{11}(t) - a_{22}(t)\}\,dt\right).$$

The problem of the global existence and uniqueness of solutions of this type of system are treated in [1, 2, 6]. However, keep in mind that we do not require the uniqueness of solutions of (1.1) throughout this paper.

In the special case that $a_{11}(t) \equiv 0$ and $a_{12}(t) \equiv 1$, (1.1) is transformed into the equation

$$\left(\phi_p(x')\right)' - a_{22}(t)\phi_p(x') - a_{21}(t)\phi_p(x) = 0. \tag{1.2}$$

If $x(t)$ is a solution of (1.2), then $cx(t)$ is also a solution of (1.2) for any $c \in \mathbb{R}$; that is, the solution space of (1.2) is homogeneous. However, even if $x_1(t)$ and $x_2(t)$ are two solutions of (1.2), the function $x_1(t) + x_2(t)$ is not always a solution of (1.2); that is, the solution space of (1.2) is not additive. For this reason, this equation is often called *'half-linear'*. For example, half-linear differential equation or half-linear differential system can be found in [1–19] and the references cited therein. Furthermore, we can confirm that the global existence and uniqueness of solutions of (1.2) are guaranteed for the initial value problem (see [1, 2]). To study half-linear differential equations is important in the field of difference equations, dynamic equations on time scales, partial differential equations and various functional equations, because the method of differential equation might be able to use for them. The reader may refer to [20–24] for example.

In the other special case, $p = 2$, (1.1) becomes the two-dimensional linear differential system

$$x' = a_{11}(t)x + a_{12}(t)y,$$
$$y' = a_{21}(t)x + a_{22}(t)y.$$

We now consider more general linear differential systems of the form

$$\mathbf{x}' = A(t)\mathbf{x}, \tag{1.3}$$

with an $n \times n$ continuous matrix $A(t)$. It is well known that the zero solution of (1.3) is uniformly asymptotically stable if and only if it is exponentially stable (see Section 2 about the precise definitions of uniform asymptotic stability and exponential stability). To be precise, the following theorem holds (for the proof, see ([25],pp.43-44), ([26], p.85), ([27], pp.499-500) and ([28], pp.29-30)).

Theorem A *If the zero solution of* (1.3) *is uniformly asymptotically stable, then it is exponentially stable.*

In general, uniform asymptotic stability is not equivalent to exponential stability in the case of nonlinear systems. Actually, we consider the scalar equation $x' = -x^3$ (see ([26], p.85) and ([29], p.49)). Clearly, this equation has the unique zero solution $x(t; t_0, 0) \equiv 0$. In the case that $x_0 \neq 0$, the solution of this equation passing through a point $x_0 \in \mathbb{R}$ at $t_0 \in I$ is given by

$$x(t) = \frac{x_0}{\sqrt{1 + 2x_0{}^2(t - t_0)}}.$$

It is known that the zero solution of this equation is uniformly asymptotically stable. However, it is not exponentially stable. In fact, we see

$$|x(t)|e^{\Lambda(t-t_0)} = \frac{x_0 e^{\Lambda(t-t_0)}}{\sqrt{1 + 2x_0{}^2(t - t_0)}} \to \infty \quad \text{as } t \to \infty$$

for any $\Lambda > 0$. It is clear that the solution $x(t)$ does not converge to the zero solution exponentially; that is, the zero solution is not exponentially stable. Here, the first question of this paper arises. Will uniform asymptotic stability guarantee exponential stability, even if the half-linear differential system (1.1) is nonlinear? We now give the answer to this question.

Theorem 1.1 *If the zero solution of* (1.1) *is uniformly asymptotically stable, then it is exponentially stable.*

Uniform asymptotic stability and exponential stability are of utmost importance for control theory. For example, we can assert that these stabilities guarantee the existence of a Lyapunov function which has good characteristics. Such results is called '*converse theorems*' on stability. The converse theorems are important in studying the properties of solutions of perturbed systems. We will consider the general nonlinear system

$$\mathbf{x}' = \mathbf{f}(t, \mathbf{x}), \tag{1.4}$$

where $\mathbf{f}(t, \mathbf{x})$ is continuous on $I \times \mathbb{R}^n$ and satisfies $\mathbf{f}(t, \mathbf{0}) = \mathbf{0}$. We define the set $S_\alpha = \{\mathbf{x} \in \mathbb{R}^n : \|\mathbf{x}\| < \alpha\}$ for $\alpha > 0$. Let $\mathbf{x}(t; t_0, \mathbf{x}_0)$ be a solution of (1.4) passing through a point $\mathbf{x}_0 \in \mathbb{R}^n$ at a time $t_0 \in I$. It is well known that if we suppose that $\mathbf{f}(t, \mathbf{x})$ satisfies locally Lipschitz condition with respect to $\mathbf{x}$ and if the zero solution of (1.4) is uniformly asymptotically stable, then there exists a *strict Lyapunov function (or strong Lyapunov function)* $V(t, \mathbf{x})$; that is, the scalar function $V(t, \mathbf{x})$ defined on $I \times S_\alpha$, where α is a suitable constant, which satisfies the following conditions:

(i) $a(\|\mathbf{x}\|) \le V(t, \mathbf{x}) \le b(\|\mathbf{x}\|)$;

(ii) $\dot{V}_{(1.4)}(t, \mathbf{x}) \le -c(\|\mathbf{x}\|)$,

where a, b, and c are continuous increasing and positive definite functions and the function $\dot{V}_{(1.4)}(t, \mathbf{x})$ is defined by

$$\dot{V}_{(1.4)}(t, \mathbf{x}) = \limsup_{h \to 0+} \frac{V(t + h, \mathbf{x}(t + h; t, \mathbf{x})) - V(t, \mathbf{x})}{h}$$

(see [25–36]).

Furthermore, global exponential stability guarantees the existence of better Lyapunov functions in the sense that it can be estimated by exact functions. The definition of global exponential stability is as follows. Let $\|\mathbf{x}\|$ be the Euclidean norm. The zero solution is said to be *globally exponentially stable* (*or globally exponentially asymptotically stable or exponentially asymptotically stable in the large*) if there exists a $\lambda > 0$ and, for any $\alpha > 0$, there exists a $\beta(\alpha) > 0$ such that $t_0 \in I$ and $\|\mathbf{x}_0\| < \alpha$ imply $\|\mathbf{x}(t; t_0, \mathbf{x}_0)\| \le \beta(\alpha) e^{-\lambda(t - t_0)} \|\mathbf{x}_0\|$ for all $t \ge t_0$. For example, we can refer to the books [28–30, 35, 37, 38] for this definition. The following result is a converse theorem on (global) exponential stability, which guarantees the existence of a Lyapunov function estimated by quadratic form $\|\mathbf{x}\|^2$ (see [30, 31, 34, 35]).

Theorem B *Suppose that for any* $\alpha > 0$, *there exists an* $L(\alpha) > 0$ *such that*

$$\left\|\mathbf{f}(t, \mathbf{x}) - \mathbf{f}(t, \mathbf{y})\right\| \le L(\alpha) \|\mathbf{x} - \mathbf{y}\|$$

for $(t, \mathbf{x}), (t, \mathbf{y}) \in I \times S_\alpha$, *where* $L(\alpha)$ *is independent of* $t \in I$. *If the zero solution of* (1.4) *is globally exponentially stable, then there exist three positive constants* $\beta_1(\alpha)$, $\beta_2(\alpha)$, β_3 *and a Lyapunov function* $V(t, \mathbf{x})$ *defined on* $I \times S_\alpha$ *which satisfies the following conditions*:

(i) $\beta_1(\alpha) \|\mathbf{x}\|^2 \le V(t, \mathbf{x}) \le \beta_2(\alpha) \|\mathbf{x}\|^2$;

(ii) $\dot{V}_{(1.4)}(t, \mathbf{x}) \le -\beta_3 \|\mathbf{x}\|^2$,

where α *is a number given in the definition of global exponential stability.*

In addition, another type converse theorem on (global) exponential stability can also be found in the classical books [28, 29].

Theorem C *Suppose that* $\mathbf{f}(t, \mathbf{x})$ *satisfy locally Lipschitz condition. If the zero solution of* (1.4) *is globally exponentially stable, then there exists a Lyapunov function* $V(t, \mathbf{x})$ *defined on* $I \times S_\alpha$ *which satisfies the following conditions*:

(i) $\|\mathbf{x}\| \le V(t,\mathbf{x}) \le \beta(\alpha)\|\mathbf{x}\|$;
(ii) $\dot{V}_{(1.4)}(t,\mathbf{x}) \le -k\lambda V(t,\mathbf{x})$, *where* $0 < k < 1$;
(iii) *there exists an* $L(t,\alpha) > 0$ *such that* $|V(t,\mathbf{x}) - V(t,\mathbf{y})| \le L(t,\alpha)\|\mathbf{x}-\mathbf{y}\|$ *for* $(t,\mathbf{x}),(t,\mathbf{y}) \in I \times S_\alpha$,

where α, β, *and* λ *are numbers given in the definition of global exponential stability.*

When restricted to the case of the linear system (1.3), the following facts are known (see ([28], pp.44-45), ([29], p.78) and ([37], p.155)).

Theorem D *If the zero solution of* (1.3) *is exponentially stable, then it is globally exponentially stable. In this case, we can find a* $\beta > 0$ *independent of* α *in the definition of globally exponentially stable.*

From Theorems A and D, for the linear differential system (1.3), uniform asymptotic stability and global exponential stability are equivalent. In the case of the linear system (1.3), the following converse theorem on (global) exponential stability is known (see ([28], pp.92-93), ([29], p.105) and ([32], p.327)).

Theorem E *Suppose that the zero solution of* (1.3) *is globally exponentially stable, i.e., there exist a* $\lambda > 0$ *and a* $\beta > 0$ *such that*

$$\left\|\mathbf{x}(t;t_0,\mathbf{x}_0)\right\| \le \beta e^{-\lambda(t-t_0)}\|\mathbf{x}_0\|$$

for all $t \ge t_0$. *Then there exists a Lyapunov function* $V(t,\mathbf{x})$ *defined on* $I \times \mathbb{R}^n$ *which satisfies the following conditions:*

(i) $\|\mathbf{x}\| \le V(t,\mathbf{x}) \le \beta\|\mathbf{x}\|$;
(ii) $\dot{V}_{(1.4)}(t,\mathbf{x}) \le -\lambda V(t,\mathbf{x})$;
(iii) $|V(t,\mathbf{x}) - V(t,\mathbf{y})| \le \beta\|\mathbf{x}-\mathbf{y}\|$ *for* $(t,\mathbf{x}),(t,\mathbf{y}) \in I \times \mathbb{R}^n$.

To present our result, we give some definitions of stability and its equivalent conditions in the next section. Also, we give a proposition which is the most important property for the proof of Theorem 1.1. In Section 3, we state the proof of Theorem 1.1. In Section 4, we give a natural generalization of Theorem D with $n = 2$. In the final section, we present the converse theorems for half-linear system (1.1), for comparison with Theorems B, C, and E.

2 Definitions and lemmas

We now give some definitions about the zero solution $\mathbf{x}(t;t_0,\mathbf{0}) \equiv \mathbf{0}$ of (1.1). The zero solution is said to be *uniformly attractive* if there exists a $\delta_0 > 0$ and, for every $\varepsilon > 0$, there exists a $T(\varepsilon) > 0$ such that $t_0 \in I$ and $\|\mathbf{x}_0\| < \delta_0$ imply $\|\mathbf{x}(t;t_0,\mathbf{x}_0)\| < \varepsilon$ for all $t \ge t_0 + T(\varepsilon)$. The zero solution of (1.1) is said to be *uniformly stable* if, for any $\varepsilon > 0$, there exists a $\delta(\varepsilon) > 0$ such that $t_0 \in I$ and $\|\mathbf{x}_0\| < \delta(\varepsilon)$ imply $\|\mathbf{x}(t;t_0,\mathbf{x}_0)\| < \varepsilon$ for all $t \ge t_0$. The zero solution is *uniformly asymptotically stable* if it is uniformly attractive and is uniformly stable. The zero solution is said to be *exponentially stable* (*or exponentially asymptotically stable*); if there exists a $\lambda > 0$ and, given any $\varepsilon > 0$, there exists a $\delta(\varepsilon) > 0$ such that $t_0 \in I$ and $\|\mathbf{x}_0\| < \delta(\varepsilon)$ imply $\|\mathbf{x}(t;t_0,\mathbf{x}_0)\| \le \varepsilon e^{-\lambda(t-t_0)}$ for all $t \ge t_0$. For example, we can refer to the books and papers [7, 18, 19, 25–53] for those definitions.

In this section, before giving the proof of Theorem 1.1, we prepare some lemmas. First, we will give the conditions which are equivalent to the above mentioned definitions. Some of these conditions are applicable to the proof of Theorem 1.1. For $\mathbf{x} = (x, y) \in \mathbb{R}^2$ and $p \geq 1$, we define a norm $\|\mathbf{x}\|_p$ by $\sqrt[p]{|x|^p + |y|^p}$. This norm is often called the '*Hölder norm*' or the '*p-norm*' (see [30, 31, 37, 54, 55]).

Lemma 2.1 *The zero solution of* (1.1) *is uniformly attractive if and only if there exists a* $\gamma_0 > 0$ *and, for every* $\rho > 0$, *there exists an* $S(\rho) > 0$ *such that* $t_0 \in I$ *and* $\|(x_0, \phi_{p^*}(y_0))\|_p < \gamma_0$ *imply*

$$\left\| \left(x(t; t_0, x_0, y_0), \phi_{p^*}\left(y(t; t_0, x_0, y_0)\right)\right) \right\|_p < \rho$$

for all $t \geq t_0 + S(\rho)$.

Proof First we prove the necessity. We suppose that the zero solution of (1.1) is uniformly attractive. That is, there exists a $\delta_0 > 0$ and, for every $\varepsilon > 0$, there exists a $T(\varepsilon) > 0$ such that $t_0 \in I$ and $\|(x_0, y_0)\| < \delta_0$ imply $\|(x(t; t_0, x_0, y_0), y(t; t_0, x_0, y_0))\| < \varepsilon$ for all $t \geq t_0 + T$. Let

$$\overline{p} = \max\{p, p^*\} \quad \text{and} \quad \gamma_0 = \min\left\{1, \left(\frac{\delta_0}{\sqrt{2}}\right)^{\frac{\overline{p}}{p}}\right\}.$$

For every $0 < \rho < 1$, we determine $\varepsilon = \rho^p/2$ and $S(\rho) = T(\rho^p/2)$. We consider the solution $(x(t; t_0, x_0, y_0), y(t; t_0, x_0, y_0))$ of (1.1) with $t_0 \in I$ and $\|(x_0, \phi_{p^*}(y_0))\|_p < \gamma_0$. From $\sqrt[p]{|x_0|^p + |y_0|^{p^*}} = \|(x_0, \phi_{p^*}(y_0))\|_p < \gamma_0$ it follows that

$$|x_0| < \gamma_0 \quad \text{and} \quad |y_0| < \gamma_0^{\frac{p}{p^*}}.$$

Hence, combining this estimation with $0 < \gamma_0 \leq 1 \leq \overline{p}/p$ and $\overline{p}/p^* \geq 1$, we obtain

$$\begin{aligned}
\left\|(x_0, y_0)\right\| &= \sqrt{x_0^2 + y_0^2} < \sqrt{\gamma_0^2 + \gamma_0^{\frac{2p}{p^*}}} \\
&= \sqrt{\min\left\{1, \left(\frac{\delta_0}{\sqrt{2}}\right)^{\frac{2\overline{p}}{p}}\right\} + \min\left\{1, \left(\frac{\delta_0}{\sqrt{2}}\right)^{\frac{2\overline{p}}{p^*}}\right\}} \\
&\leq \sqrt{2\min\left\{1, \left(\frac{\delta_0}{\sqrt{2}}\right)^2\right\}} \leq \delta_0,
\end{aligned}$$

and, therefore,

$$\left\| \left(x(t; t_0, x_0, y_0), y(t; t_0, x_0, y_0)\right) \right\| < \varepsilon = \frac{\rho^p}{2}$$

for $t \geq t_0 + S$. Taking into account that this inequality and $0 < \rho^p/2 < 1 < p$ and $p^* > 1$ hold, we have

$$\left|x(t; t_0, x_0, y_0)\right|^p < \left(\frac{\rho^p}{2}\right)^p < \frac{\rho^p}{2} \quad \text{and} \quad \left|y(t; t_0, x_0, y_0)\right|^{p^*} < \left(\frac{\rho^p}{2}\right)^{p^*} < \frac{\rho^p}{2}$$

for $t \geq t_0 + S$. From these inequalities, we see that

$$\left\| \left(x(t;t_0,x_0,y_0), \phi_{p^*}\left(y(t;t_0,x_0,y_0)\right)\right) \right\|_p < \sqrt[p]{\frac{\rho^p}{2} + \frac{\rho^p}{2}} = \rho$$

for $t \geq t_0 + S$.

Conversely, we prove the sufficiency. We suppose that there exists a $\gamma_0 > 0$ and, for every $\rho > 0$, there exists an $S(\rho) > 0$ such that $t_0 \in I$ and $\|(x_0, \phi_{p^*}(y_0))\|_p < \gamma_0$ imply $\|(x(t;t_0,x_0,y_0), \phi_{p^*}(y(t;t_0,x_0,y_0)))\|_p < \rho$ for all $t \geq t_0 + S$. Let

$$\delta_0 = \min\left\{1, \frac{{\gamma_0}^p}{2}\right\}.$$

For every $0 < \varepsilon < 1$, we determine $\rho = (\varepsilon/\sqrt{2})^{\overline{p}/p}$ and $T(\varepsilon) = S((\varepsilon/\sqrt{2})^{\overline{p}/p})$. We consider the solution $(x(t;t_0,x_0,y_0), y(t;t_0,x_0,y_0))$ of (1.1) with $t_0 \in I$ and $\|(x_0,y_0)\| < \delta_0$. From $\|(x_0,y_0)\| < \delta_0$ it follows that

$$|x_0| < \delta_0 \quad \text{and} \quad |y_0| < \delta_0.$$

Hence, combining this estimation with $0 < \delta_0 \leq 1 < p$ and $p^* > 1$, we obtain

$$\begin{aligned} \left\| \left(x_0, \phi_{p^*}(y_0)\right) \right\|_p &< \sqrt[p]{{\delta_0}^p + {\delta_0}^{p^*}} = \sqrt[p]{\min\left\{1, \left(\frac{{\gamma_0}^p}{2}\right)^p\right\} + \min\left\{1, \left(\frac{{\gamma_0}^p}{2}\right)^{p^*}\right\}} \\ &\leq \sqrt[p]{2\min\left\{1, \frac{{\gamma_0}^p}{2}\right\}} \leq \gamma_0, \end{aligned}$$

and, therefore,

$$\left\| \left(x(t;t_0,x_0,y_0), \phi_{p^*}\left(y(t;t_0,x_0,y_0)\right)\right) \right\|_p < \rho = \left(\frac{\varepsilon}{\sqrt{2}}\right)^{\frac{\overline{p}}{p}}$$

for $t \geq t_0 + T$. Taking into account that this inequality and $0 < \varepsilon^2/2 < 1 \leq \overline{p}/p$, and $\overline{p}/p^* \geq 1$ hold, we have

$$x^2(t;t_0,x_0,y_0) < \left(\frac{\varepsilon^2}{2}\right)^{\frac{\overline{p}}{p}} \leq \frac{\varepsilon^2}{2} \quad \text{and} \quad y^2(t;t_0,x_0,y_0) < \left(\frac{\varepsilon^2}{2}\right)^{\frac{\overline{p}}{p^*}} \leq \frac{\varepsilon^2}{2}$$

for $t \geq t_0 + T$. From these inequalities, we see that

$$\left\| \left(x(t;t_0,x_0,y_0), y(t;t_0,x_0,y_0)\right) \right\| < \varepsilon$$

for $t \geq t_0 + T$. This completes the proof of Lemma 2.1. □

By using the same arguments as in Lemma 2.1, we can prove the following lemma.

Lemma 2.2 *The zero solution of* (1.1) *is uniformly stable if and only if, for any $\rho > 0$, there exists a $\gamma(\rho) > 0$ such that $t_0 \in I$ and $\|(x_0, \phi_{p^*}(y_0))\|_p < \gamma(\rho)$ imply*

$$\left\| \left(x(t;t_0,x_0,y_0), \phi_{p^*}\left(y(t;t_0,x_0,y_0)\right)\right) \right\|_p < \rho$$

for all $t \geq t_0$.

Proof First we prove the necessity. We suppose that the zero solution of (1.1) is uniformly stable. That is, for any $\varepsilon > 0$, there exists a $\delta(\varepsilon) > 0$ such that $t_0 \in I$ and $\|(x_0, y_0)\| < \delta(\varepsilon)$ imply $\|(x(t; t_0, x_0, y_0), y(t; t_0, x_0, y_0))\| < \varepsilon$ for all $t \geq t_0$. For every $0 < \rho < 1$, we determine an $\varepsilon = \rho^p/2$. Let

$$\overline{p} = \max\{p, p^*\} \quad \text{and} \quad \gamma(\rho) = \min\left\{1, \left(\frac{1}{\sqrt{2}}\delta\left(\frac{\rho^p}{2}\right)\right)^{\frac{\overline{p}}{p}}\right\}.$$

We consider the solution $(x(t; t_0, x_0, y_0), y(t; t_0, x_0, y_0))$ of (1.1) with $t_0 \in I$ and $\|(x_0, \phi_{p^*}(y_0))\|_p < \gamma$. From $\sqrt[p]{|x_0|^p + |y_0|^{p^*}} = \|(x_0, \phi_{p^*}(y_0))\|_p < \gamma$ it follows that

$$|x_0| < \gamma \quad \text{and} \quad |y_0| < \gamma^{\frac{p}{p^*}}.$$

Hence, combining this estimation with $0 < \gamma \leq 1 \leq \overline{p}/p$ and $\overline{p}/p^* \geq 1$, we obtain

$$\begin{aligned}
\|(x_0, y_0)\| &= \sqrt{x_0{}^2 + y_0{}^2} < \sqrt{\gamma^2 + \gamma^{\frac{2p}{p^*}}} \\
&= \sqrt{\min\left\{1, \left(\frac{\delta}{\sqrt{2}}\right)^{\frac{2\overline{p}}{p}}\right\} + \min\left\{1, \left(\frac{\delta}{\sqrt{2}}\right)^{\frac{2\overline{p}}{p^*}}\right\}} \\
&\leq \sqrt{2\min\left\{1, \left(\frac{\delta}{\sqrt{2}}\right)^2\right\}} \leq \delta,
\end{aligned}$$

and, therefore,

$$\left\|(x(t; t_0, x_0, y_0), y(t; t_0, x_0, y_0))\right\| < \varepsilon = \frac{\rho^p}{2}$$

for $t \geq t_0$. Taking into account that this inequality and $0 < \rho^p/2 < 1 < p$ and $p^* > 1$ hold, we have

$$|x(t; t_0, x_0, y_0)|^p < \left(\frac{\rho^p}{2}\right)^p < \frac{\rho^p}{2} \quad \text{and} \quad |y(t; t_0, x_0, y_0)|^{p^*} < \left(\frac{\rho^p}{2}\right)^{p^*} < \frac{\rho^p}{2}$$

for $t \geq t_0$. From these inequalities, we see that

$$\left\|(x(t; t_0, x_0, y_0), \phi_{p^*}(y(t; t_0, x_0, y_0)))\right\|_p < \sqrt[p]{\frac{\rho^p}{2} + \frac{\rho^p}{2}} = \rho$$

for $t \geq t_0$.

Conversely, we prove the sufficiency. We suppose that for any $\rho > 0$, there exists a $\gamma(\rho) > 0$ such that $t_0 \in I$ and $\|(x_0, \phi_{p^*}(y_0))\|_p < \gamma(\rho)$ imply

$$\left\|(x(t; t_0, x_0, y_0), \phi_{p^*}(y(t; t_0, x_0, y_0)))\right\|_p < \rho$$

for all $t \geq t_0$. For every $0 < \varepsilon < 1$, we determine a $\rho = (\varepsilon/\sqrt{2})^{\overline{p}/p}$. Let

$$\delta(\varepsilon) = \min\left\{1, \frac{1}{2}\gamma^p\left(\left(\frac{\varepsilon}{\sqrt{2}}\right)^{\frac{\overline{p}}{p}}\right)\right\}.$$

We consider the solution $(x(t;t_0,x_0,y_0), y(t;t_0,x_0,y_0))$ of (1.1) with $t_0 \in I$ and $\|(x_0,y_0)\| < \delta$. From $\|(x_0,y_0)\| < \delta$ it follows that

$$|x_0| < \delta \quad \text{and} \quad |y_0| < \delta.$$

Hence, combining this estimation with $0 < \delta \leq 1 < p$ and $p^* > 1$, we obtain

$$\left\| \left(x_0, \phi_{p^*}(y_0)\right) \right\|_p < \sqrt[p]{\delta^p + \delta^{p^*}} = \sqrt[p]{\min\left\{1, \left(\frac{\gamma^p}{2}\right)^p\right\} + \min\left\{1, \left(\frac{\gamma^p}{2}\right)^{p^*}\right\}}$$
$$\leq \sqrt[p]{2\min\left\{1, \frac{\gamma^p}{2}\right\}} \leq \gamma,$$

and, therefore,

$$\left\| \left(x(t;t_0,x_0,y_0), \phi_{p^*}\left(y(t;t_0,x_0,y_0)\right)\right) \right\|_p < \rho = \left(\frac{\varepsilon}{\sqrt{2}}\right)^{\frac{\overline{p}}{p}}$$

for $t \geq t_0$. Taking into account that this inequality and $0 < \varepsilon^2/2 < 1 \leq \overline{p}/p$ and $\overline{p}/p^* \geq 1$ hold, we have

$$x^2(t;t_0,x_0,y_0) < \left(\frac{\varepsilon^2}{2}\right)^{\frac{\overline{p}}{p}} \leq \frac{\varepsilon^2}{2} \quad \text{and} \quad y^2(t;t_0,x_0,y_0) < \left(\frac{\varepsilon^2}{2}\right)^{\frac{\overline{p}}{p^*}} \leq \frac{\varepsilon^2}{2}$$

for $t \geq t_0$. From these inequalities, we see that

$$\left\| \left(x(t;t_0,x_0,y_0), y(t;t_0,x_0,y_0)\right) \right\| < \varepsilon$$

for $t \geq t_0$. This completes the proof of Lemma 2.2. □

Furthermore, by using the same arguments as in Lemmas 2.1 and 2.2, we have the following result.

Lemma 2.3 *The zero solution of* (1.1) *is exponentially stable if and only if there exists a* $\mu > 0$ *and, given any* $\rho > 0$*, there exists a* $\gamma(\rho) > 0$ *such that* $t_0 \in I$ *and* $\|(x_0, \phi_{p^*}(y_0))\|_p < \gamma(\rho)$ *imply*

$$\left\| \left(x(t;t_0,x_0,y_0), \phi_{p^*}\left(y(t;t_0,x_0,y_0)\right)\right) \right\|_p \leq \rho e^{-\mu(t-t_0)}$$

for all $t \geq t_0$.

Proof First we prove the necessity. We suppose that the zero solution of (1.1) is exponentially stable. That is, there exists a $\lambda > 0$ and, given any $\varepsilon > 0$, there exists a $\delta(\varepsilon) > 0$ such that $t_0 \in I$ and $\|(x_0,y_0)\| < \delta(\varepsilon)$ imply $\|(x(t;t_0,x_0,y_0), y(t;t_0,x_0,y_0))\| \leq \varepsilon e^{-\lambda(t-t_0)}$ for all $t \geq t_0$. Let $\mu = \lambda/p$. For every $0 < \rho < 1$, we determine an $\varepsilon = \rho^p/2$. Let

$$\overline{p} = \max\{p, p^*\} \quad \text{and} \quad \gamma(\rho) = \min\left\{1, \left(\frac{1}{\sqrt{2}}\delta\left(\frac{\rho^p}{2}\right)\right)^{\frac{\overline{p}}{p}}\right\}.$$

We consider the solution $(x(t;t_0,x_0,y_0),y(t;t_0,x_0,y_0))$ of (1.1) with $t_0 \in I$ and $\|(x_0, \phi_{p^*}(y_0))\|_p < \gamma$. From $\sqrt[p]{|x_0|^p + |y_0|^{p^*}} = \|(x_0,\phi_{p^*}(y_0))\|_p < \gamma$ it follows that

$$|x_0| < \gamma \quad \text{and} \quad |y_0| < \gamma^{\frac{p}{p^*}}.$$

Hence, combining this estimation with $0 < \gamma \le 1 \le \overline{p}/p$ and $\overline{p}/p^* \ge 1$ we obtain

$$\begin{aligned}\|(x_0,y_0)\| &= \sqrt{x_0{}^2 + y_0{}^2} < \sqrt{\gamma^2 + \gamma^{\frac{2p}{p^*}}} \\ &= \sqrt{\min\left\{1, \left(\frac{\delta}{\sqrt{2}}\right)^{\frac{2\overline{p}}{p}}\right\} + \min\left\{1, \left(\frac{\delta}{\sqrt{2}}\right)^{\frac{2\overline{p}}{p^*}}\right\}} \\ &\le \sqrt{2\min\left\{1, \left(\frac{\delta}{\sqrt{2}}\right)^{2}\right\}} \le \delta,\end{aligned}$$

and, therefore,

$$\left\|\left(x(t;t_0,x_0,y_0),y(t;t_0,x_0,y_0)\right)\right\| \le \varepsilon e^{-\lambda(t-t_0)} = \frac{\rho^p}{2}e^{-p\mu(t-t_0)}$$

for $t \ge t_0$. Taking into account that this inequality and

$$0 < \frac{\rho^p}{2}e^{-p\mu(t-t_0)} \le \frac{\rho^p}{2} < 1 < p \quad \text{and} \quad 1 < p^*$$

hold, we have

$$\left|x(t;t_0,x_0,y_0)\right|^p \le \left(\frac{\rho^p}{2}e^{-p\mu(t-t_0)}\right)^p < \frac{\rho^p}{2}e^{-p\mu(t-t_0)}$$

and

$$\left|y(t;t_0,x_0,y_0)\right|^{p^*} \le \left(\frac{\rho^p}{2}e^{-p\mu(t-t_0)}\right)^{p^*} < \frac{\rho^p}{2}e^{-p\mu(t-t_0)}$$

for $t \ge t_0$. From these inequalities, we see that

$$\left\|\left(x(t;t_0,x_0,y_0),\phi_{p^*}\left(y(t;t_0,x_0,y_0)\right)\right)\right\|_p < \rho e^{-\mu(t-t_0)}$$

for $t \ge t_0$.

Conversely, we prove the sufficiency. We suppose that there exists a $\mu > 0$ and, given any $\rho > 0$, there exists a $\gamma(\rho) > 0$ such that $t_0 \in I$ and $\|(x_0,\phi_{p^*}(y_0))\|_p < \gamma(\rho)$ imply

$$\left\|\left(x(t;t_0,x_0,y_0),\phi_{p^*}\left(y(t;t_0,x_0,y_0)\right)\right)\right\|_p \le \rho e^{-\mu(t-t_0)}$$

for all $t \ge t_0$. Let $\lambda = \mu p/\overline{p}$. For every $0 < \varepsilon < 1$, we determine a $\rho = (\varepsilon/\sqrt{2})^{\overline{p}/p}$. Let

$$\delta(\varepsilon) = \min\left\{1, \frac{1}{2}\gamma^p\left(\left(\frac{\varepsilon}{\sqrt{2}}\right)^{\frac{\overline{p}}{p}}\right)\right\}.$$

We consider the solution $(x(t;t_0,x_0,y_0),y(t;t_0,x_0,y_0))$ of (1.1) with $t_0 \in I$ and $\|(x_0,y_0)\| < \delta$. From $\|(x_0,y_0)\| < \delta$ it follows that

$$|x_0| < \delta \quad \text{and} \quad |y_0| < \delta.$$

Hence, combining this estimation with $0 < \delta \le 1 < p$ and $p^* > 1$, we obtain

$$\begin{aligned}\|(x_0,\phi_{p^*}(y_0))\|_p &< \sqrt[p]{\delta^p + \delta^{p^*}} = \sqrt[p]{\min\left\{1,\left(\frac{\gamma^p}{2}\right)^p\right\} + \min\left\{1,\left(\frac{\gamma^p}{2}\right)^{p^*}\right\}}\\ &\le \sqrt[p]{2\min\left\{1,\frac{\gamma^p}{2}\right\}} \le \gamma,\end{aligned}$$

and, therefore,

$$\|(x(t;t_0,x_0,y_0),\phi_{p^*}(y(t;t_0,x_0,y_0)))\|_p \le \rho e^{-\mu(t-t_0)} = \left(\frac{\varepsilon}{\sqrt{2}}\right)^{\frac{\overline{p}}{p}} e^{-\frac{\overline{p}\lambda}{p}(t-t_0)}$$

for $t \ge t_0$. Taking into account that this inequality and

$$0 < \frac{\varepsilon}{\sqrt{2}} e^{-\lambda(t-t_0)} \le \frac{\varepsilon}{\sqrt{2}} < 1 \le \frac{\overline{p}}{p} \quad \text{and} \quad 1 \le \frac{\overline{p}}{p^*}$$

hold, we have

$$|x(t;t_0,x_0,y_0)| \le \left(\frac{\varepsilon}{\sqrt{2}} e^{-\lambda(t-t_0)}\right)^{\frac{\overline{p}}{p}} \le \frac{\varepsilon}{\sqrt{2}} e^{-\lambda(t-t_0)}$$

and

$$|y(t;t_0,x_0,y_0)| \le \left(\frac{\varepsilon}{\sqrt{2}} e^{-\lambda(t-t_0)}\right)^{\frac{\overline{p}}{p^*}} \le \frac{\varepsilon}{\sqrt{2}} e^{-\lambda(t-t_0)}$$

for $t \ge t_0$. From these inequalities, we see that

$$\|(x(t;t_0,x_0,y_0),y(t;t_0,x_0,y_0))\| \le \varepsilon e^{-\lambda(t-t_0)}$$

for $t \ge t_0$. This completes the proof of Lemma 2.3. □

In the special case in which $p = 2$, (1.1) becomes the linear system. As is well known, the solution space of the linear system is homogeneous and additive. On the other hand, in the general case in which $p \neq 2$, the solution space of (1.1) is not homogeneous or additive. However, we can show that (1.1) has a homogeneous-like property on the solution space.

Lemma 2.4 *If $(x(t),y(t))$ is a solution of (1.1) passing through a point $(x_0,y_0) \in \mathbb{R}^2$ at $t = t_0 \in I$, then $(cx(t),\phi_p(c)y(t))$ is also a solution of (1.1) passing through a point $(cx_0,\phi_p(c)y_0) \in \mathbb{R}^2$ at $t = t_0$ for any $c \in \mathbb{R}$.*

Proof We consider the solution $(x(t), y(t))$ of (1.1) passing through a point (x_0, y_0) at $t = t_0$. Let $\tilde{x}(t) = cx(t)$ and $\tilde{y}(t) = \phi_p(c)y(t)$ with $c \in \mathbb{R}$. It is clear that $(\tilde{x}(t_0), \tilde{y}(t_0)) = (cx_0, \phi_p(c)y_0)$ is satisfied. Since ϕ_{p^*} is the inverse function of ϕ_p, we have

$$\tilde{x}'(t) = a_{11}(t)cx(t) + a_{12}(t)\phi_{p^*}\left(\phi_p(c)y(t)\right) = a_{11}(t)\tilde{x}(t) + a_{12}(t)\phi_{p^*}\left(\tilde{y}(t)\right)$$

and

$$\tilde{y}'(t) = a_{21}(t)\phi_p\left(cx(t)\right) + a_{22}(t)\phi_p(c)y(t) = a_{21}(t)\phi_p\left(\tilde{x}(t)\right) + a_{22}(t)\tilde{y}(t).$$

We therefore conclude that $(cx(t), \phi_p(c)y(t))$ is also a solution of (1.1) passing through a point $(cx_0, \phi_p(c)y_0)$ at $t = t_0$. □

We state the following proposition, which is the most important property for the proof of Theorem 1.1.

Proposition 2.5 *If the zero solution of* (1.1) *is uniformly attractive, then there exists a* $\gamma_0 >$ 0 *and, for every* $\nu > 1$, *there exists a* $T(\nu) > 0$ *such that* $t_0 \in I$ *and* $\|(x_0, \phi_{p^*}(y_0))\|_p < \gamma_0 \nu^{-(k-1)}$ *imply*

$$\left\|\left(x\left(t; t_0 + (k-1)T(\nu), x_0, y_0\right), \phi_{p^*}\left(y\left(t; t_0 + (k-1)T(\nu), x_0, y_0\right)\right)\right)\right\|_p < \gamma_0 \nu^{-k}$$

for all $t \geq t_0 + kT(\nu)$ *and* $k \in \mathbb{N}$.

Proof By using the assumption and Lemma 2.1, there exists a $\gamma_0 > 0$ and, for every $\nu > 1$, there exists an $S(\gamma_0/\nu) > 0$ such that $\tau \geq 0$ and $\|(\xi, \phi_{p^*}(\eta))\|_p < \gamma_0$ imply

$$\left\|\left(x(t; \tau, \xi, \eta), \phi_{p^*}\left(y(t; \tau, \xi, \eta)\right)\right)\right\|_p < \frac{\gamma_0}{\nu}$$

for all $t \geq \tau + S$.

Let $T(\nu) = S(\gamma_0/\nu)$. We consider the solution

$$\left(x\left(t; t_0 + (k-1)T, x_0, y_0\right), y\left(t; t_0 + (k-1)T, x_0, y_0\right)\right)$$

of (1.1) with $t_0 \in I$ and $\|(x_0, \phi_{p^*}(y_0))\|_p < \gamma_0 \nu^{-(k-1)}$. From Lemma 2.4, we conclude that

$$\left(\nu^{k-1}x\left(t; t_0 + (k-1)T, x_0, y_0\right), \phi_p\left(\nu^{k-1}\right)y\left(t; t_0 + (k-1)T, x_0, y_0\right)\right)$$

is also a solution of (1.1) passing through a point $(\nu^{k-1}x_0, \phi_p(\nu^{k-1})y_0)$ at $t = t_0 + (k-1)T$. Since

$$\left\|\left(\nu^{k-1}x_0, \nu^{k-1}\phi_{p^*}(y_0)\right)\right\|_p = \nu^{k-1}\left\|\left(x_0, \phi_{p^*}(y_0)\right)\right\|_p < \gamma_0$$

holds, we have

$$\begin{aligned}\frac{\gamma_0}{\nu} &> \left\|\left(\nu^{k-1}x\left(t; t_0 + (k-1)T, x_0, y_0\right), \nu^{k-1}\phi_{p^*}\left(y\left(t; t_0 + (k-1)T, x_0, y_0\right)\right)\right)\right\|_p \\ &= \nu^{k-1}\left\|\left(x\left(t; t_0 + (k-1)T, x_0, y_0\right), \phi_{p^*}\left(y\left(t; t_0 + (k-1)T, x_0, y_0\right)\right)\right)\right\|_p\end{aligned}$$

for all $t \geq t_0 + (k-1)T + T = t_0 + kT$. That is, we obtain

$$\left\| \left(x\left(t; t_0 + (k-1)T, x_0, y_0\right), \phi_{p^*}\left(y\left(t; t_0 + (k-1)T, x_0, y_0\right)\right)\right) \right\|_p < \gamma_0 \nu^{-k}$$

for all $t \geq t_0 + kT$. This completes the proof of Proposition 2.5. □

3 Exponential asymptotic stability

In this section, we give the proof of the main theorem.

Proof of Theorem 1.1 By using uniform attractivity of (1.1) and Proposition 2.5, there exist a $\gamma_0 > 0$ and a $T(e) > 0$ such that $t_0 \in I$ and $\|(\xi, \phi_{p^*}(\eta))\|_p < \gamma_0 e^{-(k-1)}$ imply

$$\left\| \left(x\left(t; t_0 + (k-1)T, \xi, \eta\right), \phi_{p^*}\left(y\left(t; t_0 + (k-1)T, \xi, \eta\right)\right)\right) \right\|_p < \gamma_0 e^{-k} \tag{3.1}$$

for all $t \geq t_0 + kT$ and $k \in \mathbb{N}$.

Because of the uniform stability of (1.1) and Lemma 2.2, there exists a $\gamma(\gamma_0) > 0$ such that $t_0 \in I$ and $\|(\xi, \phi_{p^*}(\eta))\|_p < \gamma$ imply

$$\left\| \left(x(t; t_0, \xi, \eta), \phi_{p^*}\left(y(t; t_0, \xi, \eta)\right)\right) \right\|_p < \gamma_0 \tag{3.2}$$

for all $t \geq t_0$. Let $\lambda = 1/T$. For every $\varepsilon > 0$, we determine a

$$\delta(\varepsilon) = \frac{\gamma \varepsilon}{\gamma_0 e} > 0.$$

We now consider the solution $(x(t; t_0, x_0, y_0), y(t; t_0, x_0, y_0))$ of (1.1) with $t_0 \in I$ and $\|(x_0, \phi_{p^*}(y_0))\|_p < \delta$. For the sake of simplicity, let

$$\left(x(t), y(t)\right) = \left(x(t; t_0, x_0, y_0), y(t; t_0, x_0, y_0)\right).$$

Using Lemma 2.4, we can find a solution

$$\left(\frac{\gamma_0 e}{\varepsilon} x(t), \phi_p\left(\frac{\gamma_0 e}{\varepsilon}\right) y(t)\right)$$

of (1.1) passing through a point $((\gamma_0 e/\varepsilon)x_0, \phi_p(\gamma_0 e/\varepsilon)y_0)$ at $t = t_0$. From $\delta = \gamma\varepsilon/(\gamma_0 e)$, we see that

$$\begin{aligned} \left\| \left(\frac{\gamma_0 e}{\varepsilon} x_0, \phi_{p^*}\left(\phi_p\left(\frac{\gamma_0 e}{\varepsilon}\right) y_0\right)\right) \right\|_p &= \left\| \left(\frac{\gamma_0 e}{\varepsilon} x_0, \frac{\gamma_0 e}{\varepsilon} \phi_{p^*}(y_0)\right) \right\|_p \\ &= \frac{\gamma_0 e}{\varepsilon} \left\| \left(x_0, \phi_{p^*}(y_0)\right) \right\|_p < \gamma \end{aligned}$$

at $t = t_0$. From this inequality and (3.2) with

$$(\xi, \eta) = \left(\frac{\gamma_0 e}{\varepsilon} x_0, \phi_p\left(\frac{\gamma_0 e}{\varepsilon}\right) y_0\right),$$

we have

$$\frac{\gamma_0 e}{\varepsilon}\|(x(t),\phi_{p^*}(y(t)))\|_p = \left\|\left(\frac{\gamma_0 e}{\varepsilon}x(t), \frac{\gamma_0 e}{\varepsilon}\phi_{p^*}(y(t))\right)\right\|_p$$

$$= \left\|\left(\frac{\gamma_0 e}{\varepsilon}x(t), \phi_{p^*}\left(\phi_p\left(\frac{\gamma_0 e}{\varepsilon}\right)y(t)\right)\right)\right\|_p < \gamma_0 \tag{3.3}$$

for $t \geq t_0$. We therefore conclude that

$$\|(x(t),\phi_{p^*}(y(t)))\|_p < \frac{\varepsilon}{e}$$

for $t_0 \leq t \leq t_0 + T$. By using (3.3), we obtain

$$\left\|\left(\frac{\gamma_0 e}{\varepsilon}x_0, \phi_{p^*}\left(\phi_p\left(\frac{\gamma_0 e}{\varepsilon}\right)y_0\right)\right)\right\|_p < \gamma_0$$

at $t = t_0$. From this inequality and (3.1) with

$$(\xi,\eta) = \left(\frac{\gamma_0 e}{\varepsilon}x_0, \phi_p\left(\frac{\gamma_0 e}{\varepsilon}\right)y_0\right), \quad k = 1,$$

we get

$$\frac{\gamma_0 e}{\varepsilon}\|(x(t),\phi_{p^*}(y(t)))\|_p = \left\|\left(\frac{\gamma_0 e}{\varepsilon}x(t), \phi_{p^*}\left(\phi_p\left(\frac{\gamma_0 e}{\varepsilon}\right)y(t)\right)\right)\right\|_p < \frac{\gamma_0}{e} \tag{3.4}$$

for $t \geq t_0 + T$. We therefore conclude that

$$\|(x(t),\phi_{p^*}(y(t)))\|_p < \frac{\varepsilon}{e^2}$$

for $t_0 + T \leq t \leq t_0 + 2T$. By using (3.4), we obtain

$$\left\|\left(\frac{\gamma_0 e}{\varepsilon}x(t_0+T), \phi_{p^*}\left(\phi_p\left(\frac{\gamma_0 e}{\varepsilon}\right)y(t_0+T)\right)\right)\right\|_p < \frac{\gamma_0}{e}$$

at $t = t_0 + T$. From this inequality and (3.1) with

$$(\xi,\eta) = \left(\frac{\gamma_0 e}{\varepsilon}x(t_0+T), \phi_p\left(\frac{\gamma_0 e}{\varepsilon}\right)y(t_0+T)\right), \quad k = 2,$$

we get

$$\frac{\gamma_0 e}{\varepsilon}\|(x(t),\phi_{p^*}(y(t)))\|_p = \left\|\left(\frac{\gamma_0 e}{\varepsilon}x(t), \phi_{p^*}\left(\phi_p\left(\frac{\gamma_0 e}{\varepsilon}\right)y(t)\right)\right)\right\|_p < \frac{\gamma_0}{e^2}$$

for $t \geq t_0 + 2T$. We therefore conclude that

$$\|(x(t),\phi_{p^*}(y(t)))\|_p < \frac{\varepsilon}{e^3}$$

for $t_0 + 2T \le t \le t_0 + 3T$. By means of the same process as in the above mentioned estimates, we see that

$$\left\| \left(x(t), \phi_{p^*}(y(t))\right) \right\|_p < \varepsilon e^{-\kappa}$$

for $t_0 + (\kappa - 1)T \le t \le t_0 + \kappa T$ and $\kappa \in \mathbb{N}$. Hence, by $t \le t_0 + \kappa T$ we have

$$-\kappa \le -\frac{1}{T}(t - t_0) = -\lambda(t - t_0),$$

and therefore

$$\left\| \left(x(t), \phi_{p^*}(y(t))\right) \right\|_p < \varepsilon e^{-\kappa} \le \varepsilon e^{-\lambda(t-t_0)}$$

for $t_0 + (\kappa - 1)T \le t \le t_0 + \kappa T$ and $\kappa \in \mathbb{N}$. Note that we can divide the interval $[t_0, t_0 + \kappa T]$ as

$$[t_0, t_0 + \kappa T] = \bigcup_{n=1}^{\kappa} \left[t_0 + (n-1)T, t_0 + nT\right]$$

for $\kappa \in \mathbb{N}$. Thus, it turns out that

$$\left\| \left(x(t), \phi_{p^*}(y(t))\right) \right\|_p \le \varepsilon e^{-\lambda(t-t_0)}$$

for $t \ge t_0$. Using Lemma 2.3, we conclude that the zero solution of (1.1) is exponentially stable. This completes the proof of Theorem 1.1. □

Note here that the proof of Theorem 1.1 does not require the uniqueness of solutions for the initial value problem.

4 Global exponential stability

Clearly concepts of above mentioned stability are local theory about the zero solution. In this section, we will discuss any initial disturbance and initial state. We consider the nonlinear scalar equation $x' = -x + x^2$ (see [27], p.508). It is easy to see that the solution of this equation is given by

$$x(t; t_0, x_0) = \frac{x_0}{x_0 - (x_0 - 1)e^{t-t_0}},$$

where $t_0 \in I$ and $x_0 \in \mathbb{R}$. Clearly, $x(t; t_0, 0) \equiv 0$ and $x(t; t_0, 1) \equiv 1$ are the trivial solution. If $x_0 < 1$, then $x(t; t_0, x_0) \to 0$ as $t \to \infty$. Moreover, for every $0 < \varepsilon < 1$, we choose $\delta(\varepsilon) = \varepsilon/2$. If $|x_0| < \delta(\varepsilon)$, then we have

$$\left|x(t; t_0, x_0)\right| \le \frac{|x_0| e^{-(t-t_0)}}{1 - |x_0|(1 - e^{-(t-t_0)})} < \frac{|x_0| e^{-(t-t_0)}}{1 - |x_0|} < \varepsilon e^{-(t-t_0)}$$

for $t \ge t_0$. This means that the zero solution is exponentially stable. On the other hand, if $1 < x_0$, then $x(t; t_0, x_0) \to \infty$ as $t \to t_0 + \log(x_0/(x_0 - 1))$; that is, in the case that $1 < x_0$, all solutions are unbounded. Therefore, we can conclude that local theory and global theory are completely different concepts. Here, the second question of this paper arises. Will exponential stability guarantee global exponential stability, even if the half-linear differential system (1.1) is nonlinear? We now give the answer to this question.

Theorem 4.1 *If the zero solution of* (1.1) *is uniformly asymptotically stable, then there exist a* $\lambda > 0$ *and a* $\beta > 0$ *such that* $t_0 \in I$ *and* $(x_0, y_0) \in \mathbb{R}^2$ *imply*

$$\left\| \left(x(t;t_0,x_0,y_0), \phi_{p^*}\left(y(t;t_0,x_0,y_0)\right)\right) \right\|_p \leq \beta e^{-\lambda(t-t_0)} \left\| \left(x_0, \phi_{p^*}(y_0)\right) \right\|_p$$

for all $t \geq t_0$*, where* $\beta > 0$ *is independent of the size of* $\|(x_0, \phi_{p^*}(y_0))\|_p$.

Proof By virtue of Theorem 1.1, it turns out that the uniform asymptotic stability of (1.1) implies exponential stability. Using Lemma 2.3, there exist a $\lambda > 0$ and a $\delta(1) > 0$ such that $t_0 \in I$ and $\|(\xi, \phi_{p^*}(\eta))\|_p < \delta$ imply

$$\left\| \left(x(t;t_0,\xi,\eta), \phi_{p^*}\left(y(t;t_0,\xi,\eta)\right)\right) \right\|_p \leq e^{-\lambda(t-t_0)}$$

for all $t \geq t_0$.

We choose $\beta = 2/\delta$. Let $t_0 \in I$ and $(x_0, y_0) \in \mathbb{R}^2$ be given. We may assume without loss of generality that $(x_0, y_0) \neq (0,0)$. Consider the solution $(x(t;t_0,x_0,y_0), y(t;t_0,x_0,y_0))$ of (1.1). For the sake of convenience, we write

$$\left(x(t), y(t)\right) = \left(x(t;t_0,x_0,y_0), y(t;t_0,x_0,y_0)\right), \qquad c = \frac{\delta}{2\|(x_0, \phi_{p^*}(y_0))\|_p}.$$

Hence, we have

$$\left\| \left(cx_0, \phi_{p^*}\left(\phi_p(c)y_0\right)\right) \right\|_p = \left\| \left(cx_0, c\phi_{p^*}(y_0)\right) \right\|_p = c\left\| \left(x_0, \phi_{p^*}(y_0)\right) \right\|_p < \delta.$$

Using Lemma 2.4, $(cx(t), \phi_p(c)y(t))$ is also a solution of (1.1) passing through a point $(cx_0, \phi_p(c)y_0)$ at $t = t_0$. Thus, we get

$$c\left\| \left(x(t), \phi_{p^*}\left(y(t)\right)\right) \right\|_p = \left\| \left(cx(t), \phi_{p^*}\left(\phi_p(c)y(t)\right)\right) \right\|_p \leq e^{-\lambda(t-t_0)}$$

for all $t \geq t_0$, and therefore

$$\left\| \left(x(t), \phi_{p^*}\left(y(t)\right)\right) \right\|_p \leq \frac{2}{\delta} e^{-\lambda(t-t_0)} \left\| \left(x_0, \phi_{p^*}(y_0)\right) \right\|_p = \beta e^{-\lambda(t-t_0)} \left\| \left(x_0, \phi_{p^*}(y_0)\right) \right\|_p$$

for all $t \geq t_0$. This completes the proof of Theorem 4.1. □

Theorem 4.1 is a natural generalization of Theorem D with $n = 2$. Actually, in the case that $p = 2$, Theorem 4.1 becomes Theorem D with $n = 2$.

Moreover, let us give some definitions. The zero solution of (1.1) is said to be *globally uniformly attractive* (*or uniformly attractive in the large*) if for any $\alpha > 0$ and any $\varepsilon > 0$, there exists a $T(\alpha, \varepsilon) > 0$ such that $t_0 \in I$ and $\|\mathbf{x}_0\| < \alpha$ imply $\|\mathbf{x}(t;t_0,\mathbf{x}_0)\| < \varepsilon$ for all $t \geq t_0 + T(\alpha, \varepsilon)$. The solutions of (1.1) are said to be *uniformly bounded* if, for any $\alpha > 0$, there exists a $B(\alpha) > 0$ such that $t_0 \in I$ and $\|\mathbf{x}_0\| < \alpha$ imply $\|\mathbf{x}(t;t_0,\mathbf{x}_0)\| < B(\alpha)$ for all $t \geq t_0$.

The zero solution of (1.1) is *globally uniformly asymptotically stable* (*or uniformly asymptotically stable in the large*) if it is globally uniformly attractive and is uniformly stable and if the solutions of (1.1) are uniformly bounded. For example, we can refer to the books and papers [7, 27–31, 33, 36, 37, 47] for those definitions. When restricted to the case of the linear system (1.3), the following facts are well known.

Theorem F *If the zero solution of* (1.3) *is uniformly asymptotically stable, then it is globally uniformly asymptotically stable.*

We can state a natural generalization of Theorem F with $n = 2$ as follows.

Theorem 4.2 *If the zero solution of* (1.1) *is uniformly asymptotically stable, then it is globally uniformly asymptotically stable.*

Proof By virtue of Theorem 4.1, if the zero solution of (1.1) is uniformly asymptotically stable, then there exist a $\lambda > 0$ and a $\beta > 0$ such that $t_0 \in I$ and $(\xi, \eta) \in \mathbb{R}^2$ imply

$$\left\| \left(x(t;t_0,\xi,\eta), \phi_{p^*}\left(y(t;t_0,\xi,\eta)\right)\right) \right\|_p \le \beta e^{-\lambda(t-t_0)} \left\| \left(\xi, \phi_{p^*}(\eta)\right) \right\|_p$$

for all $t \ge t_0$. We have only to show that the zero solution of (1.1) is globally uniformly attractive and the solutions of (1.1) are uniformly bounded.

First we will prove the global uniform attractivity. Let $\varepsilon > 0$ and $\alpha > 0$ be given. We now consider the solution $(x(t;t_0,x_0,y_0), y(t;t_0,x_0,y_0))$ of (1.1) with $t_0 \in I$ and $\|(x_0,y_0)\| < \alpha$. For the sake of convenience, we write

$$\left(x(t), y(t)\right) = \left(x(t;t_0,x_0,y_0), y(t;t_0,x_0,y_0)\right), \qquad c(\alpha) = \left\| \left(\alpha, \phi_{p^*}(\alpha)\right) \right\|_p.$$

We choose a $T(\varepsilon,\alpha)$ such that

$$T(\varepsilon,\alpha) = \frac{1}{\min\{1,p-1\}\lambda} \log \frac{\sqrt{(\beta c(\alpha))^2 + (\beta c(\alpha))^{2(p-1)}}}{\varepsilon}.$$

Since $|x_0| < \alpha$ and $|y_0| < \alpha$, we get

$$\left\| \left(x_0, \phi_{p^*}(y_0)\right) \right\|_p < c(\alpha).$$

We therefore conclude that

$$|x(t)| \le \left\| \left(x(t), \phi_{p^*}(y(t))\right) \right\|_p < \beta c(\alpha) e^{-\lambda(t-t_0)}$$

and

$$|y(t)| \le \phi_p\left(\left\| \left(x(t), \phi_{p^*}(y(t))\right) \right\|_p\right) < \left(\beta c(\alpha) e^{-\lambda(t-t_0)}\right)^{p-1}$$

for $t \ge t_0$. From these inequalities, we see that

$$\left\| \left(x(t), y(t)\right) \right\| < e^{-\min\{1,p-1\}\lambda(t-t_0)} \sqrt{\left(\beta c(\alpha)\right)^2 + \left(\beta c(\alpha)\right)^{2(p-1)}}$$

for $t \geq t_0$. Hence, we obtain

$$\left\| \big(x(t), y(t)\big) \right\| < \varepsilon$$

for $t \geq t_0 + T$.

We next prove the uniform boundedness. Let $\alpha > 0$ be given. As mentioned in the proof of global uniform attractivity, we see that

$$\left\| \big(x(t), y(t)\big) \right\| < e^{-\min\{1,p-1\}\lambda(t-t_0)} \sqrt{\big(\beta c(\alpha)\big)^2 + \big(\beta c(\alpha)\big)^{2(p-1)}}$$

for $t \geq t_0$ and $\|(x_0, y_0)\| < \alpha$, where

$$\big(x(t), y(t)\big) = \big(x(t; t_0, x_0, y_0), y(t; t_0, x_0, y_0)\big), \qquad c(\alpha) = \left\| \big(\alpha, \phi_{p^*}(\alpha)\big) \right\|_p.$$

We choose a $B(\alpha) > 0$ such that

$$B(\alpha) = \sqrt{\big(\beta c(\alpha)\big)^2 + \big(\beta c(\alpha)\big)^{2(p-1)}}.$$

Hence, we obtain

$$\left\| \big(x(t; t_0, x_0, y_0), y(t; t_0, x_0, y_0)\big) \right\| < B(\alpha)$$

for $t \geq t_0$. This completes the proof of Theorem 4.2. □

The claim of Theorem 4.2 contributes to clear that uniform asymptotic stability and global uniform asymptotic stability are completely same concept for the half-linear differential system (1.1). Moreover, we can conclude that uniform asymptotic stability, exponential stability and global exponential stability are equivalent for the half-linear differential system (1.1) from the results of Theorems 1.1 and 4.1.

5 Converse theorems on exponential stability

In this section, for comparison with Theorems B, C and E, we will discuss the converse theorems for half-linear system (1.1). Recall that the right-hand side of (1.1) does not satisfy Lipschitz condition at the origin. For this reason, unfortunately, Theorems B, C and E cannot apply to (1.1). First, let us consider the existence of a Lyapunov function estimated by the form $\|\mathbf{x}\|_p^p$. For this purpose, we give a lemma as follows.

Lemma 5.1 *If* $(x(t), y(t))$ *is a solution of* (1.1) *passing through a point* $(x_0, y_0) \in \mathbb{R}^2$ *at* $t = t_0 \in I$*, then the following inequality holds*:

$$\left\| \big(x(t), \phi_{p^*}(y(t))\big) \right\|_p^p \geq \left\| \big(x_0, \phi_{p^*}(y_0)\big) \right\|_p^p \exp\left(\int_{t_0}^t \psi(s)\, ds \right)$$

for $t \geq t_0$*, where the continuous function* $\psi(t)$ *defined by*

$$\psi(t) = \min\big\{p a_{11}(t) - \big|(p-1)a_{12}(t) + a_{21}(t)\big|, p^* a_{22}(t) - \big|a_{12}(t) + (p^*-1)a_{21}(t)\big|\big\}.$$

Proof Define

$$w(t) = \| (x(t), \phi_{p^*}(y(t))) \|_p^p = |x(t)|^p + |y(t)|^{p^*}$$

for $t \geq t_0$. From the right-hand side of (1.1), we have

$$w'(t) = pa_{11}(t)|x(t)|^p + p^* a_{22}(t)|y(t)|^{p^*} + (pa_{12}(t) + p^* a_{21}(t))\phi_p(x(t))\phi_{p^*}(y(t))$$

for $t \geq t_0$. Using the classical Young inequality, we obtain

$$\begin{aligned} w'(t) &\geq pa_{11}(t)|x(t)|^p + p^* a_{22}(t)|y(t)|^{p^*} - |pa_{12}(t) + p^* a_{21}(t)||x(t)|^{p-1}|y(t)|^{p^*-1} \\ &\geq pa_{11}(t)|x(t)|^p + p^* a_{22}(t)|y(t)|^{p^*} \\ &\quad - |pa_{12}(t) + p^* a_{21}(t)| \left(\frac{|x(t)|^{(p-1)p^*}}{p^*} + \frac{|y(t)|^{(p^*-1)p}}{p} \right) \\ &= (pa_{11}(t) - |(p-1)a_{12}(t) + a_{21}(t)|)|x(t)|^p \\ &\quad + (p^* a_{22}(t) - |a_{12}(t) + (p^*-1)a_{21}(t)|)|y(t)|^{p^*} \\ &\geq \psi(t)w(t) \end{aligned}$$

for $t \geq t_0$. Therefore, we get

$$\left(\exp\left(-\int_{t_0}^t \psi(s)\, ds \right) w(t) \right)' \geq 0$$

for $t \geq t_0$. Integrate this inequality from t_0 to t to obtain

$$w(t) \geq w(t_0) \exp\left(\int_{t_0}^t \psi(s)\, ds \right) = \| (x_0, \phi_{p^*}(y_0)) \|_p^p \exp\left(\int_{t_0}^t \psi(s)\, ds \right)$$

for $t \geq t_0$. □

The first converse theorem of this section is as follows. We can prove this theorem without requiring the uniqueness of solutions of (1.1) for the initial value problem.

Theorem 5.2 *Suppose that all coefficients of* (1.1) *are bounded on I and that there exist a* $\lambda > 0$ *and a* $\beta > 0$ *such that*

$$\| (x(t; t_0, x_0, y_0), \phi_{p^*}(y(t; t_0, x_0, y_0))) \|_p \leq \beta e^{-\lambda(t-t_0)} \| (x_0, \phi_{p^*}(y_0)) \|_p$$

for all $t \geq t_0 \geq 0$, *where* $(x(t; t_0, x_0, y_0), y(t; t_0, x_0, y_0))$ *is a solution of* (1.1). *Then there exist three positive constants* β_1, β_2, β_3 *and a Lyapunov function* $V(t, x, y)$ *defined on* $I \times \mathbb{R}^2$ *which satisfies the following conditions*:

(i) $\beta_1 \| (x, \phi_{p^*}(y)) \|_p^p \leq V(t, x, y) \leq \beta_2 \| (x, \phi_{p^*}(y)) \|_p^p$;

(ii) $\dot{V}_{(1.1)}(t, x, y) \leq -\beta_3 \| (x, \phi_{p^*}(y)) \|_p^p$.

Proof From the assumption, all solutions $(x(s;t,x,y), y(s;t,x,y))$ of (1.1) passing through a point $(x,y) \in \mathbb{R}^2$ at $t \in I$ satisfy

$$\left\| \left(x(s;t,x,y), \phi_{p^*}\left(y(s;t,x,y)\right)\right) \right\|_p \le \beta \left\| \left(x, \phi_{p^*}(y)\right) \right\|_p$$

for $s \ge t$. Therefore, we can consider the function

$$\overline{v}(s;t,x,y) = \sup \left\| \left(x(s;t,x,y), \phi_{p^*}\left(y(s;t,x,y)\right)\right) \right\|_p$$

for $s \ge t$. Note that if we suppose the uniqueness of solutions of (1.1) for the initial value problem, then $\overline{v}(s;t,x,y) = \|(x(s;t,x,y), \phi_{p^*}(y(s;t,x,y)))\|_p$ holds for $s \ge t$. However, we can prove this theorem without requiring the uniqueness of solutions. Let $V(t,x,y)$ be defined by

$$V(t,x,y) = \int_t^{t+T} \overline{v}^p(s;t,x,y)\, ds,$$

where $T = (1/\lambda)\log(\beta \sqrt[p]{2})$ is a constant. From the assumption, we obtain the following estimate:

$$\begin{aligned} V(t,x,y) &\le \int_t^{t+T} e^{-p\lambda(s-t)}\, ds \left(\beta \left\| \left(x, \phi_{p^*}(y)\right) \right\|_p\right)^p \\ &= \frac{\beta^p(1-e^{-p\lambda T})}{p\lambda} \left\| \left(x, \phi_{p^*}(y)\right) \right\|_p^p = \beta_2 \left\| \left(x, \phi_{p^*}(y)\right) \right\|_p^p. \end{aligned}$$

We will show that $\beta_1 \|(x, \phi_{p^*}(y))\|_p^p \le V(t,x,y)$. Since all coefficients of (1.1) are bounded on I, there exists an $L > 0$ such that $|\psi(s)| \le L$ for all $s \in I$, where ψ is the continuous function given in Lemma 5.1. From Lemma 5.1, we have

$$\overline{v}^p(s;t,x,y) \ge \exp\left(\int_t^s \psi(\tau)\, d\tau\right) \left\| \left(x, \phi_{p^*}(y)\right) \right\|_p^p \ge e^{-L(s-t)} \left\| \left(x, \phi_{p^*}(y)\right) \right\|_p^p$$

for $s \ge t$. Thus, we get

$$\begin{aligned} V(t,x,y) &\ge \int_t^{t+T} e^{-L(s-t)}\, ds \left\| \left(x, \phi_{p^*}(y)\right) \right\|_p^p \\ &= \frac{1-e^{-LT}}{L} \left\| \left(x, \phi_{p^*}(y)\right) \right\|_p^p = \beta_1 \left\| \left(x, \phi_{p^*}(y)\right) \right\|_p^p. \end{aligned}$$

Therefore, condition (i) is satisfied.

We next prove the condition (ii). Let $h > 0$ and

$$\left(x(s), y(s)\right) = \left(x(s;t,x,y), y(s;t,x,y)\right) \quad \text{for } s \ge t.$$

From the definition of $\overline{v}$, we see that

$$\overline{v}\left(s;u,x(u),y(u)\right) \le \overline{v}(s;t,x,y)$$

for $t \leq u \leq s$. Then we get

$$\begin{aligned}
V\big(t+h, x(t+h), y(t+h)\big) &= \int_{t+h}^{t+h+T} \overline{v}^p\big(s; t+h, x(t+h), y(t+h)\big)\, ds \\
&\leq \int_{t+h}^{t+h+T} \overline{v}^p(s; t, x, y)\, ds \\
&= V(t,x,y) + \int_{t+T}^{t+h+T} \overline{v}^p(s; t, x, y)\, ds - \int_{t}^{t+h} \overline{v}^p(s; t, x, y)\, ds \\
&\leq V(t,x,y) + \int_{t+T}^{t+h+T} \beta^p e^{-p\lambda(s-t)} \big\| (x, \phi_{p^*}(y)) \big\|_p^p\, ds \\
&\quad - \int_{t}^{t+h} e^{-L(s-t)} \big\| (x, \phi_{p^*}(y)) \big\|_p^p\, ds \\
&= V(t,x,y) \\
&\quad + \left(\frac{\beta^p e^{-p\lambda T}(1 - e^{-p\lambda h})}{p\lambda} - \frac{1 - e^{-Lh}}{L} \right) \big\| (x, \phi_{p^*}(y)) \big\|_p^p
\end{aligned}$$

from the assumption and Lemma 5.1. Therefore, we can estimate that

$$\begin{aligned}
&\frac{1}{h}\big(V\big(t+h, x(t+h), y(t+h)\big) - V(t,x,y)\big) \\
&\leq \left(\frac{\beta^p e^{-p\lambda T}(1 - e^{-p\lambda h})}{p\lambda h} - \frac{1 - e^{-Lh}}{Lh} \right) \big\| (x, \phi_{p^*}(y)) \big\|_p^p.
\end{aligned}$$

From this inequality and

$$\lim_{h \to 0} \left(\frac{\beta^p e^{-p\lambda T}(1 - e^{-p\lambda h})}{p\lambda h} - \frac{1 - e^{-Lh}}{Lh} \right) = \beta^p e^{-p\lambda T} - 1 = -\frac{1}{2},$$

we obtain

$$\dot{V}_{(1.1)}(t,x,y) \leq -\frac{1}{2} \big\| (x, \phi_{p^*}(y)) \big\|_p^p = -\beta_3 \big\| (x, \phi_{p^*}(y)) \big\|_p^p.$$

This completes the proof of Theorem 5.2. □

Note that three positive constants β_1, β_2, β_3 in Theorem 5.2 are independent of the size of $\|(x_0, \phi_{p^*}(y_0))\|_p$. The second converse theorem is as follows.

Theorem 5.3 *Suppose that there exist a $\lambda > 0$ and a $\beta > 0$ such that*

$$\big\| \big(x(t; t_0, x_0, y_0), \phi_{p^*}\big(y(t; t_0, x_0, y_0)\big)\big) \big\|_p \leq \beta e^{-\lambda(t-t_0)} \big\| \big(x_0, \phi_{p^*}(y_0)\big) \big\|_p$$

for all $t \geq t_0 \geq 0$, where $(x(t; t_0, x_0, y_0), y(t; t_0, x_0, y_0))$ is a solution of (1.1). *Then there exists a Lyapunov function $V(t,x,y)$ defined on $I \times \mathbb{R}^2$ which satisfies the following conditions:*

(i) $\|(x, \phi_{p^*}(y))\|_p \leq V(t,x,y) \leq \beta \|(x, \phi_{p^*}(y))\|_p$;

(ii) $\dot{V}_{(1.1)}(t,x,y) \leq -\lambda V(t,x,y)$.

Proof Let $V(t,x,y)$ be defined by

$$V(t,x,y) = \sup_{\sigma \geq 0} \overline{v}(t+\sigma; t, x, y) e^{\lambda\sigma},$$

where $\overline{v}$ is the function given in the proof of Theorem 5.2. It is clear that $\|(x,\phi_{p^*}(y))\|_p \le V(t,x,y)$. On the other hand, we can easy to see that

$$V(t,x,y) \le \sup_{\sigma\ge 0} \beta e^{-\lambda\sigma} \left\|(x,\phi_{p^*}(y))\right\|_p e^{\lambda\sigma} = \beta \left\|(x,\phi_{p^*}(y))\right\|_p,$$

by the assumption.

We next prove the condition (ii). Let $h > 0$ and

$$\bigl(x(s),y(s)\bigr) = \bigl(x(s;t,x,y),y(s;t,x,y)\bigr) \quad \text{for } s \ge t.$$

Then we get

$$\begin{aligned} V\bigl(t+h,x(t+h),y(t+h)\bigr) &= \sup_{\sigma\ge 0} \overline{v}\bigl(t+h+\sigma;t+h,x(t+h),y(t+h)\bigr)e^{\lambda\sigma} \\ &= \sup_{\tau\ge h} \overline{v}\bigl(t+\tau;t+h,x(t+h),y(t+h)\bigr)e^{\lambda\tau}e^{-\lambda h} \\ &\le \sup_{\tau\ge h} \overline{v}(t+\tau;t,x,y)e^{\lambda\tau}e^{-\lambda h} \\ &\le \sup_{\tau\ge 0} \overline{v}(t+\tau;t,x,y)e^{\lambda\tau}e^{-\lambda h} = V(t,x,y)e^{-\lambda h} \end{aligned}$$

from $\overline{v}(s;u,x(u),y(u)) \le \overline{v}(s;t,x,y)$ for $t \le u \le s$. Therefore, we can estimate that

$$\frac{V(t+h,x(t+h),y(t+h)) - V(t,x,y)}{h} \le \frac{e^{-\lambda h}-1}{h} V(t,x,y).$$

From this inequality and

$$\lim_{h\to 0} \frac{e^{-\lambda h}-1}{h} = -\lambda,$$

we obtain

$$\dot{V}_{(1.1)}(t,x,y) \le -\lambda V(t,x,y).$$

This completes the proof of Theorem 5.3. □

Competing interests
The authors declare that they have no competing interests.

Authors' contributions
Both authors contributed equally to the writing of this paper. Both authors read and approved the final manuscript.

Acknowledgements
The first author's work was supported in part by Grant-in-Aid for Young Scientists (B), No. 23740115 from the Japan Society for the Promotion of Science (JSPS). The authors thank the referee for useful and valuable comments.

References

1. Došlý, O: Half-linear differential equations. In: Handbook of Differential Equations, pp. 161-357. Elsevier, Amsterdam (2004)
2. Došlý, O, Řehák, P: Half-Linear Differential Equations. North-Holland Mathematics Studies, vol. 202. Elsevier, Amsterdam (2005)

3. Elbert, Á: Asymptotic behaviour of autonomous half-linear differential systems on the plane. Studia Sci. Math. Hung. **19**(2-4), 447-464 (1984)
4. Mirzov, JD: On some analogs of Sturm's and Kneser's theorems for nonlinear systems. J. Math. Anal. Appl. **53**(2), 418-425 (1976)
5. Mirzov, JD: Principal and nonprincipal solutions of a nonlinear system. Tbil. Gos. Univ. Inst. Prikl. Mat. Tr. **31**, 100-117 (1988) (Russian)
6. Mirzov, JD: Asymptotic Properties of Solutions of Systems of Nonlinear Nonautonomous Ordinary Differential Equations. Folia Facultatis Scientiarium Naturalium Universitatis Masarykianae Brunensis. Mathematica, vol. 14. Masaryk University, Brno (2004)
7. Onitsuka, M, Sugie, J: Uniform global asymptotic stability for half-linear differential systems with time-varying coefficients. Proc. R. Soc. Edinb., Sect. A **141**(5), 1083-1101 (2011)
8. Sugie, J, Hata, S, Onitsuka, M: Global attractivity for half-linear differential systems with periodic coefficients. J. Math. Anal. Appl. **371**(1), 95-112 (2010)
9. Sugie, J, Onitsuka, M: Global asymptotic stability for half-linear differential systems with coefficients of indefinite sign. Arch. Math. **44**(4), 317-334 (2008)
10. Agarwal, RP, Grace, SR, O'Regan, D: Oscillation Theory for Second Order Linear, Half-Linear, Superlinear and Sublinear Dynamic Equations. Kluwer Academic, Dordrecht (2002)
11. Elbert, Á: A half-linear second order differential equation. In: Qualitative Theory of Differential Equations (Szeged, 1979), vols. I, II. Colloq. Math. Soc. János Bolyai, vol. 30, pp. 153-180. North-Holland, Amsterdam (1981)
12. Hasil, P, Mařík, R, Veselý, M: Conditional oscillation of half-linear differential equations with coefficients having mean values. Abstr. Appl. Anal. **2014**, Article ID 258159 (2014)
13. Hatvani, L: On stability properties of solutions of second order differential equations. Proceedings of the 6th Colloquium on the Qualitative Theory of Differential Equations (Szeged, 1999). Electron. J. Qual. Theory Differ. Equ. **2000**, 11 (2000)
14. Pašić, M, Wong, JSW: Rectifiable oscillations in second-order half-linear differential equations. Ann. Mat. Pura Appl. (4) **188**(3), 517-541 (2009)
15. Řehák, P: De Haan type increasing solutions of half-linear differential equations. J. Math. Anal. Appl. **412**(1), 236-243 (2014)
16. Sugie, J, Matsumura, K: A nonoscillation theorem for half-linear differential equations with periodic coefficients. Appl. Math. Comput. **199**(2), 447-455 (2008)
17. Sugie, J, Onitsuka, M: Global asymptotic stability for damped half-linear differential equations. Acta Sci. Math. **73**(3-4), 613-636 (2007)
18. Sugie, J, Onitsuka, M: Integral conditions on the uniform asymptotic stability for two-dimensional linear systems with time-varying coefficients. Proc. Am. Math. Soc. **138**(7), 2493-2503 (2010)
19. Sugie, J, Onitsuka, M: Growth conditions for uniform asymptotic stability of damped oscillators. Nonlinear Anal. **98**, 83-103 (2014)
20. Agarwal, RP, Bohner, M, Li, T: Oscillatory behavior of second-order half-linear damped dynamic equations. Appl. Math. Comput. **254**, 408-418 (2015)
21. Došlý, O, Fišnarová, S: Linearized Riccati technique and (non-)oscillation criteria for half-linear difference equations. Adv. Differ. Equ. **2008**, Article ID 438130 (2008)
22. Došlý, O, Yamaoka, N: Oscillation constants for second-order ordinary differential equations related to elliptic equations with p-Laplacian. Nonlinear Anal. **113**, 115-136 (2015)
23. Fišnarová, S: Oscillatory properties of half-linear difference equations: two-term perturbations. Adv. Differ. Equ. **2012**, 101 (2012)
24. Hasil, P, Mařík, R, Veselý, M: Oscillation and nonoscillation of asymptotically almost periodic half-linear difference equations. Abstr. Appl. Anal. **2013**, Article ID 432936 (2014)
25. Halanay, A: Differential Equations: Stability, Oscillations, Time Lags. Academic Press, New York (1966)
26. Lefschetz, S: Differential Equations: Geometric Theory, 2nd edn. Pure and Applied Mathematics, vol. 6. Interscience, New York (1963)
27. Sansone, G, Conti, R: Non-linear Differential Equations. International Series of Monographs in Pure and Applied Mathematics, vol. 67. Macmillan Co., New York (1964). Revised ed., Translated from the Italian by Ainsley H Diamond
28. Yoshizawa, T: Stability Theory by Liapunov's Second Method. The Mathematical Society of Japan, Tokyo (1966)
29. Yoshizawa, T: Stability Theory and the Existence of Periodic Solutions and Almost Periodic Solutions. Applied Mathematical Sciences, vol. 14. Springer, New York (1975)
30. Bacciotti, A, Rosier, L: Liapunov Functions and Stability in Control Theory, 2nd edn. Communications and Control Engineering Series. Springer, Berlin (2005)
31. Hahn, W: Stability of Motion. Die Grundlehren der Mathematischen Wissenschaften, vol. 138. Springer, New York (1967). Translated from the German manuscript by Arne P Baartz
32. Hale, JK: Ordinary Differential Equations. Pure and Applied Mathematics, vol. 11. Wiley-Interscience, New York (1969)
33. Loría, A, Panteley, E: Stability, told by its developers. In: Advanced Topics in Control Systems Theory. Lecture Notes in Control and Inform. Sci., vol. 328, pp. 199-258. Springer, London (2006)
34. Malisoff, M, Mazenc, F: Constructions of Strict Lyapunov Functions. Communications and Control Engineering Series. Springer, London (2009)
35. Michel, AN, Hou, L, Liu, D: Stability of Dynamical Systems: Continuous, Discontinuous, and Discrete Systems. Systems and Control: Foundations and Applications. Birkhäuser Boston, Boston (2008)
36. Rouche, N, Habets, P, Laloy, M: Stability Theory by Liapunov's Direct Method. Applied Mathematical Sciences, vol. 22. Springer, New York (1977)
37. Harris, CJ, Miles, JF: Stability of Linear Systems. Mathematics in Science and Engineering, vol. 153. Academic Press, New York (1980)
38. Logemann, H, Ryan, EP: Ordinary Differential Equations: Analysis, Qualitative Theory and Control. Springer Undergraduate Mathematics Series. Springer, London (2014)
39. Barreira, L, Pesin, Y: Introduction to Smooth Ergodic Theory. Graduate Studies in Mathematics, vol. 148. Am. Math. Soc., Providence (2013)

40. Burton, TA: Stability and Periodic Solutions of Ordinary and Functional Differential Equations. Mathematics in Science and Engineering, vol. 178. Academic Press, San Diego (1985)
41. Cesari, L: Asymptotic Behavior and Stability Problems in Ordinary Differential Equations, 3rd edn. Ergebnisse der Mathematik und ihrer Grenzgebiete, vol. 16. Springer, New York (1971)
42. Coppel, WA: Stability and Asymptotic Behavior of Differential Equations. Heath, Boston (1965)
43. Halanay, A, Răsvan, V: Applications of Liapunov Methods in Stability. Mathematics and Its Applications, vol. 245. Kluwer Academic, Dordrecht (1993)
44. Hatvani, L: On the uniform attractivity of solutions of ordinary differential equations by two Lyapunov functions. Proc. Jpn. Acad., Ser. A, Math. Sci. **67**(5), 162-167 (1991)
45. Ignatyev, AO: Stability of a linear oscillator with variable parameters. Electron. J. Differ. Equ. **1997**, 17 (1997)
46. LaSalle, JP, Lefschetz, S: Stability by Liapunov's Direct Method with Applications. Mathematics in Science and Engineering, vol. 4. Academic Press, New York (1961)
47. Malkin, IG: On the question of the reciprocal of Lyapunov's theorem on asymptotic stability. Prikl. Mat. Meh. **18**, 129-138 (1954)
48. Massera, JL: Contributions to stability theory. Ann. Math. (2) **64**, 182-206 (1956)
49. Onitsuka, M: Non-uniform asymptotic stability for the damped linear oscillator. Nonlinear Anal. **72**(3-4), 1266-1274 (2010)
50. Onitsuka, M: Uniform asymptotic stability for damped linear oscillators with variable parameters. Appl. Math. Comput. **218**(4), 1436-1442 (2011)
51. Strauss, A, Yorke, JA: Perturbing uniform asymptotically stable nonlinear systems. J. Differ. Equ. **6**, 452-483 (1969)
52. Sugie, J, Ogami, Y, Onitsuka, M: Asymptotic stability for quasi-linear systems whose linear approximation is not assumed to be uniformly attractive. Ann. Mat. Pura Appl. (4) **190**(3), 409-425 (2011)
53. Yang, T: Impulsive Control Theory. Lecture Notes in Control and Information Sciences, vol. 272. Springer, Berlin (2001)
54. Agarwal, RP, O'Regan, D: An Introduction to Ordinary Differential Equations. Springer, New York (2008)
55. Mahmudov, E: Single Variable Differential and Integral Calculus: Mathematical Analysis. Atlantis Press, Paris (2013)

16

Periodic solutions for nonautonomous first order delay differential systems via Hamiltonian systems

Qiong Meng*

*Correspondence:
mengqiong@qq.com
School of Mathematical Science,
Shanxi University, Wucheng Road,
Taiyuan, Shanxi, China

Abstract

The existence of the nontrivial periodic solutions for the nonautonomous first order delay differential equation $x'(t) = -[f(t,x(t-1)) + f(t,x(t-2)) + \cdots + f(t,x(t-(2N-1)))]$ is investigated, where $f \in C(\mathbf{R} \times \mathbf{R}, \mathbf{R})$ is $2N$-periodic in t and odd in x, N is a positive integer. We prove several new existence results by some recent critical point theorems.

MSC: 34K13; 34K18; 58E50

Keywords: first order delay differential equation; nonautonomous; periodic solution; critical point theorems

1 Introduction

It is well known that critical point theory is a powerful tool that deals with the multiplicity of periodic solutions to ordinary differential systems as well as partial differential equations (see [1–6]). In 1998, Li and He [7] firstly applied the critical point theory to study the multiplicity of periodic solutions to delay differential equations. Especially, in 2005, Guo and Yu [8] established a variational framework for autonomous systems. In the past several years, some results on the existence of periodic solutions for the functional differential equation were obtained by critical point theory (see [7–21]).

In 2006, Fei [11, 12] considered the following autonomous functional differential equation:

$$(*) \quad x'(t) = -[f(x(t-1)) + f(x(t-2)) + \cdots + f(x(t-(n-1)))],$$

where f is odd and $n \geq 2$ is an integer. Using the Yorke-Kaplan technique, $(*)$ is changed into a Hamiltonian system. Periodic solutions of the Hamiltonian systems are still obtained as critical points of a function φ over a Hilbert space E. However, instead of finding critical points of φ over E directly, the author worked on a subspace of E which has a symmetric structure. When n is even, the function φ is invariant and φ is equivariant about a compact group. This allows one to find critical points of φ on a subspace of E which is invariant under the mentioned group action. Then one can apply the pseudo-index theory [2] to obtain periodic solutions in this subspace, which surely have the required symmetric structure and give solutions to $(*)$. When n is odd, the function φ is still invariant about a similar compact group. However, φ is not equivariant about this compact group action anymore. Therefore one cannot directly apply the same idea as in the case when n is even. In order to overcome the difficulty, one has to construct equivariant pseudo-gradient vector fields and prove a new deformation theorem. Then one can combine a Galerkin approximation with the S^1-index theory ([4], Chapter 6) to obtain critical points of φ with the required symmetric structure (for details see [12]).

But most of these functional differential equation are autonomous, and the results on the nonautonomous functional differential equation are relatively few (see [9, 15–19]).

Motivated by the work of [10–12, 15–18], we consider a class of nonautonomous first order delay differential equations,

$$x'(t) = -[f(t,x(t-1)) + f(t,x(t-2)) + \cdots + f(t,x(t-(2N-1)))], \tag{1.1}$$

where $f \in C(\mathbf{R} \times \mathbf{R}, \mathbf{R})$ is $2N$-periodic in t and odd in $x, N \in \mathbf{Z}^+$ and $\mathbf{Z}^+$ is the set of all positive integers.

More precisely, if the solution $x(t)$ of (1.1) satisfies $x(t) = -x(t-2N)$, let

$$x_1(t) = x(t), \quad x_2(t) = x(t-1), \quad \ldots, \quad x_{2N}(t) = x(t-(2N-1)), \tag{1.2}$$

then $X(t) = (x_1(t), x_2(t), \ldots, x_{2N}(t))^T$ satisfies

$$\frac{d}{dt}X(t) = A_{2N}G(t,X), \quad \text{where } A_{2N} = \begin{pmatrix} 0 & -1 & \cdots & -1 \\ 1 & 0 & \ddots & \vdots \\ \vdots & \ddots & \ddots & -1 \\ 1 & \cdots & 1 & 0 \end{pmatrix}, \tag{1.3}$$

i.e., A_{2N} is a $2N \times 2N$ skew symmetric matrix, and

$$G(t,X) = (f(t,x_1), f(t,x_2), \ldots, f(t,x_{2N}))^T.$$

Also

$$X(t) = (x_1(t), x_2(t), \ldots, x_{2N}(t))^T$$

satisfies the following symmetric structure:

$$\begin{aligned} &x_1(t) = -x_{2N}(t-1), \quad x_2(t) = x_1(t-1), \quad x_3(t) = x_2(t-1), \quad \ldots, \\ &x_{2N}(t) = x_{2N-1}(t-1). \end{aligned} \tag{1.4}$$

In this paper, we have the following conditions on f.

(f1) There exist two $2N$-periodic and continuous functions α, β such that

$$\alpha(t) = \lim_{x\to 0}\frac{f(t,x)}{x}, \qquad \beta(t) = \lim_{x\to\infty}\frac{f(t,x)}{x}$$

uniformly for $t \in [0, 2N]$.

(f2$^{\pm}$) There exist $d_1, d_2 > 0$ such that

$$\pm\left[2F(t,x) - xf(t,x)\right] \geq d_1|x| - d_2, \quad (t,x) \in [0,2N] \times \mathbf{R},$$

where $F(t,x) = \int_0^x f(t,s)\,ds$.

Denote $m^-(t) = 1$ if all $t < 0$; $m^-(t) = 0$ otherwise.

Theorem 1.1 *Assume that f satisfies* (f1)-(f2$^-$) *with*

$$\beta(t) > \frac{\pi}{2N}\tan\frac{\pi}{4N}, \quad \textit{for } t \in [0,2N].$$

Then (1.1) *possesses at least* $2m$ *pairs* $4N$*-periodic solutions with* $x(t) = -x(t-2N)$, *where*

$$m = \sum_{j=1}^{\infty} m^-\left[\frac{\pi(2j-1)}{2N}\tan\frac{\pi(2j-1)}{4N} - \beta(t)\right],$$

for

$$\tan\frac{\pi(2j-1)}{4N} > 0, \quad j \in \mathbf{Z}^+ \textit{ and for all } t \in [0,2N].$$

Theorem 1.2 *Assume that f satisfies* (f1)-(f2$^+$) *with*

$$\beta(t) < \frac{\pi(4N-1)}{2N}\tan\frac{\pi(4N-1)}{4N}, \quad \textit{for } t \in [0,2N].$$

Then (1.1) *possesses at least* $2m$ *pairs* $4N$*-periodic solutions with* $x(t) = -x(t-2N)$, *where*

$$m = \sum_{j=1}^{\infty} m^-\left[\beta(t) - \frac{\pi(2j-1)}{2N}\tan\frac{\pi(2j-1)}{4N}\right],$$

for

$$\tan\frac{\pi(2j-1)}{4N} < 0, \quad j \in \mathbf{Z}^+ \textit{ and for all } t \in [0,2N].$$

In this paper, the main purpose is to study the multiplicity of periodic solutions for the systems (1.1) via some recent critical point theorems for strongly indefinite functionals. In order to achieve this, some preliminaries are necessary. Let X and Y be Banach spaces with X being separable and reflexive, and set $E = X \oplus Y$. Let $\mathcal{S} \subset X^*$ be a dense subset. For each $s \in \mathcal{S}$, there is a semi-norm on E defined by

$$p_s : E \to \mathbf{R}, \quad p_s(u) = |s(x)| + \|y\| \quad \text{for } u = x + y \in X \oplus Y.$$

We denote by $\mathcal{T}_\mathcal{S}$ the topology on E induced by a semi-norm family $\{p_s\}$, and let ω and ω^* denote the weak topology and weak* topology, respectively. Clearly, the $\mathcal{T}_\mathcal{S}$ topology contains the product topology on $E = X \oplus Y$ produced by the weak topology on X and the strong topology on Y.

For a functional $\Phi \in C^1(E, \mathbf{R})$ we write $\Phi_a = \{u \in E : \Phi(u) \geq a\}$. Recall that Φ' is weakly sequentially continuous if $u_k \rightharpoonup u$ in E, and one has $\lim_{k\to\infty} \Phi'(u_k)v \to \Phi'(u_k)v$ for each

$v \in E$, i.e. $\Phi' : (E, \omega) \to (E^*, \omega^*)$ is sequentially continuous. For $c \in \mathbf{R}$ we say that Φ satisfies the $(C)_c$ condition if any sequence $\{u_k\} \subset E$ such that $\Phi(u_k) \to c$ and $(1 + \|u_k\|)\Phi'(u_k) \to 0$ as $k \to \infty$ contains a convergent subsequence.

Suppose that

(Φ_0) $\Phi \in C^1(E, \mathbf{R})$, Φ_c is $\mathcal{T}_S$-closed for every $c \in \mathbf{R}$, and $\Phi' : (\Phi_c, \mathcal{T}_S) \to (E^*, \omega^*)$ is continuous.

(Φ_1) There exists a $\rho > 0$ such that $\kappa := \inf \Phi(B_\rho \cap Y) > 0 = \Phi(0)$, where $B_\rho = \{u \in E : \|u\| = \rho\}$.

(Φ_2) There exist a finite dimensional subspace $Y_0 \subset Y$ and $R > \rho$ such that $\bar{c} := \sup \Phi(E_0) < \infty$ and $\sup \Phi(E_0 \backslash S_0) < \inf \Phi(B_\rho \cap Y)$, where $E_0 := X \oplus Y_0$, and $S_0 = \{u \in E_0 : \|u\| \leq R\}$.

The following critical point theorem will be used later (see [1, 5, 15]).

Theorem A *Assume that Φ is even and (Φ_0)-(Φ_2) are satisfied. Then Φ has at least $m = \dim Y_0$ pairs of critical points with critical values less than or equal to $\bar{c}$ provided Φ satisfies the $(C)_c$ condition for all $c \in [\kappa, \bar{c}]$.*

2 Preliminaries

It is easy to see that A_{2N} is a nonsingular skew symmetric Hamiltonian matrix, and (1.3) becomes a classical Hamiltonian system,

$$z'(t) = A_{2N} \nabla H(t, z), \tag{2.1}$$

where $\nabla H(t, z)$ denotes the gradient of $H(t, z)$ with respect to the z variable, and

$$H(t, z) = F(t, z_1) + F(t, z_2) + \cdots + F(t, z_{2N}), \tag{2.2}$$

for any $z = (z_1, z_2, \ldots, z_{2N})^T \in \mathbf{R}^{2N}$.

For $S^1 = \mathbf{R}/(4N\mathbf{Z})$, let $H^{1/2} = W^{1/2,2}(S^1, \mathbf{R}^{2N})$. Then $H^{1/2}$ is a Hilbert space with norm $\| \cdot \|$ and inner product $\langle , \rangle$, and $H^{1/2}$ consists of those $z(t)$ in $L^2(S^1, \mathbf{R}^{2N})$ whose Fourier series,

$$z(t) = a_0 + \sum_{j=1}^{\infty} \left[a_j \cos\left(\frac{\pi}{2N} jt\right) + a_j \cos\left(\frac{\pi}{2N} jt\right) \right],$$

satisfies

$$\|z\| = 4N|a_0|^2 + 2N \sum_{j=1}^{\infty} j\left(|a_j|^2 + |b_j|^2\right) < \infty,$$

where $a_j, b_j \in \mathbf{R}^{2N}$. By Proposition 6.6 [4], we know that $H^{1/2}$ is compactly embedded in $L^s(S^1, \mathbf{R}^{2N})$, where $s \in [1, \infty)$.

Since A_{2N} is a nonsingular skew, so is A_{2N}^{-1}. We define an operator L by extending the bilinear form

$$\langle Lz, y \rangle = \int_0^{4N} \left(-A_{2N}^{-1} \dot{z}(t), y(t)\right) dt. \tag{2.3}$$

By direct computation, L is a bounded self-adjoint linear operator on $H^{1/2}$.

By (f1), one can show that $H(t,z) \in C^1(\mathbf{R} \times \mathbf{R}^{2N}, \mathbf{R})$ and there exist two continuous functions $a(t) > 0$, $b(t) \geq 0$ such that

$$H(t,z) \leq a(t)|z|^2 + b(t), \quad \forall z \in \mathbf{R}^{2N}.$$

By using similar arguments as in [4], Chapter 6, let

$$\varphi(z) = \frac{1}{2}\langle Lz, z\rangle + \psi(z), \tag{2.4}$$

where

$$\psi(z) = \int_0^{4N} H(t, z(t))\, dt. \tag{2.5}$$

Thus the critical points of $\varphi(z)$ in $H^{1/2}$ are classical solutions of (2.1).

Let T_{2N} be the $2N \times 2N$ matrix given

$$T_{2N} = \begin{pmatrix} 0 & 0 & \cdots & 0 & -1 \\ 1 & 0 & \ddots & 0 & 0 \\ 0 & 1 & \ddots & \vdots & \vdots \\ \vdots & \ddots & \ddots & 0 & 0 \\ 0 & \ldots & 0 & 1 & 0 \end{pmatrix}.$$

For $z(t) \in H^{1/2}$, define by

$$\delta z(t) = T_{2N} z(t-1). \tag{2.6}$$

Then $\delta^{2N} z(t) = -z(t-2N)$, $\delta^{4N} z(t) = z(t)$, and $G = \{\delta, \delta^2, \ldots, \delta^{4N}\}$ is a compact group action over $H^{1/2}$. Moreover, if $\delta z(t) = z(t)$ holds, $z(t)$ has the symmetric structure (1.4).

Lemma 2.1 [11] *Denote $E = \{z(t) \in H^{1/2} : \delta z(t) = z(t)\}$. Then*

$$E = \left\{ z(t) \in H^{1/2}(S^1, \mathbf{R}^{2N}) : z(t) = \sum_{j=1}^{\infty} \left[a_j \cos\left(\frac{\pi}{2N}(2j-1)t\right) + b_j \sin\left(\frac{\pi}{2N}(2j-1)t\right) \right] : \right.$$
$$\left. \begin{pmatrix} a_j \\ b_j \end{pmatrix} \in \operatorname{span}\left\{ \begin{pmatrix} u_j \\ w_j \end{pmatrix}, \begin{pmatrix} -w_j \\ u_j \end{pmatrix} \right\} \right\},$$

where $\beta_j = \frac{\pi}{2N}(2j-1)$ and

$$u_j = \left(1, \cos\beta_j, \cos(2\beta_j), \ldots, \cos(2N-1)\beta_j\right)^T,$$
$$w_j = \left(0, \sin\beta_j, \sin(2\beta_j), \ldots, \sin(2N-1)\beta_j\right)^T.$$

Then we have the following lemma by using similar arguments [11].

Lemma 2.2 *If $z(t)$ is a critical point of φ in E, then $z(t)$ is a critical point of φ in $H^{1/2}$.*

Proof By (2.2) and direct computation, we have

$$A_{2N}T_{2N} = T_{2N}A_{2N}, \qquad H(t, T_{2N}z) = H(t,z), \qquad \nabla H(t, T_{2N}z) = T_{2N}\nabla H(t,z).$$

Combing these with (2.6) and the fact that any $z(t) \in E$ is $4N$-periodic, one can easily verify that

$$\varphi(\delta z) = \varphi(z), \qquad \varphi'(\delta z) = \delta\varphi'(z),$$

i.e., φ is G-invariant, and φ' is G-equivariant. The conclusion follows directly. □

Remark 2.1 By Lemma 2.2, the function φ is invariant and φ is equivariant about the compact group G. This allows to find critical points of φ on a subspace of E which is invariant under the group G. Then we can apply Theorem A to obtain periodic solutions in this subspace, which surely have the required symmetric structure and give solutions to (1.1).

Moreover, we also denote by $M^+(\cdot)$, $M^-(\cdot)$ and $M^0(\cdot)$ the positive definite, negative definite, and null subspaces of the self-adjoint linear operator defining it, respectively.

Then E has an orthogonal decomposition

$$E = M^+(L) \oplus M^-(L) \oplus M^0(L).$$

By Lemma 2.3 [11], for $z_j(t) = a_j\cos(\frac{\pi}{2N}(2j-1)t) + b_j\sin(\frac{\pi}{2N}(2j-1)t)$ $(j = 1,2,\ldots)$, a_j and b_j are defined in Lemma 2.1, we have

$$\langle Lz_j, z_j\rangle = -\frac{\pi(2j-1)}{2N}\tan\frac{\pi(2j-1)}{4N}\big[(a_j,a_j) + (b_j,b_j)\big]. \tag{2.7}$$

Let

$$\sigma = \min\left\{\left|\frac{\pi(2j-1)}{2N}\tan\frac{\pi(2j-1)}{4N}\right| : j \in Z^+\right\} > 0.$$

So we have

$$\langle Lz_j, z_j\rangle \geq \sigma\|z_j\|^2, \quad \text{for } \gamma_j > 0,$$

$$\langle Lz_j, z_j\rangle \leq -\sigma\|z_j\|^2, \quad \text{for } \gamma_j < 0,$$

where $\gamma_j = -\tan\frac{\pi(2j-1)}{4N}$. Therefore, we get

$$\langle Lz, z\rangle \geq \sigma\|z\|^2, \quad \text{for } z \in M^+(L), \tag{2.8}$$

$$\langle Lz, z\rangle \leq -\sigma\|z\|^2, \quad \text{for } z \in M^-(L). \tag{2.9}$$

3 Proofs of theorems

In this section, c_i stand for different positive constants for $i \in Z^+$.

By direct computation, (f1) implies that $H(t,z)$ is even and satisfies

$$\nabla H(t,z) = \alpha(t)z + o(|z|) \quad \text{as } |z| \to 0, \tag{3.1}$$

$$\nabla H(t,z) = \beta(t)z + o(|z|) \quad \text{as } |z| \to \infty, \tag{3.2}$$

uniformly for $t \in [0,2N]$.

($f2^{\pm}$) implies that

$$2H(t,z) - (\nabla H(t,z),z) \geq d_1|z| - 2Nd_2, \quad \forall(t,z) \in [0,2N] \times \mathbf{R}^{2N} \tag{3.3}$$

and

$$-[2H(t,z) - (\nabla H(t,z),z)] \geq d_1|z| - 2Nd_2, \quad \forall(t,z) \in [0,2N] \times \mathbf{R}^{2N}. \tag{3.4}$$

Remark 3.1 Here we prove that (3.3) and (3.4) hold. By (2.2) and ($f2^{\pm}$), we have

$$\begin{aligned}\pm[2H(t,z) - (\nabla H(t,z),z)] &= \pm\sum_{i=1}^{2N}[2F(t,z_i) - z_i f(t,z_i)] \\ &\geq d_1\sum_{i=1}^{2N}|z_i| - 2Nd_2 \geq d_1|z| - 2Nd_2.\end{aligned}$$

Lemma 3.1 *Suppose that f satisfies* (f1)-($f2^{\pm}$). *Then the function φ satisfies the* $(C)_c$ *condition for any $c \in \mathbf{R}$.*

Proof First, we need some notations. For any $z,y \in E$, define an operator B by extending the bilinear form

$$\langle Bz,y\rangle = \int_0^{4N} (\beta(t)z(t),y(t))\,dt.$$

By direct computation, B is a bounded self-adjoint linear operator on E. Thus $L+B$ is also a self-adjoint linear operator on E.

Let

$$E = M^+(L+B) \oplus M^-(L+B) \oplus M^0(L+B).$$

Also there exists $\sigma_1 > 0$ such that

$$\langle (L+B)z,z\rangle \geq \sigma_1\|z\|^2, \quad \text{for } z \in M^+(L+B), \tag{3.5}$$

$$\langle (L+B)z,z\rangle \leq -\sigma_1\|z\|^2, \quad \text{for } z \in M^-(L+B). \tag{3.6}$$

So we have

$$\Phi(z) = -\varphi(z) = -\frac{1}{2}\langle (L+B)z,z\rangle - \int_0^{4N}\left[H(t,z) - \frac{1}{2}(\beta(t)z,z)\right]dt$$

and

$$\langle \Phi'(z),y\rangle = -\langle (L+B)z,y\rangle - \int_0^{4N}(\nabla H(t,z) - \beta(t)z,y)\,dt.$$

Let $\{z_k\} \subset E$ be any sequence such that

$$\Phi(z_k) \to c, \quad \left(1 + \|z_k\|\right)\Phi'(z_k) \to 0 \quad \text{as } k \to \infty. \tag{3.7}$$

We first prove that $\{z_k\}$ is bounded. Since $z_k \in E$, we have

$$z_k = z_k^+ + z_k^- + z^0 \in M^+(L+B) \oplus M^-(L+B) \oplus M^0(L+B).$$

Let

$$h(z) = \int_0^{4N} \tilde{H}(t,z)\,dt, \qquad \tilde{H}(t,z) = H(t,z) - \frac{1}{2}\left(\beta(t)z, z\right), \quad z \in E.$$

By (3.2), we have

$$\frac{|\nabla \tilde{H}(t,z)|}{|z|} \to 0 \quad \text{as } |z| \to \infty,$$

uniformly for $t \in [0, 2N]$. This means that, for any $\varepsilon > 0$, there exists $M > 0$ such that

$$\left|\nabla \tilde{H}(t,z)\right| \le \varepsilon |z| + M, \quad \text{for all } (t,z) \in [0,2N] \times \mathbf{R}^{2N}. \tag{3.8}$$

Therefore, for any $y \in E$,

$$\begin{aligned}
\left|\langle h'(z_k), y\rangle\right| &= \left|\int_0^{4N} \left(\nabla \tilde{H}(t, z_k(t)), y(t)\right) dt\right| \\
&\le \int_0^{4N} \left|\nabla \tilde{H}(t, z_k(t))\right| \left|y(t)\right| dt \\
&\le \left(\int_0^{4N} \left|\nabla \tilde{H}(t, z_k(t))\right|^2 dt\right)^{1/2} \|y\|_{L^2} \\
&\le \left(2\varepsilon^2 \|z_k\|_{L^2}^2 + 8NM^2\right)^{1/2} \|y\|_{L^2} \\
&\le \left(2\varepsilon^2 \|z_k\|^2 + 8NM^2\right)^{1/2} \|y\|.
\end{aligned}$$

This implies

$$\lim_{k\to\infty} \frac{\|h'(z_k)\|}{\|z_k\|} \le \sqrt{2}\varepsilon, \quad \text{for any } \varepsilon > 0,$$

i.e.

$$\frac{\|h'(z_k)\|}{\|z_k\|} \to 0 \quad \text{as } k \to \infty. \tag{3.9}$$

Then

$$\begin{aligned}
\left\langle -\Phi'(z_k), z_k^+\right\rangle &= \left\langle (L+B)z_k^+, z_k^+\right\rangle + \left\langle h'(z_k), z_k^+\right\rangle \\
&\ge \sigma_1 \left\|z_k^+\right\|^2 - \left\|h'(z_k)\right\| \left\|z_k^+\right\|.
\end{aligned}$$

By (3.7) and (3.9), we have

$$\frac{\|z_k^+\|}{\|z_k\|} \to 0 \quad \text{as } k \to \infty. \tag{3.10}$$

Similarly,

$$\frac{\|z_k^-\|}{\|z_k\|} \to 0 \quad \text{as } k \to \infty. \tag{3.11}$$

Case (i): (3.3) holds. We have

$$\begin{aligned}
&\langle \Phi'(z_k), z_k \rangle - 2\Phi(z_k) \\
&= \int_0^{4N} \left[2H(t, z_k) - (\nabla H(t, z_k), z_k)\right] dt \\
&\geq d_1 \int_0^{4N} |z_k|\, dt - 8N^2 d_2 \\
&\geq d_1 \int_0^{4N} |z_k^0|\, dt - d_1 \int_0^{4N} (|z_k^+| + |z_k^-|)\, dt - 8N^2 d_2 \\
&\geq d_1 \|z_k^0\| - d_1 (\|z_k^+\| + \|z_k^-\|) - 8N^2 d_2.
\end{aligned}$$

Here we used the fact that $M^0(L + B)$ is finite dimensional. By (3.8), (3.10), and (3.11), we have

$$\frac{\|z_k^0\|}{\|z_k\|} \to 0 \quad \text{as } k \to \infty. \tag{3.12}$$

But this implies the following contradiction:

$$1 = \frac{\|z_k\|}{\|z_k\|} \leq \frac{\|z_k^0\| + \|z_k^-\| + \|z_k^+\|}{\|z_k\|} \to 0 \quad \text{as } k \to \infty. \tag{3.13}$$

Therefore $\{\|z_k\|\}$ must be bounded.

Case (ii): (3.4) holds. Similar to Case (i), we have

$$\begin{aligned}
&\langle 2\Phi(z_k) - \Phi'(z_k), z_k \rangle \\
&= \int_0^{4N} \left[(\nabla H(t, z_k), z_k) - 2H(t, z_k)\right] dt \\
&\geq d_1 \int_0^{4N} |z_k|\, dt - 8N^2 d_2 \\
&\geq d_1 \int_0^{4N} |z_k^0|\, dt - d_1 \int_0^{4N} (|z_k^+| + |z_k^-|)\, dt - 8N^2 d_2 \\
&\geq d_1 \|z_k^0\| - d_1 (\|z_k^+\| + \|z_k^-\|) - 8N^2 d_2.
\end{aligned}$$

This implies (3.12) and (3.13). Thus $\{\|z_k\|\}$ must be bounded.

So we see that $\{z_k\}$ is bounded, and going if necessary to a subsequence, we can assume that $z_k \rightharpoonup z$ in E and $z_k \to z$ in $L(S^1, \mathbf{R}^{2N})$. Write $z_k = z_k^+ + z_k^- + z_k^-$ and $z = z^+ + z^- + z^0$, then $z_k^\pm \rightharpoonup z^\pm$ in E, $z_k^0 \rightharpoonup z^0$ in E, and $z_k^\pm \to z^\pm$ in $L(S^1, \mathbf{R}^{2N})$.

In view of (3.8) and $z_k^- \to z^-$ in $L(S^1, \mathbf{R}^{2N})$, it is easy to verify

$$\int_0^{4N} \left(\nabla H(t, z_k) - \beta(t) z_k, z_k^- - z^-\right) dt \to 0$$

and

$$\int_0^{4N} \left(\nabla H(t, z) - \beta(t) z, z_k^- - z^-\right) dt \to 0.$$

But then $\langle \Phi'(z_k) - \Phi'(z), z_k^- - z^- \rangle \to 0$ as $k \to \infty$, and

$$\begin{aligned}
&\left\langle \Phi'(z_k) - \Phi'(z), z_k^- - z^- \right\rangle \\
&\quad = -\left\langle (L + B)\left(z_k^- - z^-\right), z_k^- - z^- \right\rangle \\
&\qquad - \int_0^{4N} \left(\nabla H(t, z_k) - \beta(t) z_k, z_k^- - z^-\right) dt + \int_0^{4N} \left(\nabla H(t, z) - \beta(t) z, z_k^- - z^-\right) dt \\
&\quad \geq \sigma_1 \left\| z_k^- - z^- \right\|^2 - \int_0^{4N} \left(\nabla H(t, z_k) - \beta(t) z_k, z_k^- - z^-\right) dt \\
&\qquad + \int_0^{4N} \left(\nabla H(t, z) - \beta(t) z, z_k^- - z^-\right) dt.
\end{aligned}$$

This yields $z_k^- \to z^-$ in E. Similarly, $z_k^+ \to z^+$ in E and hence $z_k \to z$ in E, that is, Φ satisfies the $(C)_c$ condition. Thus φ satisfies the $(C)_c$ condition. The proof of Lemma 3.1 is complete. □

Proof of Theorem 1.1 Let $X = M^+(L) \oplus M^0(L)$, $Y = M^-(L)$, $E = X \oplus Y$, and

$$\Phi(z) = -\varphi(z), \qquad \Psi(z) = \psi(z),$$

where $\varphi(z)$ and $\psi(z)$ are defined in Section 2.

In order to obtain this theorem, we will apply Theorem A to the functional $\Phi(z)$. Let $\mathcal{S} = X^*$, $\mathcal{T}_{\mathcal{S}} = E = X \oplus Y$. The proof of this theorem is divided into the following three steps.

Step 1. Φ satisfies (Φ_0).

We first check that Φ_c is $\mathcal{T}_{\mathcal{S}}$-closed for any $c \in \mathbf{R}$. Let $\{z_k\}$ be any sequence $\mathcal{T}_{\mathcal{S}}$-converging to some $z \in E$. Write $z_k = z_k^+ + z_k^0 + z_k^-$ and $z = z^+ + z^0 + z^-$, then $z_k^- \to u^-$ in E and hence $\{z^-\}$ is bounded in the norm topology.

Note that $\beta(t) > \frac{\pi}{2N} \tan \frac{\pi}{4N}$ for $t \in [0, 2N]$, then for any $z \in E$ and small $\varepsilon > 0$, by (3.8), we have

$$\begin{aligned}
\Psi(z) &= \int_0^{4N} H(t, z)\, dt = \int_0^{4N} \int_0^1 \left(\nabla H(t, sz), z\right) ds\, dt \\
&\geq \int_0^{4N} \int_0^1 \left(\beta(t) sz, z\right) ds\, dt
\end{aligned}$$

$$
\begin{aligned}
&-\int_0^{4N}\int_0^1 \left|\nabla H(t,sz)-\big(\beta(t)sz,z\big)\right||z|\,ds\,dt \\
&\geq \frac{1}{2}\int_0^{4N}\big(\beta(t)z,z\big)\,dt-\int_0^{4N}\int_0^1\big(\varepsilon|sz|+M\big)|z|\,ds\,dt \\
&\geq \frac{1}{2}\left(\frac{\pi}{2N}\tan\frac{\pi}{4N}-\varepsilon\right)\|z\|_{L^2}^2-M\|z\|_{L^2}\geq c_1\|z\|_{L^2}^2-c_2\|z\|_{L^2},
\end{aligned}
$$

which implies that Ψ is bounded from below on E. Consequently, combining $z_k\in\Phi_c$ and $\frac{1}{2}\langle Lz_k^+,z_k^+\rangle=-\frac{1}{2}\langle Lz_k^-,z_k^-\rangle-\Phi(z_k)-\Psi(z_k)$ shows that $\{z_k^+\}$ is bounded in E by (2.7) and hence

$$
\begin{aligned}
\Psi(z_k)&=-\frac{1}{2}\langle Lz_k^+,z_k^+\rangle-\frac{1}{2}\langle Lz_k^-,z_k^-\rangle-\Phi(z_k)\\
&\leq-\frac{1}{2}\langle Lz_k^+,z_k^+\rangle-\frac{1}{2}\langle Lz_k^-,u_k^-\rangle-c\leq c_3.
\end{aligned}
\tag{3.14}
$$

Moreover, since $\|z_k\|_{L^2}^2=\|z_k^+\|_{L^2}^2+\|z_k^0\|_{L^2}^2+\|z_k^-\|_{L^2}^2$, we have

$$
\Psi(z_k)\geq c_1\|z_k\|_{L^2}^2-c_2\|z_k\|_{L^2}\geq\|z_k\|_{L^2}-c_4\geq\left\|z_k^0\right\|_{L^2}-c_4. \tag{3.15}
$$

It follows from (3.14) and (3.15) that $\{z_k^0\}$ is also bounded in E since all norms are equivalent in a finite dimensional space. Then $\|z_k\|^2$ is bounded, and hence we can assume that $\{z_k\}$ converges weakly to $z=z^++z^0+z^-$ in E. Thus we have $\Psi(z_k)\to\Psi(z)$. Note that $\langle Ly,y\rangle^{1/2}$ is an equivalent norm on H^+. By the lower semi-continuity of the norm, we get

$$
\begin{aligned}
c\leq\limsup_{k\to\infty}\Phi(z_k)&=\limsup_{k\to\infty}\left(-\frac{1}{2}\langle Lz_k^+,z_k^+\rangle-\frac{1}{2}\langle Lz_k^-,z_k^-\rangle-\Psi(z_k)\right)\\
&=-\liminf_{k\to\infty}\frac{1}{2}\langle Lz_k^+,z_k^+\rangle-\frac{1}{2}\langle Lz^-,z^-\rangle-\Psi(z)\leq\Phi(z),
\end{aligned}
$$

that is, $z\in\Phi_c$ and hence Φ_c is $\mathcal{T}_S$-closed.

Next, we prove that $\Phi':(\Phi_c,\mathcal{T}_S)\to(E^*,\omega^*)$ is continuous. To achieve this, it is sufficient to demonstrate that Ψ' has the same property. Suppose $z_k\rightharpoonup u$ in E, then $\{z_k\}$ converges uniformly to z on $[0,4N]$. Hence, for every given $y\in E$, we see that $(\nabla H(t,z_k(t)),y(t))$ converges to $(\nabla H(t,z(t)),y(t))$ in measure on $[0,4N]$. Moreover, by (3.8), one has

$$
\left|\big(\nabla H\big(t,z_k(t)\big),y(t)\big)\right|\leq\big(\varepsilon\|z_k\|_\infty+\bar{\beta}\|z_k\|_\infty+M\big)\|y\|_\infty\leq c_5
$$

for all k and $t\in[0,4N]$, where $\bar{\beta}=\max_{t\in[0,2N]}\{\beta(t)\}$ and $\|\cdot\|_\infty$ denotes the natural norm of $C(S^1,\mathbf{R}^{2N})$. Thus, the Vitali theorem is applicable and

$$
\langle\Psi'(z_k),y\rangle=\int_0^{4N}\big(\nabla H(t,z_k),y\big)\,dt\to\int_0^{4N}\big(\nabla H(t,z),y\big)\,dt=\langle\Psi'(z),y\rangle
$$

for any $y\in E$. So Φ satisfies (Φ_0).

Step 2. Φ satisfies (Φ_1).

By (3.1) and (3.2), for any $\varepsilon>0$, there exist $C_\varepsilon>0$ and $p>2$ such that

$$
H(t,z)\leq\varepsilon|z|^2+C_\varepsilon|z|^p,\quad\text{for all }(t,z)\in[0,4N]\times\mathbf{R}^{2N}.
$$

Hence, for $z \in Y$ and small ε, we have

$$\begin{aligned}\Phi(z) &= -\frac{1}{2}\langle Lz, z\rangle - \int_0^{4N} H(t,z)\,dt \\ &\geq -\frac{1}{2}\langle Lz, z\rangle - \int_0^{4N} \left(\varepsilon|z|^2 + C_\varepsilon |z|^p\right) dt \\ &\geq \frac{\sigma}{2}\|z\|^2 - \varepsilon\|z\|_{L^2}^2 - C_\varepsilon\|z\|_{L^p}^p \\ &\geq \frac{\sigma}{2}\|z\|^2 - \varepsilon\xi_1\|z\|^2 - C_\varepsilon\xi_2\|z\|^p \\ &\geq \frac{\sigma}{4}\|z\|^2 - c_5\|z\|^p,\end{aligned}$$

where $\xi_1, \xi_2 > 0$ satisfy $\|z\|_{L^2} \leq \xi_1\|z\|$, $\|z\|_{L^p} \leq \xi_2\|z\|$; see [4], Proposition 6.6.

Since $p > 2$, there exists a small $\rho > 0$ such that $\frac{1}{8}\rho^2 \geq c_5\rho^p$. Therefore,

$$\kappa := \inf \Phi(B_\rho \cap Y) \geq \frac{1}{8}\rho^2 > 0,$$

and hence (Φ_1) holds.

Step 3. Φ satisfies (Φ_2).

Let

$$\begin{aligned}Y_0 = \operatorname{span}\Big\{ & a_j \cos\left(\frac{\pi}{2N}(2j-1)t\right), b_j \sin\left(\frac{\pi}{2N}(2j-1)t\right) : \\ & \beta(t) > \frac{\pi(2j-1)}{2N}\tan\frac{\pi(2j-1)}{4N} > 0, \text{for all } t \in [0, 2N]\Big\},\end{aligned}$$

where a_j and b_j are defined in Lemma 2.1.

Obviously, $Y_0 \subset Y$ and $\dim Y_0 = 2m$. In order to obtain the desired conclusion, it is sufficient to prove that $\Phi(z) \to -\infty$ as $\|z\| \to \infty$ on $E_0 := X \oplus Y_0$.

Let

$$m_1 = \max\left\{ j \in \mathbf{Z}^+ : \beta(t) > \frac{\pi(2j-1)}{2N}\tan\frac{\pi(2j-1)}{4N} > 0, \text{for all } t \in [0, 2N]\right\}$$

and

$$\Gamma = -\frac{\pi(2m_1-1)}{2N}\tan\frac{\pi(2m_1-1)}{4N}.$$

By the definition of Γ, there exists a constant $0 < \delta < \sigma$; σ is defined in (2.8), such that

$$\beta(t) \geq -\Gamma + \delta \tag{3.16}$$

for $t \in [0, 2N]$. Let

$$J = \left\{ j \in \mathbf{Z}^+ : \beta(t) > \frac{\pi(2j-1)}{2N}\tan\frac{\pi(2j-1)}{4N} > 0, \text{for all } t \in [0, 2N]\right\}.$$

For any $y \in Y_0$, by (2.7),

$$\begin{aligned}\langle Ly, y\rangle &= \sum_{j\in J}\left(-\frac{\pi(2j-1)}{2N}\tan\frac{\pi(2j-1)}{4N}\left[(a_j,a_j)+(b_j,b_j)\right]\right)\\ &\geq \Gamma\|y\|_{L^2}^2. \end{aligned} \tag{3.17}$$

Let $\widetilde{H}(t,z) = H(t,z) - \frac{1}{2}(\beta(t)z,z)$. We claim that, for $0 \neq z \in E$,

$$\|z\|^{-2}\int_0^{4N}\widetilde{H}(t,z)\,dt \to 0 \quad \text{as } \|z\| \to \infty. \tag{3.18}$$

Indeed, for $0 \neq z \in E$, by (3.8), one has

$$\begin{aligned}&\|z\|^{-2}\left|\int_0^{4N}\widetilde{H}(t,z)\,dt\right|\\ &= \|z\|^{-2}\left|\int_0^{4N}\int_0^1\big(\nabla H(t,sz)-\beta(t)sz,z\big)\,ds\,dt\right|\\ &\leq \|z\|^{-2}\int_0^{4N}\big(\varepsilon|z|+M\big)|z|\,dt\\ &\leq \|z\|^{-2}\big(\varepsilon\|z\|_{L^2}^2+M\|z\|_{L^2}\big)\\ &\leq \varepsilon+\frac{M}{\|z\|},\end{aligned}$$

which implies that (3.18) is true by the arbitrariness of ε. Then, for $z = z^+ + z^0 + z^- \in E_0$, by (3.16)-(3.17) and (2.8), one has

$$\begin{aligned}\Phi(z) &= -\frac{1}{2}\langle Lz^-,z^-\rangle - \frac{1}{2}\langle Lz^+,z^+\rangle - \int_0^{4N}H(t,z)\,dt\\ &\leq -\frac{\Gamma}{2}\|z^-\|_{L^2}^2 - \frac{\delta}{2}\|z^+\|^2 - \frac{1}{2}\int_0^{4N}(\beta(t)z,z)\,dt - \int_0^{4N}\widetilde{H}(t,z)\,dt\\ &\leq -\frac{\Gamma}{2}\|z^-\|_{L^2}^2 - \frac{\delta}{2}\|z^+\|^2 - \frac{1}{2}(-\Gamma+\delta)\|z\|_{L^2}^2 - \int_0^{4N}\widetilde{H}(t,z)\,dt\\ &\leq -\frac{\Gamma}{2}\|z^-\|_{L^2}^2 - \frac{\delta}{2}\|z^+\|^2 - \frac{1}{2}(-\Gamma+\delta)\big(\|z^-\|_{L^2}^2+\|z^0\|_{L^2}^2\big)\\ &\quad - \int_0^{4N}\widetilde{H}(t,z)\,dt\\ &\leq -\frac{\delta}{2}\big(\|z^-\|_{L^2}^2+\|z^+\|^2+\|z^0\|_{L^2}^2\big) - \int_0^{4N}\widetilde{H}(t,z)\,dt.\end{aligned}$$

Since M^0 and Y_0 are finitely dimensional, (3.18) and the above estimate imply that $\Phi(z) \to -\infty$ as $\|z\| \to \infty$, $z \in E_0 := X \oplus Y_0$. Hence (Φ_2) holds.

The proof of Theorem 1.1 is complete. □

Proof of Theorem 1.2 Let $X = M^-(L) \oplus M^0(L)$, $Y = M^+(L)$, $E = X \oplus Y$, and

$$\Phi(z) = \varphi(z), \qquad \Psi(z) = -\psi(z)$$

and

$$Y_0 = \operatorname{span}\left\{a_j \cos\left(\frac{\pi}{2N}(2j-1)t\right), b_j \sin\left(\frac{\pi}{2N}(2j-1)t\right) : \right.$$
$$\left. \beta(t) < \frac{\pi(2j-1)}{2N} \tan\frac{\pi(2j-1)}{4N} < 0, \text{for } t \in [0,2N]\right\},$$

where a_j and b_j are defined in Lemma 2.1.

Then the conclusion will be obtained by the same argument as in the proof of Theorem 1.1. The proof of Theorem 1.2 is complete. □

Consider the nonautonomous delay equation

$$x'(t) = -f\big(t, x(t-1)\big). \tag{3.19}$$

Example 3.1 Consider the nonautonomous delay equation (3.19), where

$$f(t,x) = a(t)x - b(t)\left[\frac{2x}{\ln(e+x^2)} - \frac{2x^3}{(e+x^2)\ln^2(e+x^2)}\right],$$

and $a(t) = \pi(2 + \tan(\frac{\pi t}{4} - \frac{\pi}{4}))$, $b(t) = \frac{3\pi}{2}$ for $t \in [0,2]$.

It is easy to see that $\alpha(t) = a(t) - 2b(t) \leq 0$, $\beta(t) = a(t) \geq \pi$ for $t \in [0,2]$, and $F(t,x) = \int_0^x f(t,s)\,ds = \frac{a(t)x^2}{2} - \frac{b(t)x^2}{\ln(e+x^2)}$. By Theorem 1.1, we get $m = 1$. Thus (3.19) possesses at least two pairs of 4-periodic solutions with $x(t) = -x(t-2)$.

Example 3.2 Consider the nonautonomous delay equation (3.19), where

$$f(t,x) = a(t)x - b(t)\left[\frac{2x}{\ln(e+x^2)} - \frac{2x^3}{(e+x^2)\ln^2(e+x^2)}\right],$$

and $a(t) = 4\pi(-2 + \tan(\frac{\pi t}{4} - \frac{\pi}{4}))$, $b(t) = -6\pi$ for $t \in [0,2]$.

It is easy to see that $\alpha(t) = a(t) - 2b(t) \geq 0$, $\beta(t) = a(t) \leq -4\pi$ for $t \in [0,2]$. By Theorem 1.2, we get $m = 2$. Thus (3.19) possesses at least four pairs of 4-periodic solutions with $x(t) = -x(t-2)$.

Competing interests
The author declares that they have no competing interests.

Acknowledgements
The author is grateful for the referee's careful reviewing and helpful comments. The research was supported by the Natural Science Foundation of Shanxi Province of China (No. 2012011004-1) and by the National Natural Science Foundation of China (No. 61473180).

References
1. Bartsch, T, Ding, Y: Deformation theorems on non-metrizable vector spaces and applications to critical point theory. Math. Nachr. **279**(12), 1267-1288 (2006)
2. Fannio, LO: Multiple periodic solutions of Hamiltonian systems with strong resonance at infinity. Discrete Contin. Dyn. Syst. **3**, 251-264 (1997)
3. Mawhin, J, Willem, M: Critical Point Theory and Hamiltonian Systems. Springer, New York (1989)
4. Rabinowits, PH: Minimax Methods in Critical Point Theory with Applications to Differential Equations. CBMS, vol. 65. Am. Math. Soc., Providence (1986)
5. Su, JB: Nontrivial periodic solutions for the asymptotically linear Hamiltonian systems with resonance at infinity. J. Differ. Equ. **145**, 252-273 (1998)

6. Zhao, F, Zhao, L, Ding, Y: Existence and multiplicity of solutions for a non-periodic Schrödinger equation. Nonlinear Anal. TMA **69**, 3671-3678 (2008)
7. Li, J, He, X: Multiple periodic solutions of differential delay equations created by asymptotically linear Hamiltonian systems. Nonlinear Anal. **31**, 45-54 (1998)
8. Guo, Z, Yu, J: Multiplicity results for periodic solutions to delay differential equations via critical point theory. J. Differ. Equ. **218**, 15-35 (2005)
9. Cheng, R, Xu, J: Periodic solutions for a class of non-autonomous differential delay equations. Nonlinear Differ. Equ. Appl. **16**, 793-809 (2009)
10. Fei, G: Nontrival periodic solutions of asymptotically linear Hamiltonian systems. Electron. J. Differ. Equ. **2001**, 69 (2001)
11. Fei, G: Multiple periodic solutions of differential delay equations via Hamiltonian systems (I). Nonlinear Anal. **65**, 25-39 (2006)
12. Fei, G: Multiple periodic solutions of differential delay equations via Hamiltonian systems (II). Nonlinear Anal. **65**, 40-58 (2006)
13. Guo, C, Guo, Z: Existence of multiple periodic solutions for a class of second-order delay differential equations. Nonlinear Anal., Real World Appl. **10**, 3285-3297 (2009)
14. Li, J, He, X, Liu, Z: Hamiltonian symmetric groups and multiple periodic solutions of differential delay equations. Nonlinear Anal. TMA **35**, 457-474 (1999)
15. Wu, K, Wu, X: Existence of periodic solutions for a class of first order delay differential equations dealing with vectors. Nonlinear Anal. **72**, 4518-4529 (2010)
16. Wu, K, Wu, X: Multiplicity results of periodic solutions for systems of first order delay differential equation. Appl. Math. Comput. **218**, 1765-1773 (2011)
17. Wu, K: Multiplicity results of periodic solutions for a class of first order delay differential equations. J. Math. Anal. Appl. **390**, 427-438 (2012)
18. Wu, K, Wu, X, Zhou, F: Multiplicity results of periodic solutions for a class of second order delay differential equations. Nonlinear Anal. **75**, 5836-5844 (2012)
19. Yu, J, Xiao, H: Multiplicity periodic solutions with minimal period 4 of delay differential equations $x'(t) = -f(t, x(t-1))$. J. Differ. Equ. **254**, 2158-2172 (2013)
20. Yu, JS: A note on periodic solutions of the delay differential equation $x'(t) = -f(x(t-1))$. Proc. Am. Math. Soc. **141**, 1267-1280 (2013)
21. Zhang, X, Meng, Q: Nontrivial periodic solutions for delay differential systems via Morse theory. Nonlinear Anal. **74**, 1960-1968 (2011)

17

A new exponential Jacobi pseudospectral method for solving high-order ordinary differential equations

Ali H Bhrawy[1,2*], Ramy M Hafez[3] and Jameel F Alzaidy[2]

*Correspondence:
alibhrawy@yahoo.co.uk
[1]Department of Mathematics, Faculty of Science, King Abdulaziz University, Jeddah, Saudi Arabia
[2]Department of Mathematics, Faculty of Science, Beni-Suef University, Beni-Suef, Egypt
Full list of author information is available at the end of the article

Abstract

This paper reports new orthogonal functions on the half line based on the definition of the classical Jacobi polynomials. We derive an operational matrix representation for the differentiation of exponential Jacobi functions which is used to create a new exponential Jacobi pseudospectral method based on the operational matrix of exponential Jacobi functions. This exponential Jacobi pseudospectral method is implemented to approximate solutions to high-order ordinary differential equations (ODEs) on semi-infinite intervals. The advantages of using the exponential Jacobi pseudospectral method over other techniques are discussed. Several numerical examples are presented to confirm the validity and applicability of the proposed method. Moreover, the obtained results are compared with those obtained using other techniques.

Keywords: high-order ODEs; exponential Jacobi functions; operational matrix of differentiation; pseudospectral method

1 Introduction

Consider the initial value problem

$$u^{(m)}(x) + \sum_{i=0}^{m-1} a_i(x)u^{(i)}(x) = f(x), \quad x \in [0,\infty[, \tag{1}$$

with initial conditions

$$u^{(i)}(0) = \alpha_i, \quad i = 0,1,\ldots,m-1, \tag{2}$$

where the functions $a_0(x), a_1(x), \ldots, a_{m-1}(x)$ are continuous on the half line $[0,\infty[$ and $\alpha_0, \alpha_1, \ldots, \alpha_{m-1}$ are constants. Eq. (1) was investigated to a wide class of deterministic and stochastic problems. This problem describes several phenomena in engineering, physics and chemical reaction, and so on. Many numerical methods are applied by many authors to study high-order differential equations. Some of more recent methods are collocation method [1–3], multiple iterative splitting method [4], modified Adomian decomposition method [5–7], spectral-Galerkin method [8–11], wavelets method [12, 13], homotopy method [14, 15], variational iteration method [16, 17], elegant harmonic numbers operational matrix of derivatives [18] and others [19–24].

In this article, we derive the operational matrix of differentiation of exponential Jacobi functions, and then we implement a new exponential Jacobi pseudospectral method in conjunction with the operational matrix of differentiation of exponential Jacobi functions to obtain numerical solutions of high-order ordinary differential equations on a semi-infinite interval. Spectral methods (see, for instance, [25–29]), based on using operational matrices, have been implemented in various problems such as fractional differential equations [30, 31], fractional optimal control problems [32], Lane-Emden equation [33, 34], and various integral equations [35]. This equation is collocated at the exponential Jacobi-Gauss quadrature nodes. Doing so, we find that we can obtain very accurate results with minimal computation. Hence, the method is rather computationally efficient compared with other numerical or analytical approaches. Finally, numerical experiments of high-order ODEs are implemented to demonstrate the validity and efficiency of the proposed algorithm. In particular, in Section 2 we design the exponential Jacobi pseudospectral method technique for solving high-order ODEs. In Section 3, several numerical examples are presented to demonstrate the efficiency of present numerical algorithm. Finally, in Section 4, a few concluding remarks and future work are included.

2 Exponential Jacobi pseudospectral method

This section presents technical details of the new exponential Jacobi functions and the exponential Jacobi pseudospectral method (EJPM). First, we outline some useful mathematical properties that we shall make use of. Then, the derivative operational matrix of exponential Jacobi functions is derived and proved. Finally, we derive the EJPM.

2.1 Mathematical preliminaries

Here we list some useful mathematical relations and identities useful in the construction of the exponential Jacobi operational matrix (EJOM). Consider the classical Jacobi polynomials $J_k^{(\theta,\vartheta)}(z)$ on the interval $[-1,1]$ with the weight function $\omega^{(\theta,\vartheta)}(x) = (1-x)^{\theta}(1+x)^{\vartheta}$, $\theta, \vartheta > -1$,

$$J_0^{(\theta,\vartheta)}(z) = 1, \qquad J_1^{(\theta,\vartheta)}(z) = \frac{1}{2}\big(\theta - \vartheta + z(\theta + \vartheta + 2)\big),$$

the set $\{J_k^{(\theta,\vartheta)}(z) : k = 0, 1, \dots\}$ forms a complete orthogonal system in the weighted Hilbert space $L^2_{\omega^{\theta,\vartheta}(x)}[-1,1]$ with the inner product

$$(u, v)_{\omega^{(\theta,\vartheta)}(x)} := \int_{-1}^{1} u(x)v(x)\omega^{(\theta,\vartheta)}(x)\,dx$$

and the norm

$$\|u\|_{\omega^{(\theta,\vartheta)}(x)} = (u, u)^{\frac{1}{2}}_{\omega^{(\theta,\vartheta)}(x)}.$$

Let us define the exponential Jacobi functions by replacing z by $1 - 2\exp(-x/L)$. Denoting the exponential Jacobi functions $J_i^{(\theta,\vartheta)}(1 - 2\exp(-x/L))$ by $EJ_i^{(\theta,\vartheta)}(x)$, $x \in [0, \infty)$. Therefore,

$EJ_i^{(\theta,\vartheta)}(x)$ may be created by the following recurrence relation:

$$EJ_{k+1}^{(\theta,\vartheta)}(x) = \frac{(2k+\theta+\vartheta+1)(2k+\theta+\vartheta+2)}{(k+1)(k+\theta+\vartheta+1)}\left[\left(\frac{((\theta+1)(\theta+\vartheta)+2k^2+2k(\theta+\vartheta+1))}{(2k+\theta+\vartheta)(2k+\theta+\vartheta+2)} - \exp(-x/L)\right)EJ_k^{(\theta,\vartheta)}(x) - \frac{(k+\theta)(k+\vartheta)}{(2k+\theta+\vartheta)(2k+\theta+\vartheta+1)}EJ_{k-1}^{(\theta,\vartheta)}(x)\right], \quad k \geq 1, \tag{3}$$

where

$$EJ_0^{(\theta,\vartheta)}(x) = 1, \qquad EJ_1^{(\theta,\vartheta)}(x) = (\theta+1) - (\theta+\vartheta+2)\exp(-x/L),$$

and

$$(k+\theta+\vartheta)EJ_i^{(\theta,\vartheta)}(x) = (k+\vartheta)EJ_i^{(\theta,\vartheta-1)}(x) + (k+\theta)EJ_i^{(\theta-1,\vartheta)}(x).$$

The exponential Jacobi functions $EJ_i^{(\theta,\vartheta)}(x)$ of degree i can be written as

$$EJ_i^{(\theta,\vartheta)}(x) = \sum_{k=0}^{i}(-1)^k \frac{\Gamma(i+\theta+1)\Gamma(i+k+\theta+\vartheta+1)}{\Gamma(\theta+k+1)\Gamma(i+\theta+\vartheta+1)(i-k)!k!}\exp(-kx/L),$$

where

$$EJ_i^{(\theta,\vartheta)}(0) = \sum_{k=0}^{i}(-1)^k \frac{\Gamma(i+\theta+1)\Gamma(i+k+\theta+\vartheta+1)}{\Gamma(\theta+k+1)\Gamma(i+\theta+\vartheta+1)(i-k)!k!}, \tag{4}$$

$$D^q EJ_i^{(\theta,\vartheta)}(0) = \sum_{k=0}^{i}(-1)^{k+q} \frac{k^q\Gamma(i+\theta+1)\Gamma(i+k+\theta+\vartheta+1)}{L^q\Gamma(\theta+k+1)\Gamma(i+\theta+\vartheta+1)(i-k)!k!}. \tag{5}$$

Let $\chi_R^{(\theta,\vartheta)}(x) = 1/L\exp(-x(\theta+1)/L)(1-\exp(-x/L))^{\vartheta}, \theta, \vartheta > -1$. The orthogonality condition of exponential Jacobi functions is

$$\int_0^{\infty} EJ_i^{(\theta,\vartheta)}(x)EJ_j^{(\theta,\vartheta)}(x)\chi_R^{(\theta,\vartheta)}(x)\,dx = \begin{cases}\gamma_i^{(\theta,\vartheta)}, & i=j,\\ 0, & i\neq j,\end{cases} \tag{6}$$

where

$$\gamma_i^{(\theta,\vartheta)} = \frac{\Gamma(i+\theta+1)\Gamma(i+\vartheta+1)}{(2i+\theta+\vartheta+1)\Gamma(i+1)\Gamma(i+\theta+\vartheta+1)}.$$

Any function $u(x) \in L^2_{\chi_R^{(\theta,\vartheta)}(x)}[0,\infty)$ can be expanded in terms of exponential Jacobi functions as

$$u(x) = \sum_{j=0}^{\infty} c_j EJ_j^{(\theta,\vartheta)}(x),$$

where the coefficients c_j are given by

$$c_j = \frac{1}{\gamma_j^{(\theta,\vartheta)}} \int_0^{\infty} u(x) EJ_j^{(\theta,\vartheta)}(x) \chi_R^{(\theta,\vartheta)}(x)\, dx, \quad j = 0, 1, \ldots.$$

Here, we outline the exponential Jacobi-Gauss quadrature. Assume that $x_{N,j}^{(\theta,\vartheta)}$, $0 \le j \le N$, are the zeros of the Jacobi-Gauss interpolation on the interval $(-1,1)$ and $\varpi_{N,j}^{(\theta,\vartheta)}$, $0 \le j \le N$, are the corresponding weights of this interpolation. The nodes of the exponential Jacobi-Gauss interpolation on the interval $(0,\infty)$ are the zeros of $EJ_{N+1}^{(\theta,\vartheta)}(x)$, which are denoted by $x_{R,N,j}^{(\theta,\vartheta)}$, $0 \le j \le N$. Clearly $x_{R,N,j}^{(\theta,\vartheta)} = -L\ln(\frac{1-x_{N,j}^{(\theta,\vartheta)}}{2})$, and weights are $\varpi_{R,N,j}^{(\theta,\vartheta)} = \frac{1}{2^{\theta+\vartheta+1}}\varpi_{N,j}^{(\theta,\vartheta)}$, $0 \le j \le N$. Let $S_N(0,\infty)$ be the set of all polynomials of degree at most N.

Let N be any positive integer, and

$$S_N(0,\infty) = \text{span}\{EJ_0^{(\theta,\vartheta)}(x), EJ_1^{(\theta,\vartheta)}(x), \ldots, EJ_N^{(\theta,\vartheta)}(x)\}. \tag{7}$$

Then, for any $\phi \in S_{2N+1}(0,\infty)$, we obtain

$$\begin{aligned}
&\int_0^{\infty} \frac{1}{L}\exp\left(\frac{-x(\theta+1)}{L}\right)\left(1-\exp\left(\frac{-x}{L}\right)\right)^{\vartheta} \phi(x)\, dx \\
&= \frac{1}{2^{\theta+\vartheta+1}} \int_{-1}^{1} (1-x)^{\theta}(1+x)^{\vartheta} \phi\left(-L\ln\left(\frac{1-x}{2}\right)\right) dx \\
&= \frac{1}{2^{\theta+\vartheta+1}} \sum_{j=0}^{N} \varpi_{N,j}^{(\theta,\vartheta)} \phi\left(-L\ln\left(\frac{1-x_{N,j}^{(\theta,\vartheta)}}{2}\right)\right) \\
&= \sum_{j=0}^{N} \varpi_{R,N,j}^{(\theta,\vartheta)} \phi(x_{R,N,j}^{(\theta,\vartheta)}).
\end{aligned} \tag{8}$$

Now, approximating $u(x)$ by $N+1$ terms of exponential Jacobi functions yields

$$u(x) \simeq \sum_{j=0}^{N} c_j EJ_j^{(\theta,\vartheta)}(x) = C^T \phi(x), \tag{9}$$

where C and $\phi(x)$ are the unknown coefficients vector and the exponential Jacobi function vector, respectively, and they are given by:

$$C = [c_0, c_1, \ldots, c_N]^T, \tag{10}$$

$$\phi(x) = [EJ_0^{(\theta,\vartheta)}(x), EJ_1^{(\theta,\vartheta)}(x), \ldots, EJ_N^{(\theta,\vartheta)}(x)]^T. \tag{11}$$

2.2 The derivative operational matrix of exponential Jacobi function

Here we shall give the derivation of a new operational matrix of derivative of the exponential Jacobi functions, which is essential to our numerical method.

Theorem 2.1 *Let $\phi(x)$ be the exponential Jacobi vector defined in* (11). *The derivative of the vector $\phi(x)$ can be expressed by*

$$\phi'(x) = \frac{d\phi(x)}{dx} \simeq \mathbf{D}\phi(x), \tag{12}$$

where $\mathbf{D}$ *is the* $(N+1)\times(N+1)$ *operational matrix of the derivative. Then the nonzero elements* d_{ij} *for* $0\le i,j\le N$ *are given as follows:*

$$d_{i+1,i}=\frac{(\theta+i+1)(\theta+\vartheta+2i+1)}{L(\theta+\vartheta+i+1)},\qquad d_{ii}=-\frac{i}{L},$$

$$d_{ij}=\frac{(-1)^{i+j+1}(2j+\theta+\vartheta+1)}{L}\prod_{k=1}^{i-j}\frac{(\theta+i-k+1)}{(\theta+\vartheta+i-k+1)},\quad j<i-1.$$

It is noticed that $\mathbf{D}$ *is a lower-Heisenberg matrix.*

Proof By differentiation with respect to x in (3) we get

$$\begin{aligned}\frac{d}{dx}EJ_{k+1}^{(\theta,\vartheta)}(x)&=\frac{(2k+\theta+\vartheta+1)(2k+\theta+\vartheta+2)}{(k+1)(k+\theta+\vartheta+1)}\\&\times\Bigg[\bigg(\frac{((\theta+1)(\theta+\vartheta)+2k^2+2k(\theta+\vartheta+1))}{(2k+\theta+\vartheta)(2k+\theta+\vartheta+2)}\\&-\exp\Big(\frac{-x}{L}\Big)\bigg)\frac{d}{dx}EJ_k^{(\theta,\vartheta)}(x)+\frac{1}{L}\exp\Big(\frac{-x}{L}\Big)EJ_k^{(\theta,\vartheta)}(x)\\&-\frac{(k+\theta)(k+\vartheta)}{(2k+\theta+\vartheta)(2k+\theta+\vartheta+1)}\frac{d}{dx}EJ_{k-1}^{(\theta,\vartheta)}(x)\Bigg],\quad k\ge 1.\end{aligned}$$

According to

$$(\theta+\vartheta+2)\exp\Big(\frac{-x}{L}\Big)=(\theta+1)EJ_0^{(\theta,\vartheta)}(x)-EJ_1^{(\theta,\vartheta)}(x)=\frac{d}{dx}EJ_1^{(\theta,\vartheta)}(x),\tag{13}$$

the elements d_{ij} of the matrix $\mathbf{D}$ may be achieved from

$$\begin{aligned}\frac{d}{dx}EJ_{k+1}^{(\theta,\vartheta)}(x)&=\frac{(2k+\theta+\vartheta+1)(2k+\theta+\vartheta+2)}{(k+1)(k+\theta+\vartheta+1)}\\&\times\Bigg[\bigg(\frac{((\theta+1)(\theta+\vartheta)+2k^2+2k(\theta+\vartheta+1))}{(2k+\theta+\vartheta)(2k+\theta+\vartheta+2)}\\&-\exp\Big(\frac{-x}{L}\Big)\bigg)\frac{d}{dx}EJ_k^{(\theta,\vartheta)}(x)+\frac{1}{(\theta+\vartheta+2)L}\frac{d}{dx}EJ_1^{(\theta,\vartheta)}(x)EJ_k^{(\theta,\vartheta)}(x)\\&-\frac{(k+\theta)(k+\vartheta)}{(2k+\theta+\vartheta)(2k+\theta+\vartheta+1)}\frac{d}{dx}EJ_{k-1}^{(\theta,\vartheta)}(x)\Bigg],\quad k\ge 1,\end{aligned}\tag{14}$$

$$\frac{d}{dx}EJ_1^{(\theta,\vartheta)}(x)=(\theta+1)EJ_0^{(\theta,\vartheta)}(x)-EJ_1^{(\theta,\vartheta)}(x),\tag{15}$$

$$\frac{d}{dx}EJ_0^{(\theta,\vartheta)}(x)=0.\tag{16}$$

A combination of Eqs. (14), (15) and (16) leads to the desired result. □

The main advantages of studying the general class of exponential Jacobi functions is that the exponential Legendre functions, exponential Chebyshev functions of all kinds, and the exponential Gegenbauer functions can be obtained as immediately special cases of the exponential Jacobi functions. Accordingly, in this article we cover all the previous mentioned functions. More specifically, exponential Legendre, exponential Chebyshev and exponential Gegenbauer operational matrices can be obtained as special cases from the derived exponential Jacobi functions. These cases are summarized in the following corollaries.

Corollary 2.2 *If $\theta = \vartheta = 0$, we obtain the exponential Legendre functions, then the nonzero elements of the operational matrix of the exponential Legendre functions d_{ij} for $0 \leq i,j \leq N$ are given as follows:*

$$d_{i+1,i} = \frac{2i+1}{L}, \qquad d_{ii} = -\frac{i}{L},$$
$$d_{ij} = (-1)^{i+j+1}\frac{(2j+1)}{L}, \quad j < i-1.$$

Corollary 2.3 *If $\theta = \vartheta = -\frac{1}{2}$, we have the exponential Chebyshev functions of the first kind, then the nonzero elements of the operational matrix of the exponential Chebyshev functions of the first kind d_{ij} for $0 \leq i,j \leq N$ are given as follows:*

$$d_{i+1,i} = \frac{(2i+1)}{L}, \qquad d_{ii} = -\frac{i}{L},$$
$$d_{ij} = \frac{2(-1)^{i+j+1}j(i-\frac{1}{2})_{i-j}}{L(1-i)_{i-j}}, \quad j < i-1.$$

Corollary 2.4 *If $\theta = \vartheta = \frac{1}{2}$, we have the exponential Chebyshev functions of the second kind, then the nonzero elements of the operational matrix of the exponential Chebyshev functions of the second kind d_{ij} for $0 \leq i,j \leq N$ are given as follows:*

$$d_{i+1,i} = \frac{(2i+3)(i+1)}{L(i+2)}, \qquad d_{ii} = -\frac{i}{L},$$
$$d_{ij} = \frac{2(-1)^{i+j+1}(j+1)(i-\frac{1}{2})_{i-j}}{L(-i-1)_{i-j}}, \quad j < i-1.$$

Corollary 2.5 *If $\theta = -\frac{1}{2}$, $\vartheta = \frac{1}{2}$, we have the exponential Chebyshev functions of the third kind, then the nonzero elements d_{ij} for $0 \leq i,j \leq N$ are given as follows:*

$$d_{i+1,i} = \frac{(2i+1)^2}{2L(i+1)}, \qquad d_{ii} = -\frac{i}{L},$$
$$d_{ij} = \frac{(2j+1)(-1)^{i+j+1}\Gamma(-i)(\frac{1}{2}-i)_{i-j}}{L\Gamma(-j)}, \quad j < i-1.$$

Corollary 2.6 *If $\theta = \frac{1}{2}$, $\vartheta = -\frac{1}{2}$, we have the exponential Chebyshev functions of the fourth kind, then the nonzero elements d_{ij} for $0 \leq i,j \leq N$ are given as follows:*

$$d_{i+1,i} = \frac{(2i+3)(2i+1)}{2L(i+1)}, \qquad d_{ii} = -\frac{i}{L},$$
$$d_{ij} = \frac{2(-1)^{i+j}\Gamma(-i)\Gamma(\frac{1}{2}-j)}{L\Gamma(-i-\frac{1}{2})\Gamma(-j)}, \quad j < i-1.$$

Remark 2.7 The operational matrix for the nth derivative can be derived as

$$\frac{d^n\phi(x)}{dx^n} = \left(\mathbf{D}^{(1)}\right)^n\phi(x), \tag{17}$$

where $n \in N$ and the superscript in $\mathbf{D}^{(1)}$ denotes matrix powers. Thus

$$\mathbf{D}^{(n)} = \left(\mathbf{D}^{(1)}\right)^n, \quad n = 1, 2, \ldots. \tag{18}$$

2.3 Derivation of the pseudospectral method

The purpose of this section is to derive a numerical algorithm for the exponential Jacobi spectral collocation method based on the operational matrix of derivative of exponential Jacobi function to solve high-order ordinary differential equations on the half line. Let us consider the high-order ordinary differential equations of the form

$$u^{(m)}(x) + \sum_{i=0}^{m-1} a_i(x) u^{(i)}(x) = f(x), \quad x \in [0, \infty[, \tag{19}$$

with initial conditions

$$u^{(i)}(0) = \alpha_i, \quad \text{for } i = 0, 1, \ldots, m-1. \tag{20}$$

We give some needed properties of exponential Jacobi functions in the preceding subsections, along with the derivation of the operational matrix of derivatives of exponential Jacobi functions.

We will obtain a system of $N + 1$ algebraic equations from: (i) applying the operational matrix of an exponential Jacobi function; (ii) collocation of the high-order ordinary differential equations at $(N - m - 1)$ exponential Jacobi-Gauss points; (iii) imposition of m initial conditions.

In order to use the exponential Jacobi operator matrix for this problem, we first approximate $u(x)$, $u^{(i)}(x)$ and $u^{(m)}(x)$ by the exponential Jacobi functions as

$$u_N(x) \simeq \sum_{j=0}^{N} c_j R_j^{(\theta,\vartheta)}(x) = C^T \phi(x), \tag{21}$$

$$u_N^{(i)}(x) \simeq \sum_{j=0}^{N} c_j \frac{d^i}{dx^i} R_j^{(\theta,\vartheta)}(x) = C^T \mathbf{D}^{(i)} \phi(x), \tag{22}$$

and

$$u_N^{(m)}(x) \simeq \sum_{j=0}^{N} c_j \frac{d^m}{dx^m} R_j^{(\theta,\vartheta)}(x) = C^T \mathbf{D}^{(m)} \phi(x). \tag{23}$$

By substituting Eqs. (21), (22) and (23) in Eq. (19), we get

$$C^T \mathbf{D}^{(m)} \phi(x) + \sum_{i=0}^{m-1} a_i(x) C^T \mathbf{D}^{(i)} \phi(x) = f(x). \tag{24}$$

Now, we satisfy (24) exactly at the collocation points of Jacobi rational Gauss quadrature. In other words, we have to collocate this operational matrix relation at the $(N - m - 1)$ exponential Jacobi roots; $x_{R,N-m,k}^{(\theta,\vartheta)}$, $k = 0, 1, \ldots, N - m$,

$$C^T\mathbf{D}^{(m)}\phi\left(x_{R,N-2,k}^{(\theta,\vartheta)}\right)+\sum_{i=0}^{m-1}a_i\left(x_{R,N-2,k}^{(\theta,\vartheta)}\right)C^T\mathbf{D}^{(i)}\phi\left(x_{R,N-2,k}^{(\theta,\vartheta)}\right)=f\left(x_{R,N-2,k}^{(\theta,\vartheta)}\right),$$
$$k=0,1,\ldots,N-m. \tag{25}$$

Furthermore, for imposing of m initial conditions, substituting Eq. (21) in Eq. (20) gives

$$u^{(i)}(0)=C^T\mathbf{D}^{(i)}\phi(0)=\alpha_i. \tag{26}$$

Finally, the relations (25)-(26) constitute a system of $(N+1)$ algebraic equations, which can be solved using any iterative technique. Consequently, the approximate solution $u_N(x)$ can be obtained (for more details, see [3, 36, 37]).

3 Numerical results

This section presents several numerical examples to demonstrate the high accuracy and applicability of the present method, and all of them were performed on the computer using a program written in Mathematica 8.0. The absolute errors in the given tables are the values of $|u(x)-u_N(x)|$ at selected points. Moreover, the obtained results are compared with those obtained using other techniques. We consider the following examples.

Example 1 Consider the nonlinear Emden-Fowler equation

$$u^{(2)}(x)+\frac{6}{(x)}u^{(1)}(x)+14u(x)=-4u\ln u,\quad x\geq 0, \tag{27}$$

subject to

$$u(0)=1,\qquad u^{(1)}(0)=0.$$

The analytical solution is $u(x)=e^{-x^2}$.

In this example, ten node points in $[0,1]$ and six corresponding weights with respect to first six exponential Jacobi functions are considered. Table 1 shows the analytical and approximation solutions of $u(x)$ obtained by Chebyshev neural network (ChNN) [38] and exponential Jacobi operational matrix (EJOM) with $\theta=\vartheta=-\frac{1}{2}$ (first kind exponential Chebyshev functions), $\theta=\vartheta=0$ (exponential Legendre functions) and $\theta=\vartheta=\frac{1}{2}$ (second kind exponential Chebyshev functions), respectively. Absolute errors obtained by EJOM with $\theta=\frac{1}{2}$, $\vartheta=-\frac{1}{2}$, $N=20$ and $L=1$ for Example 1 are plotted in Figure 1. The graph of analytical solution and approximate solution for $\theta=\frac{1}{2}$, $\vartheta=-\frac{1}{2}$, $N=20$ and $L=1$ is displayed in Figure 2 to make it easer to compare with analytical solution. Moreover the resulting graph of Eq. (27) for the presented method and the analytic solution are shown in Figure 3.

Example 2 Consider the second-order nonlinear differential equation

$$u^{(2)}(x)+\frac{8}{(x)}u^{(1)}(x)+u^2(x)=\operatorname{erf}(x)^2-\frac{4e^{-x^2}(x^2-4)}{\sqrt{\pi}x},\quad x\geq 0, \tag{28}$$

subject to

$$u(0)=0,\qquad u^{(1)}(0)=\frac{2}{\sqrt{\pi}}.$$

The analytical solution is $u(x)=\frac{2}{\sqrt{\pi}}\int_0^x e^{-t^2}\,dt$.

Table 1 Comparison of the absolute errors for Example 1

x	Analytical	ChNN [38]	EJOM ($N = 20, L = 1$)		
			$\theta = \vartheta = -\frac{1}{2}$	$\theta = \vartheta = 0$	$\theta = \vartheta = \frac{1}{2}$
0.0	1.0000000	1.0000000	1.0000000	1.0000000	1.0000000
0.1	0.99004983	0.99004883	0.99004988	0.99004994	0.99004993
0.2	0.96078943	0.96077941	0.96078946	0.96078942	0.96078934
0.3	0.91393118	0.9139317	0.91393112	0.91393112	0.91393120
0.4	0.85214378	0.85224279	0.85214382	0.85214381	0.85214376
0.5	0.77880078	0.77870077	0.77880085	0.77880096	0.77880099
0.6	0.69767632	0.69767719	0.69767618	0.69767594	0.69767596
0.7	0.61262639	0.61272838	0.61262644	0.61262662	0.61262661
0.8	0.527292424	0.52729340	0.52729252	0.52729272	0.52729269
0.9	0.44485806	0.44490806	0.44485807	0.44485772	0.44485769
1.0	0.36787944	0.36782729	0.36787939	0.36787915	0.36787921

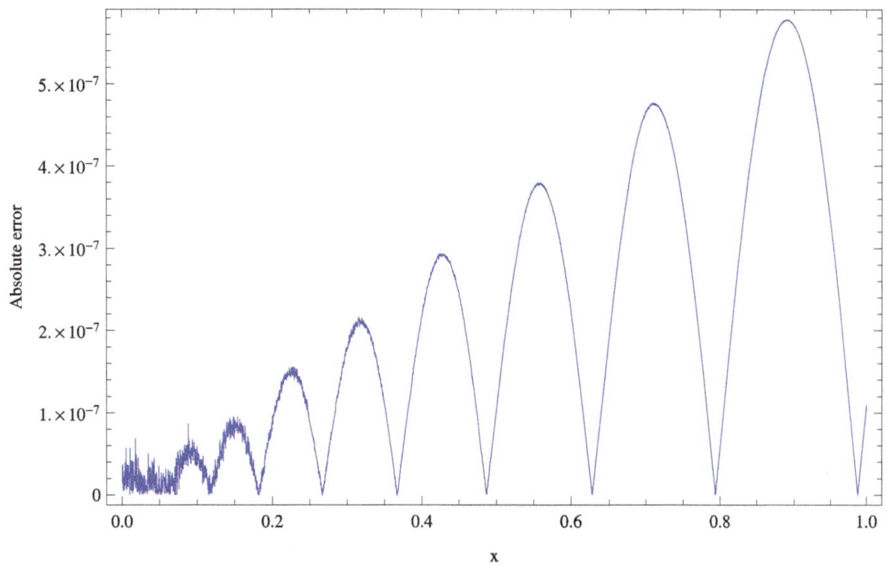

Figure 1 Absolute residual error functions with $\theta = \frac{1}{2}$, $\vartheta = -\frac{1}{2}$, $N = 20$ and $L = 1$ for Example 1.

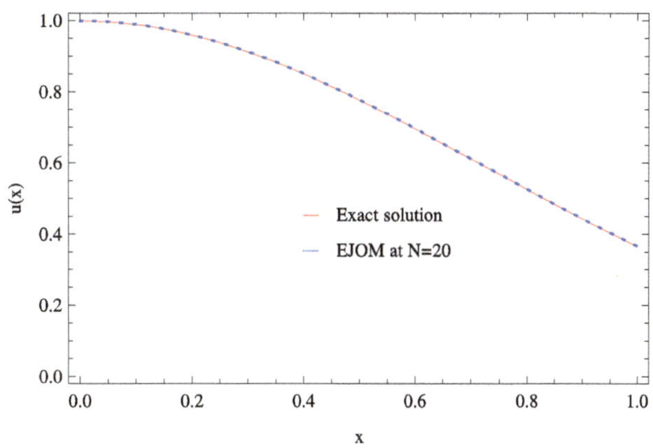

Figure 2 Exact solution and approximate solution at $\theta = \frac{1}{2}$, $\vartheta = -\frac{1}{2}$, $N = 20$ and $L = 1$ for Example 1.

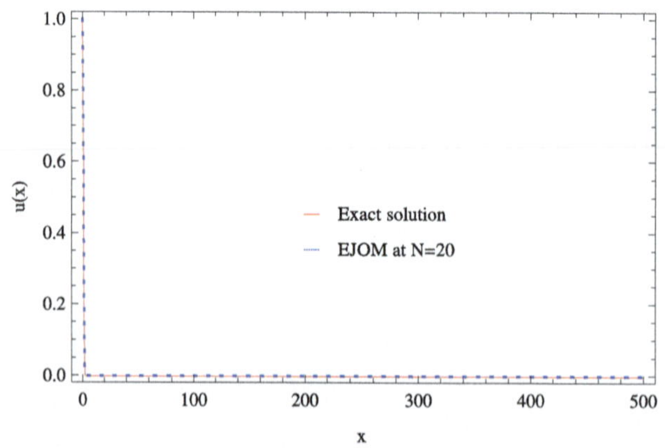

Figure 3 Exact solution and approximate solution with $\theta = \frac{1}{2}$, $\vartheta = -\frac{1}{2}$, $N = 20$ and $L = 1$ for Example 1.

Table 2 Comparison of the absolute errors for Example 2

x	$\theta = \vartheta = \frac{-1}{2}$	$\theta = \vartheta = 0$	$\theta = \vartheta = \frac{1}{2}$
0.0	0	0	0
1.0	$3.007.10^{-8}$	$2.107.10^{-8}$	$2.707.10^{-8}$
2.0	$2.353.10^{-8}$	$4.976.10^{-8}$	$5.067.10^{-8}$
3.0	$3.180.10^{-8}$	$5.733.10^{-8}$	$5.028.10^{-8}$
4.0	$9.386.10^{-8}$	$2.368.10^{-8}$	$6.252.10^{-9}$
5.0	$9.514.10^{-9}$	$6.890.10^{-9}$	$1.667.10^{-8}$
6.0	$1.007.10^{-9}$	$4.581.10^{-9}$	$3.294.10^{-9}$
7.0	$6.591.10^{-10}$	$1.593.10^{-10}$	$9.366.10^{-9}$
8.0	$1.070.10^{-9}$	$3.543.10^{-9}$	$1.079.10^{-8}$
9.0	$1.375.10^{-9}$	$5.259.10^{-9}$	$1.136.10^{-8}$
10.0	$6.492.10^{-10}$	$5.976.10^{-9}$	$1.153.10^{-8}$

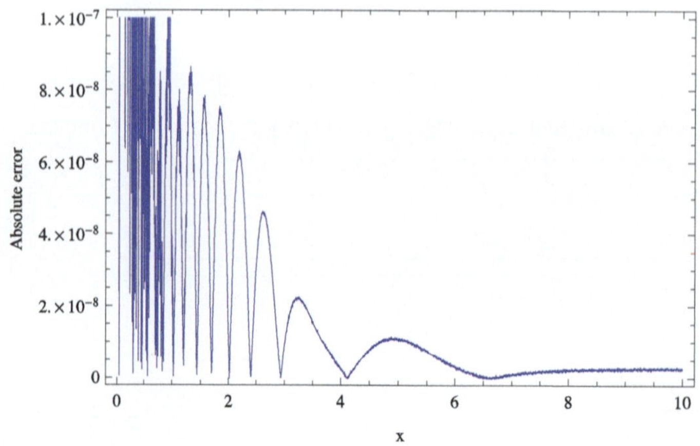

Figure 4 Absolute residual error functions with $\theta = -\frac{1}{2}$, $\vartheta = \frac{1}{2}$, $N = 24$ and $L = 1$ for Example 2.

In Table 2, we introduce the absolute errors with different values of θ, ϑ, $N = 24$ and $L = 1$. In Figure 4, we plot the absolute errors obtained by EJOM with $\theta = -\frac{1}{2}$, $\vartheta = \frac{1}{2}$, $N = 24$ and $L = 1$, while Figure 5 presents the analytic solution with the approximate solution with $\theta = -\frac{1}{2}$, $\vartheta = \frac{1}{2}$, $N = 24$ and $L = 1$.

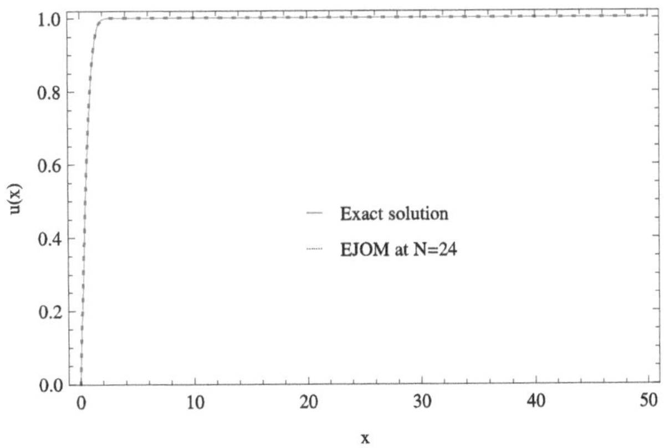

Figure 5 **Exact solution and approximate solution with $\theta = -\frac{1}{2}$, $\vartheta = \frac{1}{2}$, $N = 24$ and $L = 1$ for Example 2.**

Table 3 Comparison of the absolute errors for Example 3

x	RCC [36] method	EJOM ($N = 7, L = 10$) $\theta = \vartheta = -\frac{1}{2}$	$\theta = \vartheta = 0$	$\theta = \vartheta = \frac{1}{2}$
0.0	0	0	0	0
0.1	$1.253.10^{-4}$	$2.915.10^{-5}$	$3.871.10^{-5}$	$3.985.10^{-5}$
0.2	$5.855.10^{-4}$	$1.836.10^{-4}$	$2.508.10^{-4}$	$2.552.10^{-4}$
0.3	$1.131.10^{-3}$	$4.957.10^{-4}$	$6.943.10^{-4}$	$6.952.10^{-4}$
0.4	$2.276.10^{-3}$	$9.589.10^{-4}$	$1.368.10^{-3}$	$1.341.10^{-3}$
0.5	$3.488.10^{-3}$	$1.564.10^{-3}$	$2.255.10^{-3}$	$2.152.10^{-3}$
0.6	$4.950.10^{-3}$	$2.320.10^{-3}$	$3.342.10^{-3}$	$3.088.10^{-3}$
0.7	$6.665.10^{-3}$	$3.255.10^{-3}$	$4.631.10^{-3}$	$4.123.10^{-3}$
0.8	$8.596.10^{-3}$	$4.416.10^{-3}$	$6.139.10^{-3}$	$5.246.10^{-3}$
0.9	$1.074.10^{-2}$	$5.872.10^{-3}$	$7.897.10^{-3}$	$6.463.10^{-3}$
1.0	$1.309.10^{-2}$	$7.701.10^{-3}$	$9.949.10^{-3}$	$7.795.10^{-3}$

Example 3 Consider the linear third-order problem

$$u^{(3)}(x) + \frac{1}{(x+1)}u^{(2)}(x) - \frac{1}{(x+1)^2}u^{(1)}(x) + \frac{x}{x+1}u(x) = \frac{x\ln(x+1)}{x+1}, \quad x \in [0,1], \tag{29}$$

subject to the conditions

$$u(0) = 0, \qquad u^{(1)}(0) = 1, \qquad u^{(2)}(0) = -1,$$

and the analytical solution $u(x) = \ln(x+1)$. We apply the proposed exponential Jacobi operational matrix (EJOM) with three choices of θ and ϑ; $\theta = \vartheta = -1/2$ (first kind exponential Chebyshev functions), $\theta = \vartheta = 0$ (exponential Legendre functions), $\theta = \vartheta = 1/2$ (second kind exponential Chebyshev functions) and the numerical results are tabulated in Table 3. In this table we compare our results with those obtained by the rational Chebyshev collocation (RCC) method [36]. Numerical results of this problem show that (EJOM) is more accurate than the presented method in [36]. The graph of analytical solution and approximate solution for $\theta = \frac{1}{2}$, $\vartheta = -\frac{1}{2}$ at $N = 16$ and $L = 10$ is displayed in Figure 6 to make it easer to compare with analytical solution. Figure 7 shows the absolute residual error functions for $\theta = \frac{1}{2}$, $\vartheta = -\frac{1}{2}$ at $N = 20$ and $L = 10$.

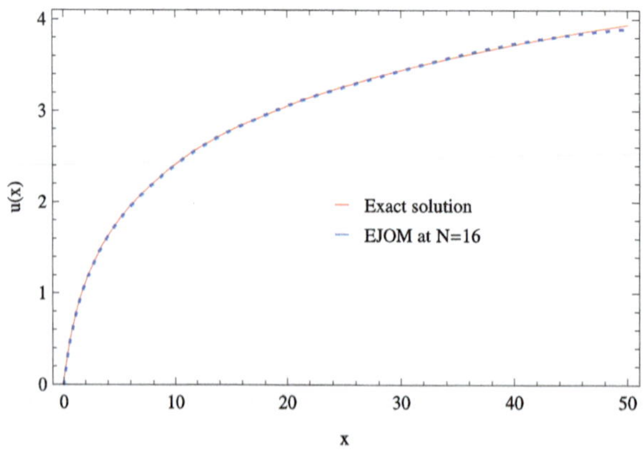

Figure 6 Exact solution and approximate solution with $\theta = \frac{1}{2}$, $\vartheta = -\frac{1}{2}$, $N = 16$ and $L = 10$ for Example 3.

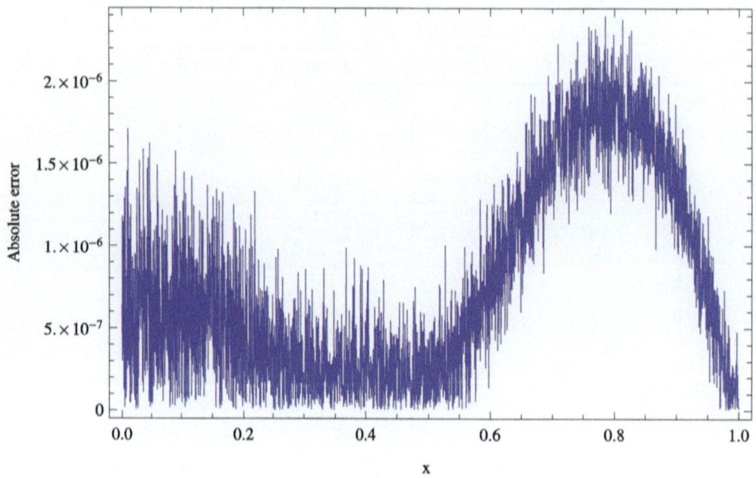

Figure 7 Absolute residual error functions with $\theta = \frac{1}{2}$, $\vartheta = -\frac{1}{2}$, $N = 20$ and $L = 10$ for Example 3.

Example 4 Consider the linear fourth-order problem

$$\begin{aligned} &u^{(4)}(x) + \cos x u^{(3)}(x) + \sin x u^{(2)}(x) \\ &\quad - 2\cos x u^{(1)}(x) - 2\sin x u(x) = f(x), \quad x \in [0, 100], \end{aligned} \tag{30}$$

subject to the conditions

$$u(0) = 0, \qquad u^{(1)}(0) = 1, \qquad u^{(2)}(0) = -2, \qquad u^{(3)}(0) = 2,$$

where f is selected such that exact solution is $u(x) = e^{-x}\sin x$. In Table 4, we list the absolute errors obtained by EJOM with different values of θ, ϑ at $N = 24$ and $L = 24$. Figure 8 is plotted to compare the analytic solution with the approximate solution with $\theta = \vartheta = -\frac{1}{2}$ at $N = 24$ and $L = 6$.

Table 4 Comparison of the absolute errors for Example 4

x	$\theta = \vartheta = 0$	$\theta = \vartheta = \frac{1}{2}$	$\theta = \vartheta = 1$
0.0	0	0	0
10.0	$5.810.10^{-8}$	$6.143.10^{-8}$	$6.856.10^{-8}$
20.0	$5.880.10^{-8}$	$1.450.10^{-6}$	$1.264.10^{-6}$
30.0	$2.953.10^{-7}$	$1.094.10^{-6}$	$5.662.10^{-8}$
40.0	$1.933.10^{-7}$	$8.744.10^{-7}$	$2.319.10^{-6}$
50.0	$1.047.10^{-7}$	$1.617.10^{-6}$	$3.126.10^{-6}$
60.0	$8.421.10^{-7}$	$1.775.10^{-6}$	$3.296.10^{-6}$
70.0	$8.012.10^{-8}$	$1.805.10^{-6}$	$3.329.10^{-6}$
80.0	$7.940.10^{-8}$	$1.811.10^{-6}$	$3.335.10^{-6}$
90.0	$7.924.10^{-8}$	$1.812.10^{-6}$	$3.336.10^{-6}$
100.0	$7.924.10^{-8}$	$1.812.10^{-6}$	$3.336.10^{-6}$

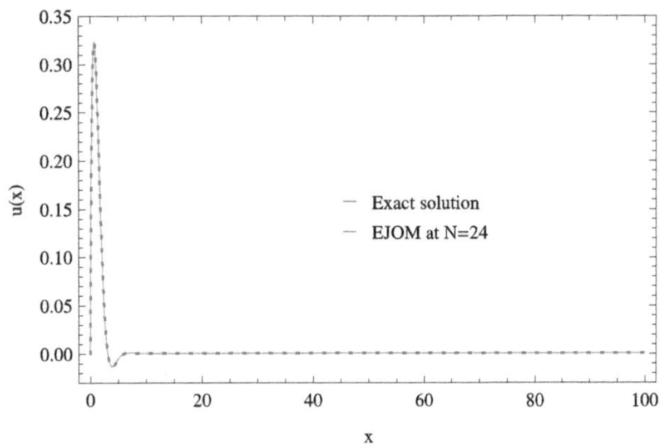

Figure 8 Exact solution and approximate solution with $\theta = \vartheta = -\frac{1}{2}$, $N = 24$ and $L = 6$ for Example 4.

4 Conclusions

In this paper, we derived the operational matrix of derivative of exponential Jacobi functions. This operational matrix in conjunction with the exponential Jacobi spectral collocation method is utilized for reducing the solution of high-order ordinary differential equations on the semi-infinite interval to that of a system of algebraic equations, which may then be solved much more easily. The operational matrices of derivatives of exponential Legendre and exponential Chebyshev functions of the first and second kinds, which often appear in conjunction with such spectral methods in the literature, may be obtained as special cases of the operational matrix of exponential Jacobi functions by taking the corresponding spacial cases of the exponential Jacobi functions parameters θ and ϑ.

Illustrative numerical examples with the satisfactory approximate solutions are achieved to demonstrate the applicability and high accuracy of the present technique. The obtained approximations of the exact solutions for the test problems make this technique very attractive and contributed to the good agreement between approximate and exact values in the numerical example. In addition, the present method could prove fruitful for those investigating not only high-order ordinary differential equations, but more broadly equations with (i) strong nonlinearity and (ii) singularities.

It can be expected that the new exponential Jacobi pseudospectral scheme coupled with a spectral element method will be an effective tool for the numerical solution of time-dependent differential equations [39]. It also may be extended to solve nonlocal boundary value problems with more complicated conditions, meanwhile its extension to the two-dimensional problems is straightforward. We assert that the proposed technique can be applied to a much larger class of fixed-order and variable-order fractional differential equations (see, for instance, [40, 41]).

Competing interests

The authors declare that they have no competing interests.

Authors' contributions

The authors have equal contributions to each part of this paper. All the authors read and approved the final manuscript.

Author details

[1]Department of Mathematics, Faculty of Science, King Abdulaziz University, Jeddah, Saudi Arabia. [2]Department of Mathematics, Faculty of Science, Beni-Suef University, Beni-Suef, Egypt. [3]Department of Basic Science, Institute of Information Technology, Modern Academy, Cairo, Egypt.

Acknowledgements

This article was funded by the Deanship of Scientific Research DSR, King Abdulaziz University, Jeddah. The authors, therefore, acknowledge with thanks DSR technical and financial support.

References

1. Mai-Duy, N: An effective spectral collocation method for the direct solution of high-order ODEs. Commun. Numer. Methods Eng. **22**, 627-642 (2006)
2. Shi, Z, Li, F: Numerical solution of high-order differential equations by using periodized Shannon wavelets. Appl. Math. Model. **38**, 2235-2248 (2014)
3. Costabile, FA, Napoli, A: Collocation for high order differential equations with two-points Hermite boundary conditions. Appl. Numer. Math. **87**, 157-167 (2015)
4. Geiser, J: A multiple iterative splitting method for higher order differential equations. J. Math. Anal. Appl. **424**, 1447-1470 (2015)
5. Hasan, YQ, Zhu, LM: Solving singular boundary value problems of higher-order ordinary differential equations by modified Adomian decomposition method. Commun. Nonlinear Sci. Numer. Simul. **14**, 2592-2596 (2009)
6. Lin, Y, Chen, C-K: Modified Adomian decomposition method for double singular boundary value problems. Rom. J. Phys. **59**(5-6), 443-453 (2014)
7. Cristescu, IA: Decomposition method for neutron transport equation. Rom. J. Phys. **60**(1-2), 179-189 (2015)
8. Doha, EH, Bhrawy, AH: A Jacobi spectral Galerkin method for the integrated forms of fourth-order elliptic differential equations. Numer. Methods Partial Differ. Equ. **25**, 712-739 (2009)
9. Doha, EH, Bhrawy, AH, Abd-Elhameed, WM: Jacobi spectral Galerkin method for elliptic Neumann problems. Numer. Algorithms **50**, 67-91 (2009)
10. Doha, EH, Bhrawy, AH, Saker, MA: On the derivatives of Bernstein polynomials: an application for the solution of high even-order differential equations. Bound. Value Probl. **2012**, Article ID 829543 (2012). doi:10.1155/2011/829543
11. Abd-Elhameed, WM: New formulae for the high-order derivatives of some Jacobi polynomials: an application to some high-order boundary value problems. Sci. World J. **2014**, Article ID 456501 (2014)
12. Fazal-i-Haq, Ali, A: Numerical solution of fourth order boundary value problems using Haar wavelets. Appl. Math. Sci. **5**, 3131-3146 (2011)
13. Abd-Elhameed, WM, Doha, EH, Youssri, YH: New spectral second kind Chebyshev wavelets algorithm for solving linear and nonlinear second-order differential equations involving singular and Bratu type equations. Abstr. Appl. Anal. **2013**, Article ID 715756 (2013)
14. Noshad, H, Bahador, SS: Numerical solution of Fokker-Planck equation for energy straggling of protons. Rom. Rep. Phys. **66**, 99-108 (2014)
15. Jafarian, A, Ghaderi, P, Golmankhaneh, AK: Construction of soliton solution to the Kadomtsev-Petviashvili-II equation using homotopy analysis method. Rom. Rep. Phys. **65**, 76-83 (2013)
16. Liu, B, Wen, Y, Zhou, X: The variational problem of fractional-order control systems. Adv. Differ. Equ. **2015**, Article ID 110 (2015)
17. Marinca, V, Ene, R-D, Marinca, B: Approximate analytic solutions of a nonlinear elastic wave equations with the anharmonic correction. Proc. Rom. Acad., Ser. A: Math. Phys. Tech. Sci. Inf. Sci. **16**, 80-86 (2015)
18. Abd-Elhameed, WM: On solving linear and nonlinear sixth-order two point boundary value problems via an elegant harmonic numbers operational matrix of derivatives. Comput. Model. Eng. Sci. **101**, 159-185 (2014)
19. Kumar, D, Singh, J, Sushila: Application of homotopy analysis transform method to fractional biological population model. Rom. Rep. Phys. **65**, 63-75 (2013)
20. Noshad, H, Bahador, SS: Numerical solution of Fokker-Planck equation for energy straggling of protons. Rom. Rep. Phys. **66**, 99-108 (2014)
21. Lakhdari, A, Boussetila, N: An iterative regularization method for an abstract ill-posed biparabolic problem. Bound. Value Probl. **2015**, Article ID 55 (2015)
22. El-Raheem, ZFA, Salama, FA: The inverse scattering problem of some Schrodinger type equation with turning point. Bound. Value Probl. **2015**, Article ID 57 (2015)
23. Al-Khaled, K: Numerical solution of time-fractional partial differential equations using Sumudu decomposition method. Rom. J. Phys. **60**, 99-110 (2015)

24. Wang, GW, Xu, TZ: The improved fractional sub-equation method and its applications to nonlinear fractional partial differential equations. Rom. Rep. Phys. **66**, 595-602 (2014)
25. Karimi Vanani, S, Soleymani, F: Tau approximate solution of weakly singular Volterra integral equations. Math. Comput. Model. **57**, 494-502 (2013)
26. Abdelkawy, MA, Ahmed, EA, Sanchez, P: A method based on Legendre pseudo-spectral approximations for solving inverse problems of parabolic types equations. Math. Sci. Lett. **4**, 81-90 (2015)
27. Abd-Elhameed, WM, Youssri, YH: New ultraspherical wavelets spectral solutions for fractional Riccati differential equations. Abstr. Appl. Anal. **2014**, Article ID 626275 (2014)
28. Gurbuz, B, Sezer, M: Laguerre polynomial approach for solving Lane-Emden type functional differential equations. Appl. Math. Comput. **242**, 255-264 (2014)
29. Akyuz-Dascioglu, A, Sezer, M: Bernoulli collocation method for high-order generalized pantograph equations. New Trends Math. Sci. **3**, 96-109 (2015)
30. Bhrawy, AH, Abdelkawy, MA: A fully spectral collocation approximation for multi-dimensional fractional Schrödinger equations. J. Comput. Phys. **294**, 462-483 (2015)
31. Bhrawy, AH, Doha, EH, Ezz-Eldien, SS, Abdelkawy, MA: A numerical technique based on the shifted Legendre polynomials for solving the time-fractional coupled KdV equation. Calcolo (2015). doi:10.1007/s10092-014-0132-x
32. Bhrawy, AH, Doha, EH, Baleanu, D, Ezz-Eldien, SS, Abdelkawy, MA: An accurate numerical technique for solving fractional optimal control problems. Proc. Rom. Acad., Ser. A: Math. Phys. Tech. Sci. Inf. Sci. **16**, 47-54 (2015)
33. Doha, EH, Bhrawy, AH, Hafez, RM, Van Gorder, RA: A Jacobi rational pseudospectral method for Lane-Emden initial value problems arising in astrophysics on a semi-infinite interval. Comput. Appl. Math. **33**, 607-619 (2014)
34. Doha, EH, Abd-Elhameed, WM, Bassuony, MA: On using third and fourth kinds Chebyshev operational matrices for solving Lane-Emden type equations. Rom. J. Phys. **60**, 281-292 (2015)
35. Bhrawy, AH, Zaky, MA, Baleanu, D: New numerical approximations for space-time fractional Burgers' equations via a Legendre spectral-collocation method. Rom. Rep. Phys. **67**, 2 (2015)
36. Sezer, M, Gulsu, M, Tanay, B: Rational Chebyshev collocation method for solving higher-order linear ordinary differential equations. Numer. Methods Partial Differ. Equ. **27**, 1130-1142 (2011)
37. Doha, EH, Bhrawy, AH, Hafez, RM: On shifted Jacobi spectral method for high-order multi-point boundary value problems. Commun. Nonlinear Sci. Numer. Simul. **17**, 3802-3810 (2012)
38. Mall, S, Chakraverty, S: Numerical solution of nonlinear singular initial value problems of Emden-Fowler type using Chebyshev neural network method. Neurocomputing **149**, 975-982 (2015)
39. Bhrawy, AH: An efficient Jacobi pseudospectral approximation for nonlinear complex generalized Zakharov system. Appl. Math. Comput. **247**, 30-46 (2014)
40. Bhrawy, AH, Zaky, MA: A method based on the Jacobi tau approximation for solving multi-term time-space fractional partial differential equations. J. Comput. Phys. **281**, 876-895 (2015)
41. Bhrawy, AH, Zaky, MA: Numerical simulation for two-dimensional variable-order fractional nonlinear cable equation. Nonlinear Dyn. **80**(1), 101-116 (2015)

Permissions

All chapters in this book were first published in ADE, by Springer; hereby published with permission under the Creative Commons Attribution License or equivalent. Every chapter published in this book has been scrutinized by our experts. Their significance has been extensively debated. The topics covered herein carry significant findings which will fuel the growth of the discipline. They may even be implemented as practical applications or may be referred to as a beginning point for another development.

The contributors of this book come from diverse backgrounds, making this book a truly international effort. This book will bring forth new frontiers with its revolutionizing research information and detailed analysis of the nascent developments around the world.

We would like to thank all the contributing authors for lending their expertise to make the book truly unique. They have played a crucial role in the development of this book. Without their invaluable contributions this book wouldn't have been possible. They have made vital efforts to compile up to date information on the varied aspects of this subject to make this book a valuable addition to the collection of many professionals and students.

This book was conceptualized with the vision of imparting up-to-date information and advanced data in this field. To ensure the same, a matchless editorial board was set up. Every individual on the board went through rigorous rounds of assessment to prove their worth. After which they invested a large part of their time researching and compiling the most relevant data for our readers.

The editorial board has been involved in producing this book since its inception. They have spent rigorous hours researching and exploring the diverse topics which have resulted in the successful publishing of this book. They have passed on their knowledge of decades through this book. To expedite this challenging task, the publisher supported the team at every step. A small team of assistant editors was also appointed to further simplify the editing procedure and attain best results for the readers.

Apart from the editorial board, the designing team has also invested a significant amount of their time in understanding the subject and creating the most relevant covers. They scrutinized every image to scout for the most suitable representation of the subject and create an appropriate cover for the book.

The publishing team has been an ardent support to the editorial, designing and production team. Their endless efforts to recruit the best for this project, has resulted in the accomplishment of this book. They are a veteran in the field of academics and their pool of knowledge is as vast as their experience in printing. Their expertise and guidance has proved useful at every step. Their uncompromising quality standards have made this book an exceptional effort. Their encouragement from time to time has been an inspiration for everyone.

The publisher and the editorial board hope that this book will prove to be a valuable piece of knowledge for researchers, students, practitioners and scholars across the globe.

List of Contributors

Xiao-Feng Yang and Yi Wei
Department of Applied Mathematics, Northwestern Polytechnical University, Xi'an, 710072, P.R. China

Zi-Chen Deng
School of Mechanics, Civil Engineering and Architecture, Northwestern Polytechnical University, Xi'an, 710072, P.R. China
State Key Laboratory of Structural Analysis of Industrial Equipment, Dalian University of Technology, Dalian, 116023, P.R. China

Chang-Wen Peng
School of Mathematics and Computer Sciences, Guizhou Normal College, Guiyang, Guizhou 550018, China

Zong-Xuan Chen
School of Mathematical Sciences, South China Normal University, Guangzhou, Guangdong 510631, China

Xingyong Zhang and Yun Wang
Department of Mathematics, Faculty of Science, Kunming University of Science and Technology, Kunming, Yunnan 650500, P.R. China

Guang Zhang
School of Science, Tianjin University of Commerce, Tianjin, 300134, P.R. China

Wenying Feng
Department of Computing & Information Systems, Department of Mathematics, Trent University, Peterborough, K9J 7B8, Canada

Yubin Yan
School of Computer Science & Mathematics, Faculty of Science & Engineering, University of Chester, Thornton Science Park, Pool Lane, Ince, Chester, CH2 4NU, UK

Shun-Zhou Wu and Xiu-Min Zheng
Institute of Mathematics and Information Science, Jiangxi Normal University, Nanchang, Jiangxi 330022, P.R. China

Shuhui Wu
School of Sciences, Zhejiang University of Science and Technology, 318 Liuhe Road, Hangzhou, 310023, China

Pargat Singh Calay and Zhanyuan Hou
Faculty of Computing, London Metropolitan University, 166-220 Holloway Road, London, N7 8DB, UK

Hao Chen
School of Mathematical Sciences, Huaibei Normal University, Huaibei, Anhui 235000, P.R. China
School of Mathematical Sciences, University of Electronic Science and Technology of China, Chengdu, Sichuan 611731, P.R. China

Jun Cheng
School of Electrical and Information Technology, Yunnan Minzu University, Kunming, Yunnan 650500, P.R. China

Shouming Zhong and Wei Kang
School of Mathematical Sciences, University of Electronic Science and Technology of China, Chengdu, Sichuan 611731, P.R. China

Jinxiang Yang
School of Computer Science and Technology, Southwest University for Nationalities, Chengdu, Sichuan 610041, P.R. China

Siqing Gan
School of Mathematics and Statistics, Central South University, Changsha, Hunan 410083, China

Shuhui Wu
School of Sciences, Zhejiang University of Science and Technology, 318 Liuhe Road, Hangzhou, 310023, China

Pargat Singh Calay and Zhanyuan Hou
Faculty of Computing, London Metropolitan University, 166–220 Holloway Road, London, N7 8DB, UK

Xuemei Zhang
Department of Mathematics and Physics, North China Electric Power University, Beijing, 102206, People's Republic of China

Meiqiang Feng
School of Applied Science, Beijing Information Science & Technology University, Beijing, 100192, People's Republic of China

Somlak Utudee
Department of Mathematics, Faculty of Science, Chiang Mai University, Chiang Mai, 50200, Thailand

Montri Maleewong
Department of Mathematics, Faculty of Science, Kasetsart University, Bangkok, 10900, Thailand

Ravi P Agarwal
Department of Mathematics, Texas A&M University, Kingsville, TX 78363-8202, USA
Department of Mathematics, Faculty of Science, King Abdulaziz University, P.O. Box 80203, Jeddah, 21589, Saudi Arabia

GuotaoWang and Lihong Zhang
School of Mathematics and Computer Science, Shanxi Normal University, Linfen, Shanxi 041004, People's Republic of China

Bashir Ahmad and Aatef Hobiny
Department of Mathematics, Faculty of Science, King Abdulaziz University, P.O. Box 80203, Jeddah, 21589, Saudi Arabia

Jessada Tariboon and Phollakrit Thiramanus
Nonlinear Dynamic Analysis Research Center, Department of Mathematics, Faculty of Applied Science, King Mongkut's University of Technology North Bangkok, Bangkok, 10800, Thailand

Sotiris K Ntouyas
Department of Mathematics, University of Ioannina, Ioannina, 451 10, Greece
Nonlinear Analysis and Applied Mathematics (NAAM)-Research Group, Department of Mathematics, Faculty of Science, King Abdulaziz University, P.O. Box 80203, Jeddah, 21589, Saudi Arabia

Fengrong Zhang, Nana Liu and Weiran Lü
College of Science, China University of Petroleum (East China), Qingdao, 266580, P.R. China

Chungchun Yang
Department of Mathematics, Nanjing University, Nanjing, 210093, P.R. China

Masakazu Onitsuka and Tomomi Soeda
Department of Applied Mathematics, Okayama University of Science, Okayama, 700-0005, Japan

Qiong Meng
School of Mathematical Science, Shanxi University, Wucheng Road, Taiyuan, Shanxi, China

Ali H Bhrawy
Department of Mathematics, Faculty of Science, King Abdulaziz University, Jeddah, Saudi Arabia
Department of Mathematics, Faculty of Science, Beni-Suef University, Beni-Suef, Egypt

Ramy M Hafez
Department of Basic Science, Institute of Information Technology, Modern Academy, Cairo, Egypt

Jameel F Alzaidy
Department of Mathematics, Faculty of Science, Beni-Suef University, Beni-Suef, Egypt